International Symposium on

Structure and Dynamics of Heterogeneous Systems

From Atoms, Molecules and Clusters in Complex Environment to Thin Films and Multilayers

International Symposium on

Structure and Dynamics of Heterogeneous Systems

From Atoms, Molecules and Clusters in Complex Environment to Thin Films and Multilayers

Duisburg, Germany 24–26 February 1999

Editors

Peter Entel & Dietrich E. Wolf

Gerhard-Mercator University, Germany

World Scientific

Singapore • New Jersey • London • Hong Kong

Published by

World Scientific Publishing Co. Pte. Ltd.

P O Box 128, Farrer Road, Singapore 912805

USA office: Suite 1B, 1060 Main Street, River Edge, NJ 07661

UK office: 57 Shelton Street, Covent Garden, London WC2H 9HE

British Library Cataloguing-in-Publication Data
A catalogue record for this book is available from the British Library.

Design of title page by R. Spettmann

International Symposium on
STRUCTURE AND DYNAMICS OF HETEROGENEOUS SYSTEMS

ISBN 981-02-4251-4

Printed in Singapore by FuIsland Offset Printing

PREFACE

This volume contains the proceedings of an interdisciplinary meeting which attracted participants from the fields of physics, chemistry, engineering and computer science. The subjects covered range from microscopic theories and experimental work of cluster and layer growth on complex substrates to large-scale dynamics of pinning and depinning of magnetic domain walls in two-dimensinal systems, with structural and dynamic aspects of large dye molecules in complex environments in between like the retinal chromophore of the human eye. The aim was to bring together experts from the various fields and to enable exchange of the latest research results stimulating further cooperative research based on the presentations at the symposium. We believe that we have succeded in doing so and that the present proceedings is an ample evidence for this.

The structure and dynamics of heterogeneous systems on microscopic and macroscopic length scales is a subject of growing interest. Increasing computer power over the last decade has especially enabled the theoreticians to perform large scale simulations of the ever growing complexity of the systems when passing from the atomic scale to larger length scales. Examplary we mention the structural changes of the retinal chromophore of the human eye brought about by the electronic excitations when hit by photons. This so-called *cis-trans* transformation of the chromophore is the first elementary step in the complex hierarchy of the vision process occuring in the picosecond range. Details of this transformation depend on how the chromophore is embedded in the protein box, which is far from being completely understood. Large-scale *ab initio* molecular dynamics of ground- and excited-state processes of molecules are nowadays at the edge of becoming feasible, although the simulation of the light-absorbing molecule in its complete protein box is still beyond present days computer capabilities.

The symposium was embedded in the research activities of the Graduate School on *Stucture and Dynamics of Heterogeneous Systems* in which eleven research groups from various fields in physics and chemistry cooperate with each other and which has been funded by the German Science Council (DFG) since 1997. We are very grateful for further financial support of the symposium from the Ministry of Science and Education Nordrhein-Westfalen, Hewlett-Packard and the Konrad-Krieger-Stiftung, which made the symposium possible. For its organization were important the efficiency of the conference secretary and coordinator of the Graduate School, Carsten Holtfort, as well as the help of all graduate students and postdocs of the Graduate School. Last not least, the main work in editing these proceedings was also done by Carsten Holtfort. We thank him for the job he has done.

Duisburg, August 23, 1999

Peter Entel & Dietrich E. Wolf

CONTENTS

ATOMS AND MOLECULES IN COMPLEX ENVIRONMENTS

AB INITIO CLUSTER CALCULATIONS FOR THE ADSORPTION OF SMALL MOLECULES ON OXIDE SURFACES - FROM SINGLE MOLECULES TO MONOLAYERS

M. PYKAVY AND V. STAEMMLER
Lehrstuhl für Theoretische Chemie
Ruhr-Universität Bochum, D-44780 Bochum, Germany

F. RITTNER
Lehrstuhl für Physikalische Chemie II
Universität Dortmund, D-44227 Dortmund, Germany

Quantum chemical ab initio cluster calculations were performed for the adsorption of small molecules on metal oxide surfaces. Two systems were studied in detail: The adsorption of N_2 on the (110) surface plane of TiO_2 (rutile) and the adsorption of CO on the polar (0001) surface of Cr_2O_3. In both cases a full five-dimensional potential for the interaction of a single molecule with the respective surface was calculated. For N_2/TiO_2 (110) the minimum was found for the end-on adsorption of N_2 atop a coordinately unsaturated surface Ti atom, with an adsorption energy of (35 $\pm$ 5) kJ/mol. In the case of CO/Cr_2O_3 (0001) the CO molecule is adsorbed strongly tilted (almost side-on) along a line connecting two Cr^{3+} ions at the surface; the calculated adsorption energy is 22 kJ/mol. In conjunction with empirical pair potentials for the N_2/N_2 and CO/CO interaction in the gas phase, Monte Carlo simulations were carried out to determine adsorption isotherms and the geometric structure of adsorbed monolayers.

1 Introduction

The three main steps in heterogeneous catalytic processes are (a) the adsorption of the reactant molecules at the solid surface, (b) the reaction of adsorbed molecules to products and (c) the desorption of the products from the surface.[14] In general, each of these steps consists of several physical and chemical elementary processes, therefore several questions have to be addressed if a detailed understanding of the underlying mechanism is to be obtained. For the adsorption process the essential points are:

a) adsorption site and adsorption energy for the interaction of a single molecule with the solid surface;

b) lateral interactions between adsorbed molecules which are of great importance for the geometric structure of adsorbed mono- and multilayers, growth modes, adsorption isotherms, sticking coefficients and so forth;

c) modification of the properties of adsorbed molecules by the interaction with the surface;

d) adsorbate-induced reconstructions and relaxations of the surface as well as possible changes of its electronic structure;

e) dynamic processes such as the diffusion at the surface.

In the present review we present some results of our recent quantum chemical ab initio calculations for the adsorption of small molecules on transition metal oxide surfaces. We will focus on two systems, $N_2/TiO_2(110)$ and $CO/Cr_2O_3(0001)$, and we will discuss in particular the transition from single adsorbed molecules to the formation of mono- and multilayers at higher coverage. Such a description is possible by a combination of quantum-chemical calculations for interaction potentials and Monte Carlo simulations for geometric structures and thermodynamic properties.

Recent comprehensive reviews covering the whole field of metal oxide surfaces can be found in Ref. 2-4.

2 Cluster Calculations

Since the adsorption of a single molecule on a solid surface is a local process, at least as long as the adsorbed molecule is not too large, it is quite natural to represent the adsorption site by a finite cluster which is embedded in the semi-infinite crystal in order to describe long-range electrostatic and electronic effects.

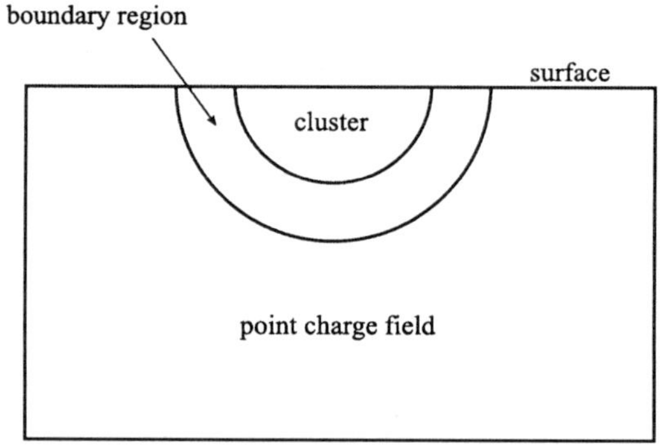

Figure 1. Schematic view of embedded cluster models.

The cluster model used in our studies for transition metal oxide surfaces is given schematically in Figure 1: The adsorption site is described by a fairly small cluster of metal cations and O^{2-} anions. This cluster can be either stoichiometric (e.g. Ti_9O_{18}, Cr_4O_6 or $Ni_{14}O_{14}$) or nonstoichiometric

(TiO_5^{6-}, CrO_3^{3-} or NiO_5^{8-}). Its shape, size and composition will depend on the geometric and electronic structure of the surface, in particular on the amount of covalency of the metal-oxygen bond, and is designed by monitoring the convergence of certain properties of the surface, e.g. ionic charges or adsorption energies.

The finite cluster has to be embedded in a semi-infinite array of point charges which describes the long-range electrostatic effects (Madelung field) of the crystal. In order to achieve a reasonable convergence rate the Evjen method of fractional charges at the boundaries of the point charge field is used. This technique has been applied more recently by Stolarczyk and Piela[1] and makes it possible to reach convergence of the Madelung energy in the bulk oxide or at the surface with 500-10000 point charges. In many oxides, e.g. in the partially covalent TiO_2, the formal ionic charges are too large and do not correctly describe the effective charges and the electrostatic potential above the surfaces. Therefore, the numerical values of the charges have to be readjusted, for instance by means of a Mulliken charge analysis for a cluster calculation of the bulk oxide. For the positions of the point charges, in general the regular lattice points of the bulk oxide are chosen.

The boundary region between the point charge field and the cluster deserves particular attention: If unscreened positive point charges are located directly adjacent to those O^{2-} anions, which are explicitly included in the cluster, the electrons of the anions will flow towards the point charges, as far as the basis set permits. This causes spurious effects in the basis set dependence of the electronic structure and changes in particular the spatial extent of the anions at the surface with the consequence that the interaction with the adsorbed molecule is strongly modified. One can avoid this problem by employing pseudopotentials with a repulsive core for the cations of the point charge field. In the present calculations, electron free effective core potentials (ECPs) were used for all cations in the first coordination shell of the O^{2-} anions in the cluster. The ECPs of Stoll and coworkers[13] were used, and it was checked by a series of test calculations that properties like the adsorption energy do not depend much on what sort of ECPs was employed explicitly (large core ECPs for the respective transition metal ion, or e.g. Mg^{2+} or Al^{3+} cores).

All the calculations presented in the next sections were performed by means of the Bochum suite of quantum chemical ab initio programs.[22,23,24,25] Most results were obtained at the Hartree-Fock level (closed-shell SCF or restricted open-shell SCF, ROHF, if necessary). Correlation effects were included in some cases at an approximate coupled cluster level,[25] but they do not contribute much to the interaction energy between closed-shell molecules such as N_2 or CO and oxide surfaces.

Gaussian basis sets of TZP (triple zeta plus one set of polarization functions) quality were used for the adsorbed N_2 and CO molecules as well as for the metal cations at the adsorption site. The basis set for the O atoms in the oxide was of TZ quality as well, but augmented by one set of semidiffuse s- and p-functions ($\eta_s = \eta_p = 0.1$) to take care of the larger spatial extent of the O^{2-} ions as compared to isolated O atoms. The reader is referred to the original papers[18,21] as far as numerical details (test of basis sets, influence of electron correlation effects, different cluster models, use of ECPs) are concerned.

The interaction energy between closed-shell molecules and oxide surfaces is generally rather small, 0-50 kJ/mol, therefore it is necessary to correct for the basis set superposition error (BSSE). The Boys-Bernardi correction[2] was used throughout.

3 The adsorption of N_2 on rutile (110)

Because of its wide range of applications, e.g. as pigment, catalyst support and photochemically reactive species, TiO_2 has attracted much recent attention, both from experimentalists and theoreticians. In particular the (110) surface plane of rutile, which is the thermodynamically most stable crystal face, has been the subject of numerous studies in which the geometric structure of the pure surface has been investigated under various experimental conditions.[17,3] The adsorption of several small molecules such as N_2, CO, NH_3 and H_2O and of reactive species as for instance K and Na atoms has been analyzed in detail as well.[17,4]

One particular problem in the context of the catalytic activity of TiO_2 is the question how much of the (110) surface face is exposed in TiO_2 powder samples. In general, the size of the active surface is measured by N_2 adsorption at 77 K, and the BET (Brunauer-Emmett-Teller) model[26] is used to estimate the monolayer capacity and the specific surface area. However, it is not clear whether or not the assumptions on which the BET model is based, namely that all adsorption sites are equivalent and no lateral interactions between the adsorbed molecules exist, are valid.

The (110) cleavage plane of rutile has a missing-row structure, as is shown in Figure 2. It consists of parallel rows of 5-fold coordinated Ti atoms (row A atoms), separated by rows of fully coordinated Ti atoms which are saturated by protruding 2-fold coordinated O atoms (row B atoms). The main adsorption site for a single N_2 molecule is certainly the atop position above the coordinately unsaturated row A Ti atoms, however, it has to be checked whether N_2 is adsorbed side-on or end-on, whether it can be adsorbed on

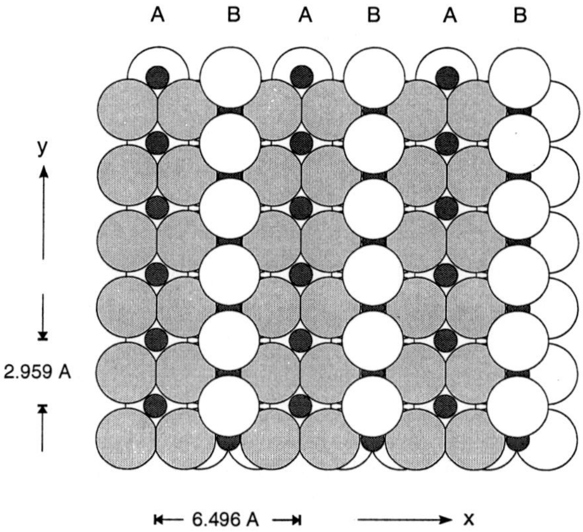

Figure 2. Top view of the (110) surface of rutile. Small dark spheres denote Ti atoms; large grey and white spheres denote O atoms in the plane and above or below the plane of Ti, respectively.

row B O atoms as well and how large the lateral interaction between two N_2 molecules adsorbed on two adjacent row A Ti atoms is.

The first step in the ab initio calculation of the $N_2/TiO_2(110)$ interaction consists of designing a cluster which allows for a correct description of the electronic structure of the $TiO_2(110)$ surface and a balanced representation of all possible adsorption sites. It turned out that it is necessary to use embedded clusters as large as Ti_9O_{18} or $Ti_{13}O_{26}$ to satisfy these requirements.[18,19] For smaller clusters, the adsorption energy depends quite sensitively on the ionicity of the embedding point charge field; only in clusters as large as Ti_9O_{18} the covalency of the $Ti-O$ bond and screening effects are satisfactorily described.

Using these clusters, we have calculated a full five-dimensional $N_2/TiO_2(110)$ potential which covers all possible geometric configurations of N_2 above the $TiO_2(110)$ surface, with the N_2 bond length fixed at its equilibrium value of 1.098 Å.[19] In Figure 3 potential curves are plotted for four different adsorption geometries, including the two configurations which exhibit an attractive interaction. Apparently, the global minimum for a single N_2 on $TiO_2(110)$ is the end-on position on top a row A Ti atom; the adsorption energy amounts to $\sim$ 46.1 kJ/mol at R$(Ti-N) = 2.39$ Å (Fig. 3a). The

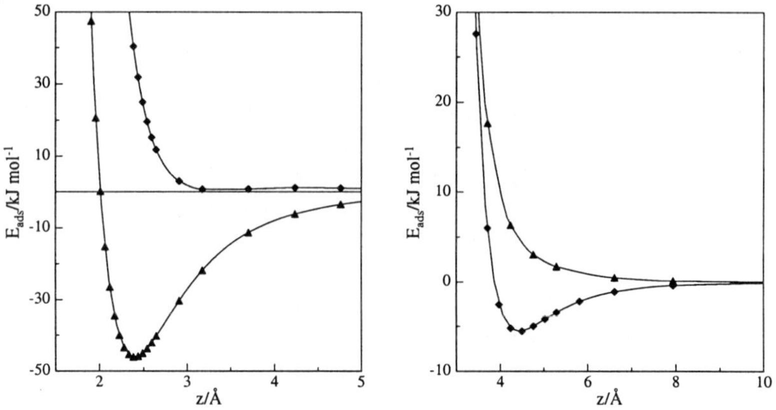

Figure 3. Potential curves for the end-on (▲) and side-on (♦) adsorption of N_2 on the different $TiO_2(110)$ surface sites: a) on-top position above a coordinately unsaturated row A Ti atom (left) and b) bridge position on row B oxygen atoms (right). z is the distance between the Ti surface plane and the closest N atom.

adsorption above row B O atoms is also attractive (Fig. 3b), with a much smaller adsorption energy of only $\sim$ 5.5 kJ/mol and with N_2 adsorbed side-on in a bridging position between two O atoms. The calculated adsorption energy compares fairly well with experimental estimates ($\sim$ 35 kJ/mol in the low coverage regime).[18] We repeated the calculation for the most favorable adsorption geometry using a hierarchy of larger clusters and applied zero point vibrational corrections to the results. We thus arrive at a final value of the adsorption energy of (35 ± 5) kJ/mol, in excellent agreement with experiment.

The main contributions to the chemical bond between N_2 and $TiO_2(110)$ are the electrostatic interaction between the quadrupole moment of N_2 and the electric field gradient above the surface and the attraction due to the polarization of N_2 by the electrostatic field above the surface. These two contributions are counterbalanced by the Pauli repulsion between N_2 and the O^{2-} anions at the surface.[18,19]

If two N_2 molecules are adsorbed end-on at adjacent row A Ti atoms, the distance between them is only 2.959 Å, as shown in Figure 2, i.e. so small that they experience a strong lateral repulsion. In order to describe the geometric structure of mono- and multilayers of N_2 on $TiO_2(110)$ we have combined our ab initio results for the $N_2/TiO_2(110)$ interaction with the semiempirical potential between two N_2 molecules in the fluid phase, as developed by Murthy et al.[27] This potential is an effective pair potential consisting of a two-site Lennard-Jones part and an electrostatic part which accounts for the quadrupole interaction between two N_2 molecules.

Based on these two potentials, canonical and grand canonical Monte Carlo (MC) simulations were performed for a temperature of 77 K which is the boiling temperature of nitrogen and is usually employed in experimental adsorption studies. Simulation boxes of 51.9 Å× 47.4 Å (128 row A Ti atoms) and a height of 30 Å above the surfaces were used in conjunction with periodic and minimum image boundary conditions in the lateral directions.[20]

The main results can be described as follows:[20]

a) At small nitrogen pressures ($p < 1$ Torr) the N_2 molecules are adsorbed exclusively at the most favorable adsorption sites, i.e. end-on atop row A Ti atoms. In order to reduce their lateral interaction, the N_2 molecules tilt against each other, perpendicular to the row A direction, thus forming nearly perfect zig-zag chains along the row A lines (Figure 4). Two of such parallel chains can be oriented in-phase or out-of-phase; since the distance between the chains is 6.5 Å, they interact only very weakly with each other and the two orientations occur with equal probability. In the out-of-phase orientation small hexagonal arrangements of end-on N_2 molecules are generated, the centers of which are above the B rows.

b) At higher nitrogen pressures, the next N_2 molecules will preferentially populate these hexagonal holes between out-of-phase zig-zag chains. If the pressure is further increased higher layers are populated without any clear orientational order (condensed phase).

c) By means of the grand canonical MC simulations adsorption isotherms at 77 K have been calculated. They show - in agreement with experiment - the typical type II behavior and can be analyzed with the BET equation.[26] Using the point B approach[5] which usually agrees quite well with the linearized version of the BET equation, one gets a monolayer capacity $n_m = 170 \pm 10$ which is substantially larger than the 128 row A Ti atoms contained in the simulation box. If the row B adsorption sites are also included in the definition of the monolayer, one expects 200 adsorption sites. Thus the value n_m is between the number of equivalent "primary" adsorption sites (row A Ti atoms) and primary plus "secondary" sites (row A plus row B). Likewise, the area occupied by one N_2 molecule would amount to 19.2 Å^2 for 128 N_2 molecules in the primary sites or to 12.3 Å^2 for all 200 row A plus row B molecules (the total surface area in the above mentioned 51.9 Å× 47.4 Å simulation box is 2460 Å^2), and not 16.1 Å^2 as usually assumed in BET analyses. All these discrepancies clearly indicate the limitations of the BET assumptions of equivalent adsorption sites without lateral interaction.

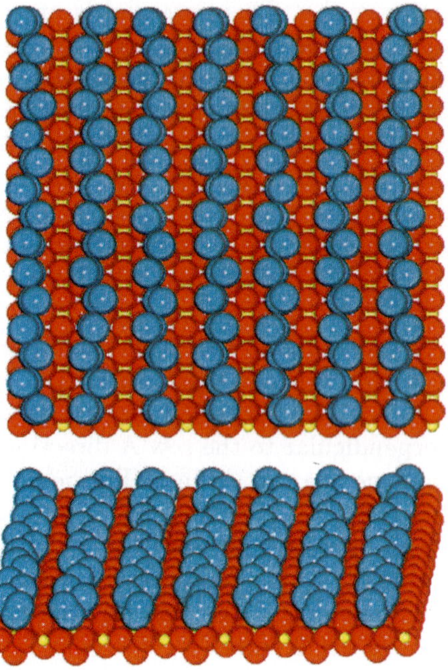

Figure 4. Adsorbed N_2 monolayers on the $TiO_2(110)$ surface.

So far, the system N_2/TiO_2 (110) has not been studied experimentally under UHV conditions at low temperatures. Therefore, the above results for the geometric structure of N_2 monolayers on rutile (110) are predictions which are waiting for an experimental confirmation.

4 The adsorption of CO on $Cr_2O_3(0001)$

Chromium oxide (Cr_2O_3) plays an important role as active component in the Phillips catalyst,[?] which is used in industrial processes for ethene polymerization. It has the same crystal structure as α-Al_2O_3 (corundum). Perpendicular to the (0001) orientation (in hexagonal notation), it possesses alternating O and Cr layers, as indicated in Figure 5. Stoichiometry is preserved by the fact that only two of the three possible octahedral positions in the Cr layer are occupied by Cr ions, one position remains empty.

The (0001) surface is the most stable surface plane of Cr_2O_3, however it is a polar surface and is thermodynamically unstable unless it undergoes sub-

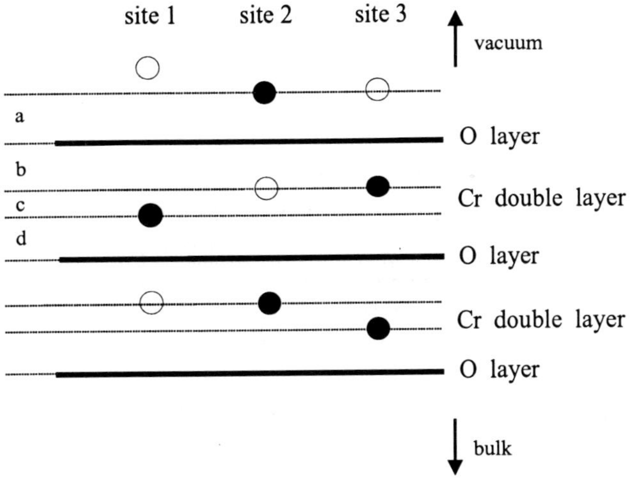

Figure 5. Side view of the $Cr_2O_3(0001)$ surface: • occupied Cr positions; ∘ empty Cr positions.

substantial reconstructions. It is by now well established,[28,6,11,29] both theoretically and experimentally, that the stabilization of the Cr-terminated $Cr_2O_3(0001)$ surface is achieved by a 50 % reduction of the number of Cr ions in the topmost Cr layer (Figure 5). At the same time, the uppermost interlayer distances perpendicular to the (0001) orientation are considerably relaxed. Table 1 shows that different theoretical and experimental methods for determining the interlayer spacings yield slightly differing results. They agree, however, that the first distance is reduced by about 50 %. This means that the Cr ions in the first (half-filled) Cr layer are pulled towards the first oxygen layer; the reason is that they are only three-fold coordinated instead of six-fold as in bulk Cr_2O_3.

It should be mentioned that it is still uncertain whether lateral relaxations do also exist in the uppermost Cr and O layers[6] and how much the ionicity of the ions in the surface layers differs from that in the bulk. Approximate values for the charges of the ions are $+2.0$ for Cr and -1.667 for O at the surface instead of $+3.0$ and -2.0 in the bulk.[31]

Contrary to the system $N_2/TiO_2(110)$, the $CO/Cr_2O_3(0001)$ adsorption has been studied repeatedly under surface science (i.e. UHV) conditions.[21,9,7,8] ARUPS and NEXAFS measurements[7] have shown that CO is weakly chemisorbed and strongly tilted, if not even planar, on $Cr_2O_3(0001)$.

Table 1. Relaxation of the first four interlayer distances of the Cr-terminated $Cr_2O_3(0001)$ surface (in %)

Method	Ref.	a	b	c	d
MD simulations	28	−58	0	36	+17
ab initio SCF	11	−50	+3.3	–	–
ab initio SCF	29	−45	−4	−24	+9
LEED analysis	28	-38 ± 4	-21 ± 6	-25 ± 16	$+11 \pm 9$
refined LEED analysis	28	−60	−3	−21	+6

Definition of the interlayer distances see Figure 5.
Interlayer distances in bulk Cr_2O_3: a, b, d: 0.94 Å; c: 0.38 Å.

LEED investigations showed that CO molecules on $Cr_2O_3(0001)$ form an ordered overlayer with a $(\sqrt{3} \times \sqrt{3})\ R\,30°$ structure.[9] Recent TDS (thermal desorption spectroscopy) measurements[21] revealed a desorption peak at about 180 K in the low coverage regime, which is consistent with a binding energy of about 45 kJ/mol for a single CO molecule at the $Cr_2O_3(0001)$ surface. Finally, IRAS measurements[21] have shown that the CO stretch frequency is blue shifted upon adsorption, from 2143 cm^{-1} (isolated CO) to 2178 cm^{-1} (low coverage regime).

Our quantum chemical ab initio SCF cluster calculations for the adsorption of a single CO molecule on $Cr_2O_3(0001)$ were performed by means of embedded CrO_3^{3-} clusters for the end-on on-top adsorption of CO on a Cr^{3+} ion and stoichiometric Cr_4O_6 clusters for CO adsorbed between two surface Cr^{3+} ions. Particular care is necessary for modeling the embedding Madelung field for this polar surface; fortunately the adsorption energy of a neutral adsorbate is not very sensitive against variations of the embedding field. Our main results are summarized in Figure 6.

A single CO molecule is bound by about 15 kJ/mol in the on-top position above a Cr^{3+} ion, end-on with the C atom towards the surface (tilt angle $\vartheta = 0°$). The interaction becomes even more attractive when CO is tilted and moved laterally in the direction towards the next Cr^{3+} ion. Since the oxygen layer underneath is not symmetric with respect to the sign of the tilt angle ϑ, two different minima with $\vartheta = -60°$ and $\vartheta = +45°$ are found. The physical origin for the bond between CO and the surface is primarily the electrostatic interaction between the quadrupole moment of CO and the electrostatic field above the surface, balanced by the Pauli repulsion between CO and the O^{2-} anions in the first O layer underneath. Figure 6 shows that the bond becomes stronger if the relaxation of the first interlayer distance

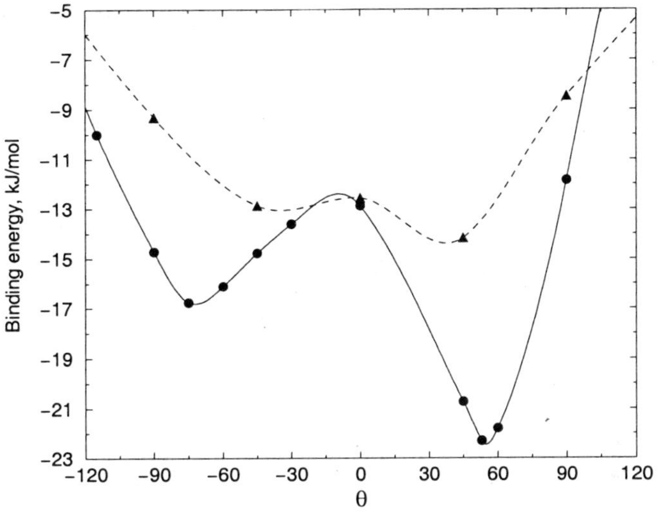

Figure 6. Potential curves for the adsorption of CO at the $Cr_2O_3(0001)$ surface for two different values of the relaxation of the topmost interlayer distance: solid line (•) – 40%, dashed line (▲) – 60%. For each tilt angle ϑ the position of the adsorbed CO molecule is optimized.

is reduced, since a smaller relaxation causes a longer $CO \cdots O$ distance and therefore a smaller Pauli repulsion.

The calculated adsorption energy of about 22 kJ/mol at the global minimum of the CO/Cr_2O_3 (0001) interaction potential amounts to about half of the experimental value of 45 kJ/mol.[21] The reason for this discrepancy is that electron correlation effects were not included in the calculations. They will lead to a smaller CO–surface distance, a larger binding energy and a larger blue shift of the CO stretch frequency. This part of our work is not yet finished, therefore we cannot present final results of how large correlation effects are.[10]

Since the $CO/Cr_2O_3(0001)$ bond is rather strong, one can expect that each $Cr^{3+} \cdots Cr^{3+}$ line at the surface is occupied by one CO molecule at monolayer coverage. However, this would lead to a very strong $CO-CO$ repulsion. The situation is much more favorable if only every second $Cr^{3+} \cdots Cr^{3+}$ line carries one CO molecule. This is indicated in Figure 7. In this arrangement only a small lateral repulsion between neighboring CO molecules is left, in particular between the more diffuse C-ends of the CO molecules. This repulsion can be further reduced by moving all CO molecules towards those Cr^{3+} ions to which they are bound with the O-end. This way one will obtain exactly the $(\sqrt{3} \times \sqrt{3})\ R\,30°$ structure that has been determined experimentally in the LEED study of Xu et al.[?]

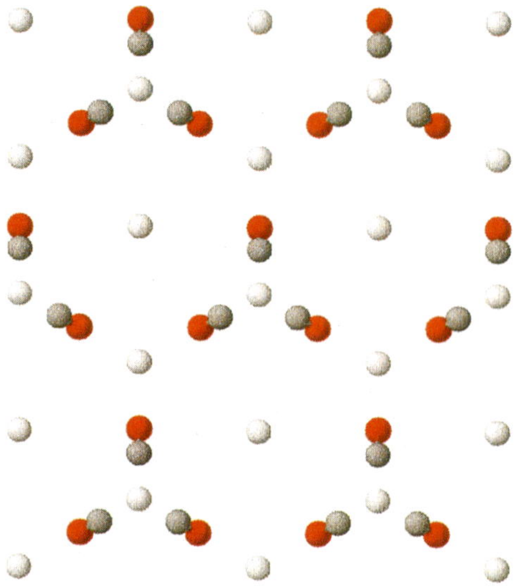

Figure 7. $(\sqrt{3} \times \sqrt{3})\ R\,30°$ structure of an adsorbed CO monolayer on $Cr_2O_3(0001)$. The topmost Cr-layer is also shown (bright circles).

MC simulations - similar to those for $N_2/TiO_2(110)$ - are currently being performed for $CO/Cr_2O_3(0001)$ at higher coverage using the $CO - CO$ pair potential of van der Pol et al.[?] Results will be published in a forthcoming paper.[?]

5 Conclusions

By combining quantum chemical cluster calculations with Monte Carlo simulations we are able to obtain a detailed understanding of the interaction between small molecules and metal oxide surfaces. For the two adsorption systems treated in this review, $N_2/TiO_2(110)$ and $CO/Cr_2O_3(0001)$, our theoretical study could rationalize

a) the mechanism of the chemical bond between a single molecule and the respective oxide surface,

b) the geometric structure of ordered monolayers of adsorbed N_2 or CO molecules, and

c) the form and behavior of the experimental adsorption isotherms.

In those cases, in which experimental information is available, the results of our studies are in very good agreement with the observed data. However, theoretical investigations of this kind are complementary to experiment since they can offer explanations for experimental observations.

Acknowledgement

Financial support of this work by Fonds der Chemischen Industrie, Ministerium für Wissenschaft und Forschung des Landes Nordrhein-Westfalen and in particular by the Deutsche Forschungsgemeinschaft through the Graduiertenkolleg "Dynamische Prozesse an Festkörperoberflächen" is gratefully acknowledged. We also thank many colleagues for their contributions, in particular B. Boddenberg (Dortmund), R. F. Fink (Bochum), M. J. Bojan and W. A. Steele (State College, PA), H.-J. Freund, H. Kuhlenbeck and O. Seiferth (Berlin).

References

1. L.Z. Stolarczyk and L. Piela, *Int. J. Quantum Chem.* **22**, 911 (1982).
2. S.F. Boys and F. Bernardi , *Mol. Phys.* **19**, 553 (1970).
3. P.J. Moller and M.C. Wu , *Surf. Sci.* **224**, 265 (1989).
4. T. Bredow, E. Aprà, M. Catti and G. Pacchioni , *Surf. Sci.* **418**, 150 (1998).
5. S.J. Gregg and K.S.W. Sing *Adsorption, Surface Area and Porosity* (Academic Press, London, 1982).
6. Th. Gloege, H.L. Meyerheim, W. Moritz and D. Wolf , to be published.
7. H. Kuhlenbeck, C. Xu, B. Dillmann, M. Hašel, B. Adam, D. Ehrlich, S. Wohlrab, H.-J. Freund, U.A. Ditzinger, H. Neddermeyer, M. Neuber and M. Neumann , *Ber. Bunsenges. Phys. Chem.* **96**, 15 (1992).
8. D. Scarano, A. Zecchina, S. Bordiga, G. Ricchiardi and G. Spoto , *Chem. Phys.* **177**, 547 (1993).
9. C. Xu, B. Dillmann, H. Kuhlenbeck and H.-J. Freund , *Phys. Rev. Lett.* **67**, 3551 (1991).
10. M. Pykavy and V. Staemmler , to be published.
11. C. Rehbein, N.M. Harrison and A. Wander , *Phys. Rev.* B **54**, 14066 (1996).

12. A. van der Pol, A. van der Avoird and P.E.S. Wormer , *J. Chem. Phys.* **92**, 7498 (1990).
13. M. Dolg, U. Wedig, H. Stoll and H. Preuss , *J. Chem. Phys.* **86** , 866 (1987).
14. M. Baerns, H. Hofmann and A. Renken , *Chemische Reaktionstechnik* (Thieme, Stuttgart, 1987).
15. C. Noguera , *Physics and Chemistry of Oxide Surfaces* (Cambridge University Press, Cambridge, 1996).
16. H.-J. Freund, H. Kuhlenbeck and V. Staemmler , *Rep. Prog. Phys.* **59** , 283 (1996).
17. V.E. Henrich and P.A. Cox , *The Surface Science of Metal Oxides*, (Cambridge University Press, Cambridge, 1994).
18. F. Rittner, R. Fink, B. Boddenberg and V. Staemmler , *Phys. Rev.* B **57**, 4160 (1998).
19. F. Rittner, B. Boddenberg, R.F. Fink and V. Staemmler , *Langmuir* **15**, 1449 (1999).
20. F. Rittner, B. Boddenberg, M.J. Bojan and W.A. Steele , *Langmuir* **15**, 1456 (1999).
21. O. Seiferth, H. Kuhlenbeck, H.-J. Freund and M. Pykavy and V. Staemmler , to be published.
22. V. Staemmler , *Theor. Chim. Acta* **45**, 89 (1977).
23. U. Meier and V. Staemmler , *Theor. Chim. Acta* **76**, 95 (1989).
24. J. Wasilewski , *Int. J. Quantum Chem.* **36**, 503 (1989).
25. R. Fink and V. Staemmler , *Theor. Chim. Acta* **87**, 129 (1993).
26. S. Brunauer, P.H. Emmett and E. Teller , *J. Am. Chem. Soc.* **60**, 309 (1938).
27. C.S. Murthy, K. Singer, M.L. Klein and I.R. McDonald , *Mol. Phys.* **41**, 1387 (1980).
28. F. Rohr, M. Bäumer, H.-J. Freund, J.A. Mejias, V. Staemmler, S. Müller, L. Hammer and K. Heinz , *Surf. Sci.* **372**, L291 (1997); *Surf. Sci.* **389**, 391 (1997).
29. A. Leitheuŝer, K. Fink and V. Staemmler , to be published.
30. S. Wang, P.J.T. Tait and C.E. Marsden , *J. Mol. Catal.* **65**, 237 (1991).
31. J.A. Mejias, V. Staemmler and H.-J. Freund , *J. Phys. C: Condensed Matter*, accepted for publication.

MOLECULES IN GROUND AND EXCITED STATES - *AB INITIO* MOLECULAR DYNAMICS AS A COMPUTATIONAL TOOL IN CHEMISTRY

S. FALZEWSKI AND V. BUSS

Institut für Physikalische und Theoretische Chemie, Gerhard-Mercator-Universität Duisburg, D 47048 Duisburg, Germany
E-mail: theobuss@uni-duisburg.de

The results of two *ab initio* molecular dynamics (AIMD) simulations of medium-sized organic molecules are described, one in the (electronic) ground state, the other in the excited state. Gradients obtained from non-empirical calculation of the electronic wave function are used to drive the motions of the nuclei by a numerical first-order approximation of Newton's equations. In the first simulation, the relaxation of a monomethine dye from the planar transition state to the helically twisted minimum energy geometry is studied, a reaction which involves only the electronic ground state of the molecule. In the second example, we follow the isomerization about the central double bond of the penta-3,5-dien-1-iminium molecular ion in an electronically excited state, a reaction which serves as a model for the initial step in the visual process of the vertebrate eye. Both simulations yield reaction trajectories that agree with classical expectations and with results previously obtained by others.

1 Introduction

The development of ever-more powerful computers and soft-ware packages have widened the scope of applied theoretical chemistry in recent years considerably. *Ab initio* molecular dynamics (AIMD) which is based on non-empirically derived force fields is one of these areas: the problem of getting a proper solution to the quantum-chemistry involved is compounded by solving the equations of motion to obtain the development of the system in time. Originally developed in physics as a tool in statistical mechanics MD has thus turned into a tool of high accuracy and predictive power.[1] In the following we apply the approach developed by E.A. Carter for the calculation of molecular trajectories on Born-Oppenheimer ground and excited state potential energy surfaces (PES)[2] to two problems of current interest in our laboratory.

The principle is straightforward: Quantum-mechanics are used to calculate the Born-Oppenheimer separated electronic wave function. Gradients are used to propagate the nuclei by integrating Newton's equations of motion. Re-optimizing the wave-function after each time-step ensures that the molecule stays on the Born-Oppenheimer PES. Coupled with the high-efficient code provided by GAUSSIAN for the calculation of analytical gradients we have shown that the method is applicable to, and yields plausible molecular trajectories for, a medium-sized (36 atoms; 170 basis functions) organic molecule.[3] In the calculations described in the following we used the GAUSSIAN94 program package[4] to obtain the electronic wave function within Restricted Hartree Fock (RHF) *ab initio* theory. For the nuclei analytical gradients were used to integrate Newton's equations of motion according to the velocity VERLET algorithm.[5] Nuclear motions occurring on a time scale faster than the time-step chosen were suppressed using the RATTLE formalism.[6]

2 Inversion of a twisted monomethine dye

Cyanine dyes are usually planar, a consequence of their highly delocalized π-electron system. The non-planar geometry of the monomethine dye shown in Fig. 1 is the result of steric interaction of the two N-methyl groups. According to the X-ray analysis of the tetrafluoroborate salt the conformation is twisted, with an angle of 43° between the aromatic end groups.[7] In solution the oppositely twisted forms equilibrate, with a barrier to inversion of 50 kJ/mol or 518.2 meV.[8] We have followed the inversion coordinate by MD. Details of the geometry optimization and the results of the MD calculation have appeared elsewhere.[9]

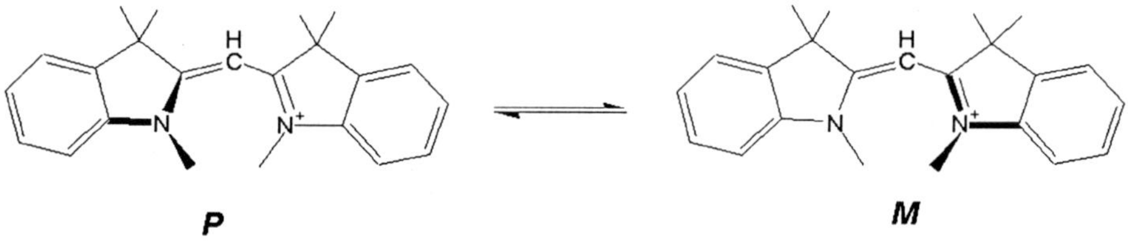

Figure 1. Dynamic equilibrium between the *P*- (left) and the *M*-helically twisted form (right) of an all-*cis*-configurated monomethine dye.

For the ground state geometry optimizations Restricted Hartree Fock (RHF) and Density Functional Theory (DFT) was used. Localizing the transition state along a reaction coordinate from which the MD calculations was to be started requires considerable computing resources because in each step of the optimization analytical frequences are needed, i.e. second derivatives of the wave function with respect to all internal degrees of freedom. The transition state for the interconverting dye conformations is shown in Fig. 2. It has C_{2v}-symmetry, with the two N-methyl groups staggered, not eclipsed. The imaginary frequency shown in the Figure involves conrotatory motion of the two methyl groups coupled with their out-of-plane displacement in order to reduce steric interaction between the methyl hydrogens. The motion shown in the Figure leads to a twisted conformation of the dye in which one of the halves has to be rotated in a clockwise direction to bring it into a co-planar arrangement with the other half; this is the definition of *P*-helicity.

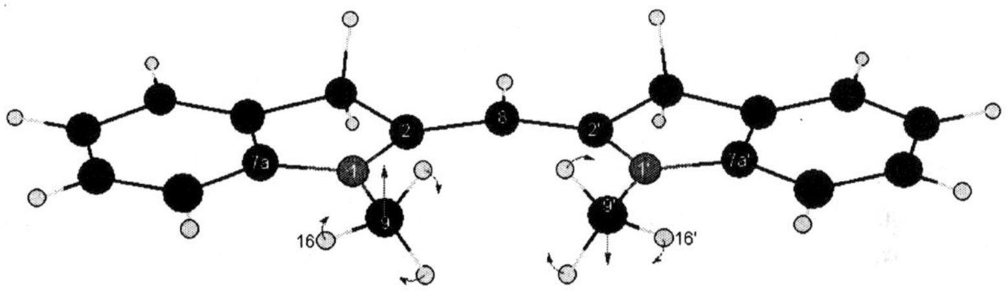

Figure 2. Transition state for the interconversion of oppositely twisted conformations of the monomethine dye. The four methyl groups at C3 and C3' have been omitted to save computer resources. The geometry was optimized with RHF and DFT theory, both methods giving essentially identical results. The arrows indicate nuclear motions corresponding to the calculated imaginary frequency of the transition state.

The molecular dynamics simulations were carried out on the basis of RHF-calculated energies and gradients with an atom centered split valence shell gaussian expansion (3-21G). The high-frequency C-H bond lengths were kept constant in order to make a time step of 30 a.u. or 0.7 fs possible. Since no significant out-of-plane forces can be expected to result from a planar transition state, we started the trajectory with a geometry twisted by 5° about both the central C2-C8 and C8-C2' bonds (C_2-symmetry), and the whole simulation was followed for 800 fs. In a second simulation which was started 200 fs into the first run, the molecule was "cooled down", i.e. the kinetic energy which builds up as a consequence of the down-hill motion of the nuclei was reduced by one third simulating the effect of vibrational and rotational energy dissipation into the environment.

In Fig. 3 is shown how the energy components and the molecular geometry change during the simulation. After some initial oscillations due to bond stretching the potential energy drops sharply after about 120 fs into the simulation and goes through a minimum some 100 fs later. The energy reached at this point is only 160 meV above the optimized minimum calculated with the same basis set which indicates that the trajectory passes this point in

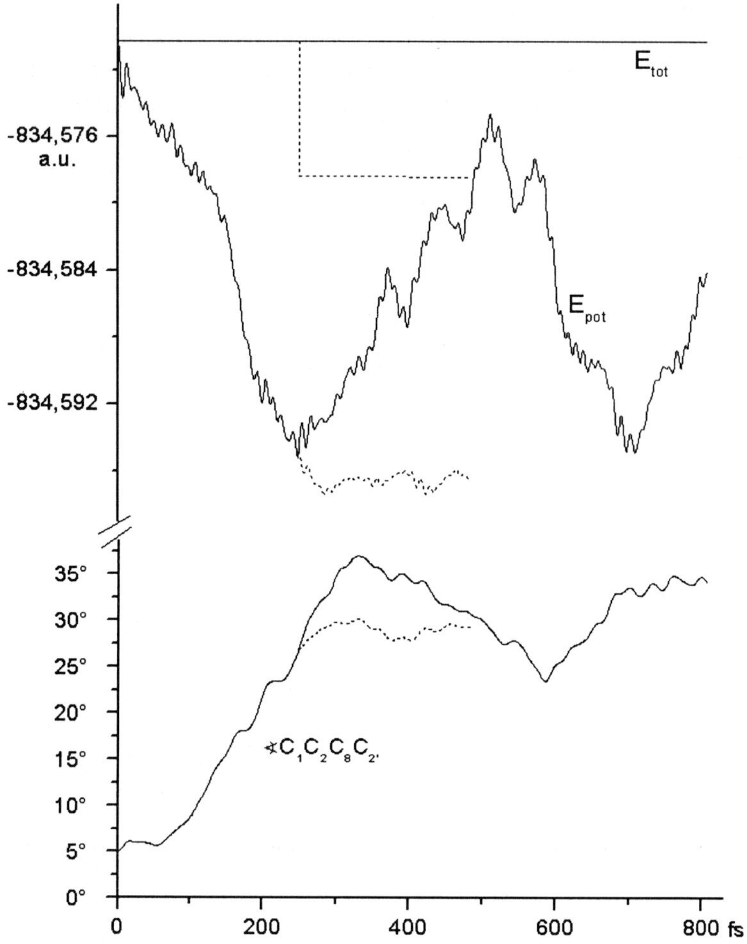

Figure 3. Results of the MD simulation starting with a geometry twisted by 5° from the transition state geometry shown in Fig. 2. From top: change of the total energy E_{tot}, of the potential energy E_{pot}, both in a.u., and of the twist twist angle (in deg) with time. The dashed lines correspond to the simulation which was started 200 fs into the simulation with the kinetic energy reduced by one third.

close proximity. Because of the kinetic energy that is building up during the downward path - note that the total energy stays constant throughout the simulation - the molecule does not stay in the (local) minimum but overshoots, with the potential energy reaching a maximum after a total simulation time of 500 fs, from where a new down-ward path begins. Driving force for the dynamics is the loss in steric repulsion forced upon the molecule in the planar geometry. This is most obviously displayed by the central dihedral angle which starting with 5° from planarity increases steadily, going through a maximum value of almost 40° before it turns back.

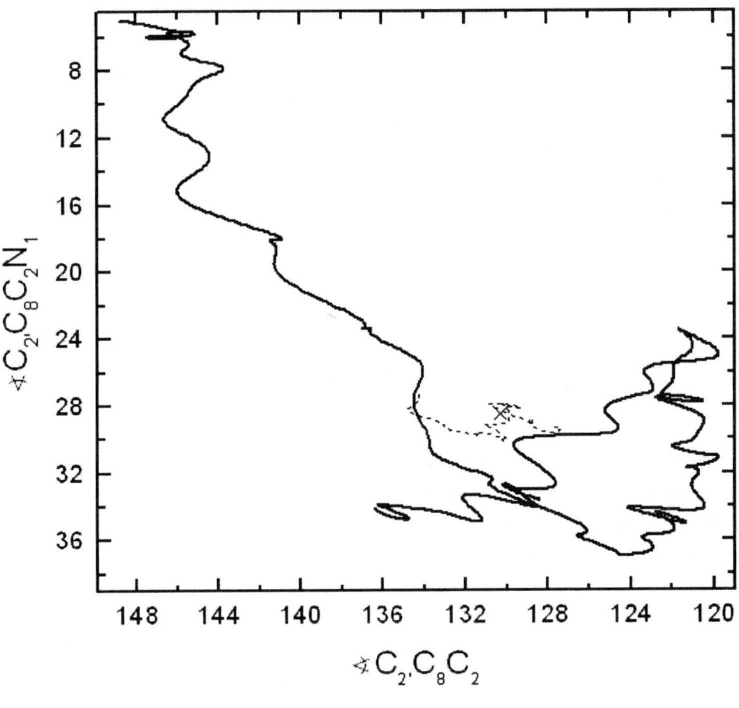

Figure 4. Another display of the MD trajectory of Fig. 3.It shows the change of the central dihedral angle (vertical axis) plotted against the central bond angle during the MD. The trajectory starts in the upper left corner, where the molecule is almost planar but, as a consequence of the steric interactions of the methyl groups, has a significantly increased central bond angle. As the MD proceeds, that angle contracts as the molecule adopts an increasingly twisted geometry. The trajectory passes the minimum of the PES (cross) and performs oscillations around that geometry. The dotted line shows the effect of reducing the kinetic energy: the oscillations around the minimum have much smaller amplitudes.

Reducing the excess kinetic energy that is building up during the simulation is a much used method in molecular dynamics to reduce large potential energy oscillations and "steering" the molecule closer to the potential energy minimum. The dotted lines in Fig. 3 show what happens when the kinetic energy is scaled by one third: the potential energy is reduced to -834.578a.u. which is now only 100meV above the computed minimum; like the potential energy the dihedral angle does not show any large fluctuations but oscillates between 28 and 30°.

A different view of the trajectory is offered in Fig. 4, where the simultaneous change of two geometric parameters is plotted as the MD proceeds.

3 MD simulation on an excited state PES

Compared to the ground state of a molecule the adequate description of electronically excited states requires tremendous computing resources. Correlation has to be dealt with explicitly which requires either the use of Configurational Interaction (CI) schemes or of perturbative schemes like Moeller-Plesset theory. With analytical gradients available from Multiconfigurational (MC) SCF wave functions in GAUSSIAN we have performed AIMD calculations of the excited S_1 state isomerization of the 3-*cis*-pentadieniminium cation (Fig. 5). Robb et al. have performed calculations on the same system using a very similar quantum-mechanical set-up. However, they either employ a minimum energy path (MEP) approach to calculate the trajectory on the S_1-surface[10] or they model the PES to second order which requires the time-consuming calculation of frequencies.[11] Since we rely on gradients only our time-steps have to be smaller (0.1 fs compared to 1 fs employed by the Robb group).

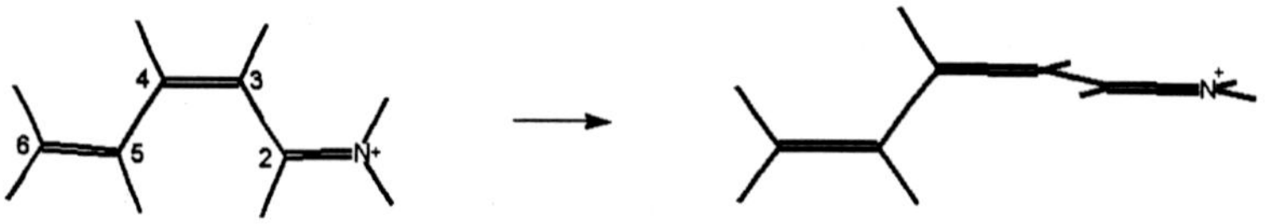

Figure 5. The photochemical conversion of the 3-*cis*-pentadieniminium cation (left) via the perpendicular Conical Intersection (CI) (right) to the 3-trans isomer (not shown) followed by MD. The starting geometry of the trajectory shown in Fig. 6 corresponds to the structure shown in the left; the geometry reached after 70 fs is essentially the one on the right. The dihedral angle shown in that Figure is the angle subtended by the two planar halves from N_1 to C_3 and from C_4-C_6.

Complete actice space (CAS) calculations with a 3-21G basis set and

6 electrons in 6 active orbitals were performed. Orbitals were chosen by visual inspection; for the initial geometry these correspond to almost pure π-electrons. The 3-*cis*-pentadieniminium cation (Fig. 5) has been called a minimal model for retinal photoisomerization.[10] The isomerization of retinal which contains 6 conjugated double bonds is the key step in the visual process of the vertebrate eye. It is an extremely fast process: the appearance of the first photoproduct takes only 200 fs.[12] The pentadieniminium cation shares with retinal the cis-configurated double bond and the positively charged terminal nitrogen atom; so the steric and the electronic conditions are comparable. The smallness of the molecule, on the other hand, allows calculations of high quality to be carried out.

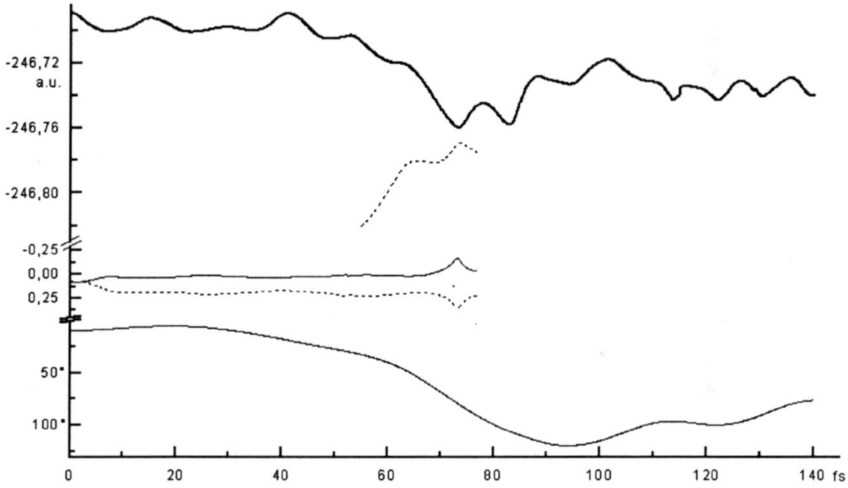

Figure 6. Results of the MD simulation of the excited 3-*cis*-pentadieniminium cation. The Figure shows (from top) the potential energy of the molecule as it moves on the S_1 excited surface; the corresponding energy on the ground state S_0-surface; the mulliken calculated charges on C_3 and C_4; and the value of the central dihedral angle. The time-step was 0.1 fs, so the whole graph took 1 400 single calculations to complete.

The simulation was started with a geometry of the cation twisted by 10° from planarity in order to generate out-of-plane forces. The S_1 energy is seen (Fig. 6) to oscillate after an initial drop; the near constancy of the dihedral angle indicates that for the first 40 fs the molecule stays essentially planar. Closer inspection of the molecular geometry reveals that during this time span strong in-plane vibrations occur which completely invert the pattern of single and double bonds in the molecule. Then, however, the S_1-energy starts to drop reaching a minimum after 68 fs; at the same time the S_0- surface rises and approaches the S_1-minimum to within 5.45 kJ/mol or 56.6meV. The dihedral angle at this point is 72°, which means that the two halves of the molecule are close to perpendicular. Calculations with a larger basis set and careful geometry optimization have revealed the presence of a Conical Intersection (CI) at this point. This is a geometry in which the S_1- and the S_0-PES cross, providing a "funnel" for the molecule to decay very efficiently from the electronically excited state to the ground state[13], probably a prerequisite for the extremely fast kinetics of the visual process. Near the CI there is a sudden charge polarization which moves electron density from the neutral half of the molecule to the positively charged one, as can be seen from the calculated charges on C_3 and C_4. These results concur closely with what Robb has observed.[10] Details of our calculation, as well as the results of more realistic methyl derivatives, are being published elsewhere.[14]

4 Conclusion

Ab initio molecular dynamics simulations present a viable alternative to the traditional methods of localizing stationary points on potential energy surfaces. The trajectories which are calculated by a combination of quantum-mechanical and classical models are unbiased and follow only the forces generated for the momentary geometry. Local maxima and transition states may be overcome with an excess of kinetic energy; reducing the kinetic energy will lead directly towards local or even global minima. The calculations show that excited state trajectories are entirely within the means of today's computational resources.

Acknowledgements

This work has been supported by the Graduate College *Structure and Dynamics of Heterogeneous Systems* at the Gerhard-Mercator-University of Duisburg. Calculations were performed at the RZ Duisburg and the RRZ Köln.

References

1. M. Parinello, *Solid State Commun.* **102**, 107 (1997).
2. D.A. Gibson, I.V. Ionova, and E.A. Carter, *Chem. Phys. Lett.* **240**, 261 (1995); Z. Liu, L.E. Carter, and E.A. Carter, *J. Phys. Chem.* **99**, 4355

(1995); D.A. Gibson and E.A. Carter, *Chem. Phys. Lett.* **271**, 266 (1997).
3. F. Terstegen, E.A. Carter, and V. Buss, *Int. J. Quant. Chem.* , in print.
4. M.J. Frisch, G.W. Trucks, H.B. Schlegel, P.M.W. Gill, B.G. Johnson, M.A. Robb, J.R. Cheeseman, T. Keith, G.A. Petersson, A. Montgomery, K. Raghavachari, M.A. Al-Laham, V.G. Zakrzewski, J.V. Ortiz, J.B. Foresman, C.Y. Peng, P.Y. Ayala, W. Chen, M.W. Wong, J.L. Andres, E.S. Replogle, R. Gomperts, R.L. Martin, D.J. Fox, J.S. Binkley, D.J. Defrees, J. Baker, J.J.P. Stewart, M. Head-Gordon, C. Gonzalez, J.A. Pople, *GAUSSIAN 94. Rev B.1 3.3* (Gaussian Inc, Pittsburgh, PA, 1995).
5. W.C. Swope, H.C. Andersen, P.H. Berends, and K.R. Wilson, *J. Chem. Phys.* **76**, 637 (1982).
6. H.C. Andersen, *J. Comput. Phys.* **52**, 24 (1983).
7. R. Allman, H.C.T. Debaerdemaeker *Cryst. Struct. Commun.* **5**, 211 (1976).
8. W. Grahn, H.H. Johannes, J. Rheinheimer, B. Knierim, and E.U. Würthwein, *Liebigs. Ann.* , 1003 (1995).
9. S.F. Falzewski, F. Terstegen, and V. Buss, *Chem. Phys.* **243**, 179 (1999).
10. M. Garavelli, P. Celani, F. Bernardi, M. A. Robb, and M. Olivucci, *J. Am. Chem. Soc.* **119**, 6891 (1997).
11. T. Vreven, F. Bernardi, M. Garavelli, M. Olivucci, M.A. Robb, and H.B. Schlegel, *J. Am. Chem. Soc.* **119**, 12687 (1997).
12. R.W. Schoenlein, L.A. Peteanu, R.A. Mathies, and C.V. Shank, *Science.* **254**, 412 (1991).
13. E. Teller, *Isr. J. Chem.* **7**, 227 (1969).
14. S.F. Falzewski, O. Weingart, and V. Buss, *Chem. Phys. Lett.* submitted.

RHODOPSIN - STRUCTURE, SPECTROSCOPY AND DYNAMICS OF A CHROMOPHORE IN A TRULY HETEROGENEOUS ENVIRONMENT

F. TERSTEGEN, K. KOLSTER, S. FALZEWSKI, AND V. BUSS
Institut für Physikalische und Theoretische Chemie, Gerhard-Mercator-Universität Duisburg, D 47048 Duisburg, Germany
E-mail: theobuss@uni-duisburg.de

We report *ab initio* calculations of the ground and several excited states of protonated 11-*cis*-and all-*trans*-retinal Schiff base, the chromophores of the visual pigment rhodopsin and its photoproduct, bathorhodopsin, respectively. A supramolecule in which a formate anion and water have been attached to the Schiff base is described as a first attempt to model part of the complex environment of the chromophore. For highly coiled geometries of the retinals CD spectral data have been calculated. The comparison with experimental data suggests that a revision of the presently accepted structures may be necessary.

1 Introduction

Protonated 11-*cis*-retinal Schiff base is the chromophore of rhodopsin, the visual pigment which is responsible for light/dark vision in the retina of the vertebrate eye (Fig. 1, left). As the first step in a series of events that finally leads to the perception of light in the brain, the molecule is isomerized by photons of suitable wave length to the all-*trans*-isomer (Fig. 1, right), a reaction which occurs with a very high quantum yield of 65 %[1] and is complete within 200 femtoseconds.[2]

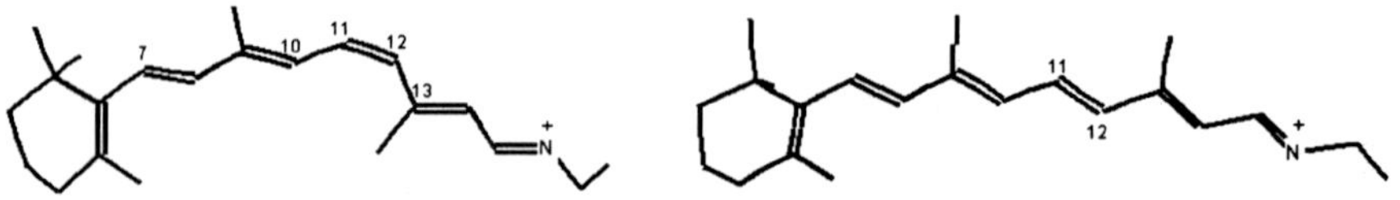

Figure 1. Configuration of protonated retinal Schiff base before (left) and after photochemical isomerization (right) of the C11-C12 double bond.

The stereospecifity of this photoreaction is remarkable: Of all double bonds in the molecule which can theoretically undergo *cis/trans* isomerization only the 11-12-bond is affected by the action of light. This, and the high efficiency of the isomerization reaction, can probably only be understood if proper account is taken of the effects of the environment on the structure and dynamics of the chromophore.

In rhodopsin the retinal chromophore is embedded in a pocket formed by seven so-called transmembrane helices which form the main part of the visual pigment. (Fig. 2). Chemically, rhodopsin is a protein, a macromolecule consisting of about 350 amino acids linearly connected into a molecular chain. While the primary structure of rhodopsin, i.e. the sequence of the constituent amino acids, is known for various animal species, knowledge about the three-dimensional structure of the protein, i.e. the atomic arrangement in space, is limited. Models of rhodopsin have been developed based on chemical experience and with reference to the structure of closely related proteins such as bacteriorhodopsin, a retinal-based proton pump of certain bacteria. In addition, experimental evidence, in particular spectroscopic studies and studies of mutant species, has increased our understanding how the interplay between the retinal molecule and its protein surrounding brings about the unique properties of the pigment.

Since the photoisomerization of the chromophore involves electronically excited states, electronic structure theory, that means quantum theory, has to be employed rather than force-field theory which describes the molecule as an ensemble of atoms interacting through mechanical forces. However, subjecting the whole molecule with thousands of heavy atoms to an ab initio treatment is out of the question for many years to come.

We have shown that the isolated chromophore can be calculated to any degree of accuracy using either Hartree-Fock or Density Functional Theory.[3] By adding what will hopefully turn out to be functionally important pieces of the environment to the chromophore at a level that is computationally feasible now, and proceeding from there with increasing complexity we hope to arrive eventually at a sufficiently sophisticated model for the protein binding pocket of rhodopsin. In the following we present, as a first step in this direction, a minimal model of rhodopsin consisting of the chromophore, water and the carboxylate group of one amino acid. We will also discuss evidence for two highly folded chromophore models of rhodopsin and bathorhodopsin based on Circular Dichroism (CD) spectroscopy. The calculated CD spectra for these systems may require the revision of the presently accepted structures.

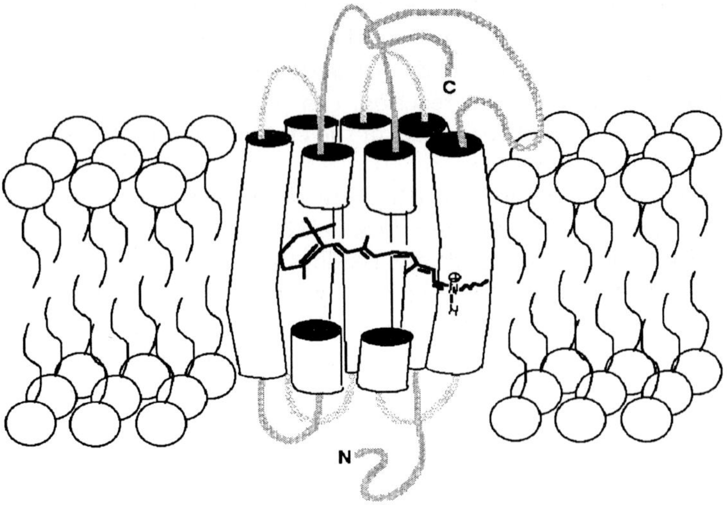

Figure 2. Cartoon of rhodopsin, with the retinal chromophore inside, embedded in a lipid bilayer membrane. The grey thread is the aminoacid chain, with the so-called N- (for nitrogen) and C- (for carbon) terminus. The chain runs through the whole protein molecule (in effect it *is* the molecule) and is coiled in seven regions into helical structures, the so-called transmembrane helices.

2 A minimal rhodopsin complex

According to density functional calculations employing a high quality atom centered basis set auf Gaussian functions (B3LYP/6-31G**) the protonated 11-*cis* retinal Schiff base chromophore is planar from C6 to N^{+3} (see Fig. 1 for the numbering of the chromophore). There is a sterically congested area, from C10 to C13, which is the most likely place for a non-planar deformation; however, the gain in steric strain when the molecule is twisted about the right bonds is obviously not enough to offset the loss in conjugation energy as a consequence of decreased π-overlap.

It is generally assumed that the retinal chromophore becomes twisted when bound to the protein and buried inside the binding pocket - a result of electronic and/or steric interactions. Apart from the most conspicuous observation which is the optical acticity of rhodopsin and which will be discussed in section 3, evidence for distortion of the chromophore comes among others from solid state magic angle spinning NMR spectroscopy. By determining with this technique the distances between the methyl group at C13 and C10 and C11, respectively, it has been possible to arrive at a quantitative estimate, 44°, of the over-all twist between the two halves of the molecule.[4] There is evidence that in rhodopsin one amino acid, Glu113, is the primary counter-ion of the Schiff base (Glu stands for glutamic acid; the index indicates that it is the 113th amino acid counting from the N-terminus). In addition 13C chemical shift differences between the free and the protein bound chromophore suggest that one of the oxygens of the glutamate ion is close to C12. We have geometry-optimized a molecular complex consisting of protonated 11-*cis* retinal Schiff base, a water molecule and a negatively charged formate ion, HCOO-, with density functional theory (B3LYP/6-311G**). The resulting geometry is shown in Fig. 3.

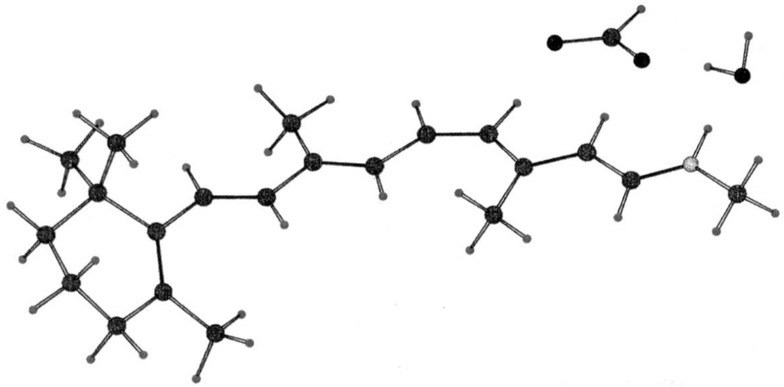

Figure 3. DFT-optimized geometry of the complex between protonated 11-*cis*-retinal Schiff base, formate ion and a water molecule. Oxygen atoms are in black.

We note close non-bonded interactions (distances 215.5 and 194.3 pm) between the formate oxygens and the hydrogens at C12 and C14, respectively. The position of the water is such that it can act as a relay for proton transport: Upon deprotonation of the Schiff base it accepts the hydrogen from N^+ and donates another one to the formate ion. To study the electronic effects of complexation, we take a closer look at the bond distances along the conjugated carbon chain of the chromophore as it changes from the free to the complexed state (Table 1). For comparison, the bond lengths of the unprotonated Schiff base are included also. Perusal of Table 1 reveals the following: Protonation of the Schiff base (last row to first row) decreases bond length alternation: All double bonds become larger, all single bonds shorter, a consequence of the

Table 1. Bond lengths along the conjugated carbon chains of 11-*cis*-retinal protonated Schiff base, its complex with water and formate ion and its deprotonated form (bond lengths in pm; first and last row entries calculated with B3LYP/6-31G**, the second row with B3LYP/6-311G**). (a) 11-*cis*-retinal, protonated SB, (b) 11-*cis*-retinal, protonated SB, complexed with water and formate, (c) unprotonated 11-*cis*-retinal SB

	C9=C10	C10–C11	C11=C12	C12–C13	C13=C14	C14–C15	C15–N
(a)	139.7	140.6	139.6	141.3	140.9	138.7	133.3
(b)	137.2	142.9	137.2	143.8	138.5	140.5	131.3
(c)	136.9	143.9	136.6	145.6	136.6	145.0	128.3

chromophore changing from a polyene-type systems, with strongly alternating double bonds, to a cyanine type system, with more equalized bond lengths. This is a well-documented fact: Protonation causes, among others, the large shift of the retinal Schiff base absorption maximum to longer wave lengths. Complexation of the protonated form with water and the formate gegenion (first row to second row) reverses this effect, though not complete. Throughout the carbon chain, bond lengths are between those of the protonated and the unprotonated Schiff base. Obviously the charge on the chromophore becomes localized due to the interaction with the negative anion and subsequent loss of conjugation.

Complexation at this level does not induce non-planarity into the chromophore, which is not surprising considering that both the formate ion and the water molecule position themselves in the plane of the chromophore.

The corresponding complex with the chromophore in the all-*trans* configuration is shown in Fig. 4. Similar to the 11-*cis* case just discussed, conjugation is decreased relative to the free chromophore, a fact which should be of significant consequences for the energies of the excited state. The stabilization of the all-trans isomer relative to 11-*cis* in the ground state increases upon complexation from 5.25 to 6.68 kcal mol^{-1}.

3 Structure and CD-spectroscopy

One of the most direct manifestations of the non-planar geometry of the protein-bound chromophore is the Circular Dichroism, or CD, observed in the absorption spectrum of rhodopsin: In a CD-spectrum the difference in absorptivity of an optically active medium for left and right circular polarized light, expressed as the so-called ellipticity θ, is plotted against the wave length. In the spectrum of rhodopsin, there is a positive CD-band with maximum at 500 nm and another stronger one at 320nm (Fig. 5). After irradiating

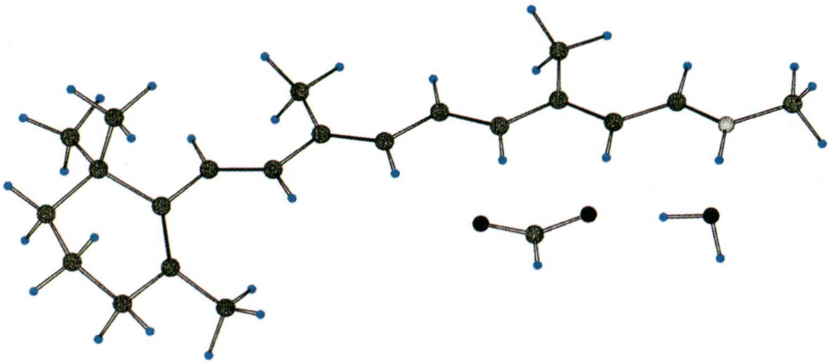

Figure 4. DFT- optimized geometry of the complex of protonated all-*trans* retinal Schiff base, formate ion and a water molecule.

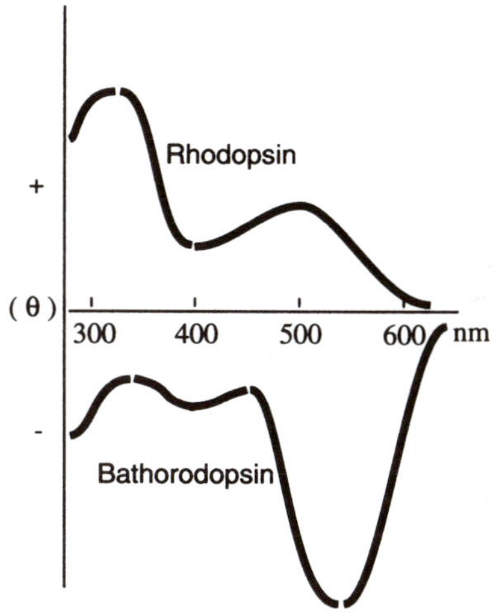

Figure 5. CD-spectra of rhodopsin and of first photoproduct (after Ref. [5]).

with blue light at 437 nm at 77 K (to slow down consecutive reactions and stabilize the intermediate) the resulting photosteady state mixture has been analyzed to give the CD spectrum of bathorhodopsin, which is rhodopsin with the retinal chromophor in the all-*trans*-geometry. This spectrum, also shown in Fig. 5, displays a prominent negative band at 520 nm followed by smaller bands towards shorter wave lengths.

The sign and the magnitude of the different CD bands allow conclusions to be drawn pertaining to the three-dimensional geometry of the chromophore on the condition that the excited states causing the bands can be identified. For the retinal protonated Schiff base in rhodopsin which represents a perturbed cyanine type chromophore this condition is easily met: the intense long wave length absorption at 500 nm which is responsible for the color of the protein corresponds to the strongly allowed $\pi\pi^*$-absorption of the conjugated double bond system.

Once the electronic states involved in the CD have been identified it remains to choose a geometry on the Born-Oppenheimer potential energy surface from which the Franck-Condon- or vertical excitation takes place. Based on this geometry electronic wave functions and transition moments are calculated from which theoretical expressions for the rotatory strengths, which is the area enclosed by a CD band, can be obtained.

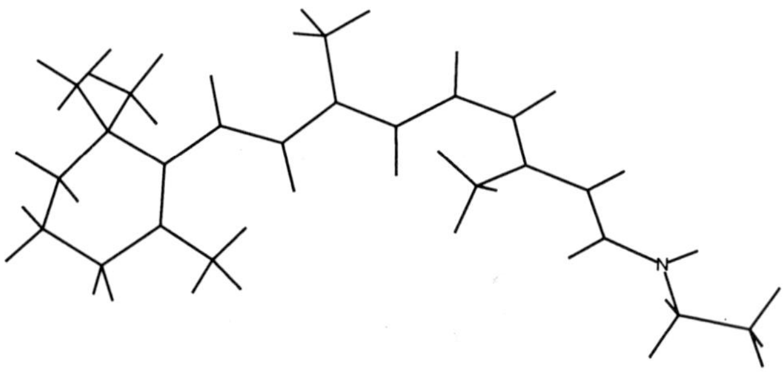

Figure 6. 11-*cis*-Retinal protonated Schiff base according to de Groot *et al.*[6]

For the geometries, we used the results of H. de Groot and co-workers who performed Car-Parrinello type molecular-dynamics calculations on the rhodopsin chromophore[6] and have obtained, as a consequence of boundary conditions imposed on the chromophores, highly coiled structures (Figs. 6 and 7). Based on these geometries we have performed excited state calculations employing the MOLCAS-4[7] and GAUSSIAN 94 program packages.[8] A 3-21G basis set was employed for the wave functions. The (restricted) active space for the CI part (RAS2) consisted of 10 electrons in 10 orbitals.

The calculated energies and oscillator strengths indicate that the excited states we have chosen correspond to the bands observed in the spectrum.

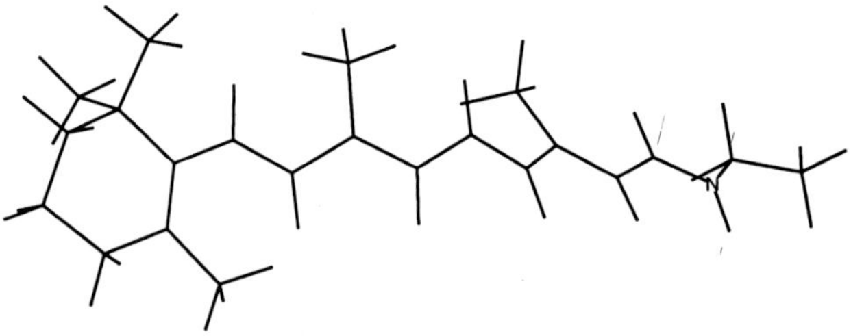

Figure 7. all-*trans*-Retinal protonated Schiff base according to de Groot *et al.*[6]

Table 2. Calculated relative energies, wave lengths λ, oscillator strengths f and rotary strengths R (in Debye-Bohr magneton) of the two lowest excited states of 11-*cis*- and all-*trans*-retinal protonated Schiff base.

	States	E (a.u.)	l (nm)	f	R ($D\mu_B$)
11-*cis*	2A–1A	0.0841	543	3.1	-0.52
	3A–1A	0.1204	379	0.4	-0.46
all-*trans*	2A–1A	0.0875	522	3.1	2.43
	3A–1A	0.1.71	426	0.1	0.43

In particular, the high oscillator strengths of the 2A and the low oscillator strengths of the 3A states which moreover decreases when the chromophore configuration changes from 11-*cis* to all-*trans* attest to this. The calculated rotatory strengths, -0.52 and +2.4 DmB, for the 2A-states of 11-*cis*- and all-*trans*-retinal protonated Schiff base, are to be compared with the experimental CD-spectra (Fig. 5), in which a positive and a negative CD band are observed for the long wave length absorptions of the corresponding species, rhodopsin and bathorhodopsin, respectively. The exact value of R depends somewhat on the method of preparation[9] but is in the range of 0.5 DmB, in good agreement with the results of our computation, except for the sign.

We are not ready at this time to include the calculated results for the next higher 3A states. The theoretical description of this state which is composed of at least two different configurations is more difficult than the lowest excited state which involves mainly the excitation from the highest occupied to the lowest empty molecular orbital. We note, however, that all calculated rotatory strengths have opposite signs compared to the experimental ones. These data suggest that the absolute configurations of the models on which the calculations of de Groot et al. are based (and which are used throughout the relevant literature) may have to berevised.

4 Conclusion

The theoretical description of the interaction between the retinal chromophore and its environment inside the rhodopsin protein binding pocket has been termed a *grand challenge* for quantum chemistry.[10] We have presented two aspects of our approach: modelling of a minimal complex consisting of the protonated chromophore, a water molecule at the (formally) positive nitrogen and Glu113 and calculation of the rotatory strengths expected for different geometries of the chromophore. The complex is planar, both in the 11-*cis* and in the all-*trans* configuration, and the delocalization of the chromophore π-system is intermediate between the protonated and the deprotonated form. The calculated CD data imply that the absolute configuration of the twisted chromophore inside the protein pocket may be opposite to the one generally assumed. This may be important for attempts to chemically model the binding site.

Acknowledgement

This work has been supported in part by the Graduate College *Structure and Dynamics of Heterogeneous Systems* at the Gerhard-Mercator-University of Duisburg. Calculations were performed at the RZ Duisburg and the RRZ Köln.

References

1. R.W. Schoenlein, L.A. Petenau, R.A. Matthies, and C.V. Shank, *Science* **254**, 412 (1991).
2. F. Boucher and R. Leblanc, *Photochem. Photobiol.* **41**, 459 (1985).
3. F. Terstegen and V. Buß, *Chem. Lett.* , 449 (1996); F. Terstegen and V. Buß, *J. Mol. Struct. (Theochem)* **369**, 53 (1996); F. Terstegen and V. Buß, *J. Mol. Struct. (Theochem)* **430**, 209 (1998).
4. X. Feng, P.J.E. Verdegem, Y.K. Lee, D. Sandström, M. Eden, P. Bovee-Geurts, W.J. de Grip, J. Lugtenburg, H.J.M. de Groot, and M.H. Levitt, *J. Am. Chem. Soc.* **119**, 6853 (1997).
5. T. Yoshizawa and S. Horiuchi in *Biochemistry and Physiology of Visual Pigments*, ed. H. Langer (Springer, Heidelberg, 1973), p. 69.
6. F. Buda, H.J.M. de Groot, and A. Bifone, *Phys. Rev. Lett.* **77**, 4474 (1996).

7. K. Andersson, M.R.A. Blomberg, M.P. Fülscher, G. Karlström, R. Lindh, P.-Å. Malmqvist, P. Neogrády, J. Olsen, B.O. Roos, A.J. Sadlej, M. Schütz, L. Seijo, L. Serrano-Andrés, P.E.M. Siegbahn, P.-O. Widmark, *MOLCAS-4*, Lund University, Sweden.
8. M.J. Frisch, G.W. Trucks, H.B. Schlegel, P.M.W. Gill, B.G. Johnson, M.A. Robb, J.R. Cheeseman, T. Keith, G.A. Petersson, A. Montgomery, K. Raghavachari, M.A. Al-Laham, V.G. Zakrzewski, J.V. Ortiz, J.B. Foresman, C.Y. Peng, P.Y. Ayala, W. Chen, M.W. Wong, J.L. Andres, E.S. Replogle, R. Gomperts, R.L. Martin, D.J. Fox, J.S. Binkley, D.J. Defrees, J. Baker, J.J.P. Stewart, M. Head-Gordon, C. Gonzalez, J.A. Pople, *GAUSSIAN 94. Rev E.1* (Gaussian Inc, Pittsburgh, PA, 1995).
9. F. Crescitelli, W.F.H.M. Mommaerts, and T.I. Shaw, *Proc. Nat. Acad. Sci. USA* **56**, 1729 (1966); A.S. Waggoner and L. Stryer, *Biochemistry* **10**, 3250 (1971).
10. M. Garavelli, T. Vreven, P. Celani, F. Bernardi, M.A. Robb, and M. Olivucci, *J. Am. Chem. Soc.* **120**, 1285 (1998).

AB INITIO MOLECULAR-DYNAMICS SIMULATIONS OF ADSORPTION OF DYE MOLECULES AT SURFACES

M. SUGIHARA, H. MEYER*, AND P. ENTEL
Theoretische Tieftemperaturphysik,
Universität Duisburg, 47048 Duisburg, Germany
E-mail: minoru@thp.uni-duisburg.de

**Max-Planck Institut für Polymerforschung, 55128 Mainz, Germany*
E-mail: hmeyer@mpip-mainz.mpg.de

Y. SAKAMOTO
Department of Physics, Osaka University, Toyonaka, Osaka 560-0043, Japan
E-mail: sakamoto@phys.sci.osaka-u.ac.jp

J. HAFNER
Institut für Materialphysik, Universtität Wien, 1090 Wien, Austria
E-mail:Juergen.Hafner@univie.ac.at

V. BUSS
Institut für Physikalische und Theoretische Chemie,
Universität Duisburg, 47048 Duisburg, Germany
E-mail:theobuss@uni-duisburg.de

We present results of *ab initio* total energy calculations and molecular-dynamics simulations of dye molecules on the NaCl(100) surface. The investigations concentrate on the flat dye molecules trimethin, $[C_{19}H_{17}N_2O_2]^+$, which form sandwich-like structures if closely packed, and the cyanine molecule monomethin, $[C_{21}H_{23}N_2]^+$, which shows a typical stereochemical deformation due to two repulsive methyl groups. The molecular-dynamics simulations are able to reproduce the experimentally observed configurations of the charged dye molecules on surfaces.

1 Introduction

Dye molecules are of particular interest, since they are a subject of intense inter-disciplinary research involving all branches of chemistry, in particular physical and theoretical chemistry, and also physics and biology. The fundamental problem is the interaction of dye molecules with light and the subsequent change of conformation and its impact on the surrounding, which is usually protein-like. In this short note we investigate whether conformational changes of the dye molecules will be altered due to the presence of the surface of an ionic crystal. In this case conformational changes are caused by the

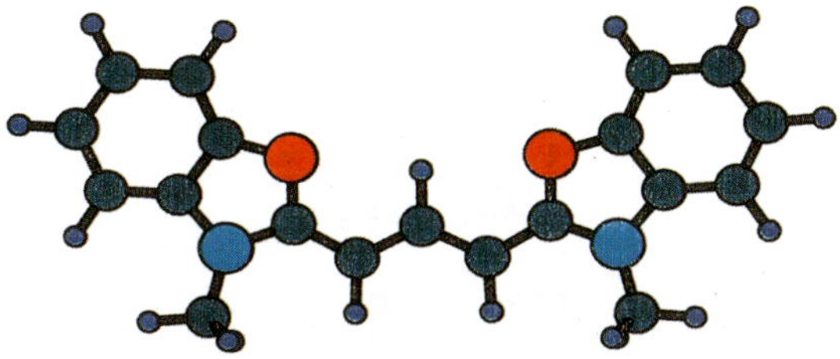

Figure 1. Conformation of trimethin.

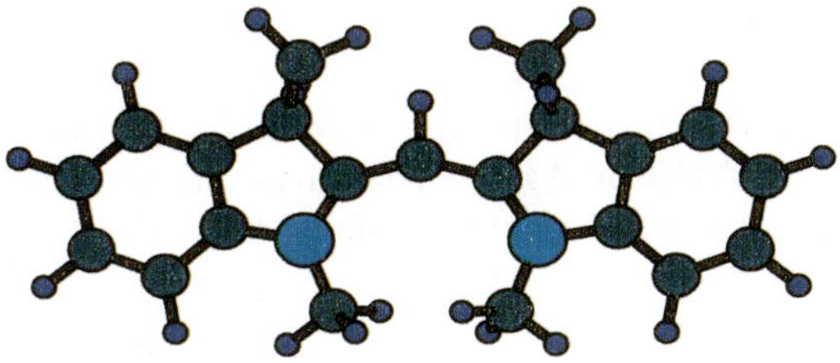

Figure 2. Conformation of monomethin.

repulsive methyl groups in the case of monomethin and not by the absorption of light leading to a cis-trans transformation.

We have used the Vienna *ab initio* simulation package (VASP) which is a program designed for molecular-dynamics (MD) simulations based on Density Functional Theory. It uses a plane wave basis set and Vanderbilt's ultrasoft pseudopotentials. All calculations are performed for a periodic supercell and the details are described elsewhere.[1]

Prior to the investigations of the dye molecules we have checked convergence properties and quality of the pseudopotentials by performing total energy calculations as a function of the bond angle and O-H bond length for a simple water molecule and for water in contact with NaCl surfaces. The results are satisfactory in good agreement with the experimental values.[2]

2 Molecular-dynamics simulations of dye molecules

Adsorption of cyanine dye molecules on surfaces have been studied experimentally for more than fifty years[3] because of the fundamental interest in the adsorption process as well as the interest in theoretical applications of their

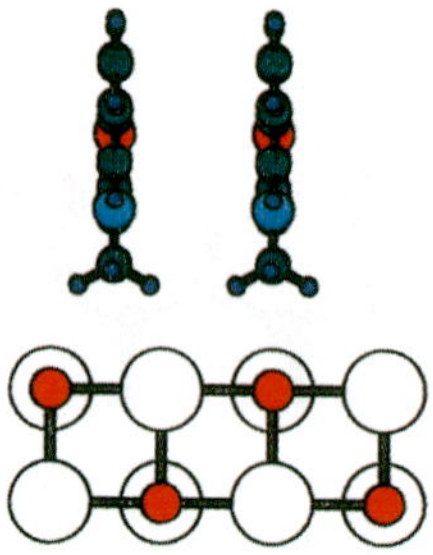

Figure 3. Initial configuration of two trimethin molecules on the NaCl(100) surface.

optical properties. With respect to photochemistry the study of the behavior of dye molecules on AgCl or AgBr surfaces would be of interest. However, because of the many valence electrons of Ag this calculation would require enormous computer capacities. Therefore, we have investigated in a preliminary study the behavior of dye molecules on the simple NaCl(100) surface, which has a quite similar lattice constant as the AgCl lattice.

In this work we discuss two kinds of dye molecules. One is the flat dye molecule trimethin, $[C_{19}H_{17}N_2O_2]^+$, which is shown in Fig. 1 (green, light blue, red and blue circles are carbon, nitrogen, oxygen and hydrogen atoms, respectively). The full name is 3,3′-dimethyloxacarbocyanine. The other dye molecule is the cyanine molecule monomethin, $[C_{21}H_{23}N_2]^+$, for which the full name is bis-(1,3,3-trimethylindolin-2′-yl)-monomethinium. This dye molecule has a twisted ground state due to the steric repulsion between the two methyl groups at the nitrogen atoms. Fig. 2 shows a metastable planar conformation of monomethin (same colors as in Fig. 1 are used for the atoms). In solution these dye molecules are positively charged and are accompanied by a negative counter ion, for example I^- in the case of monomethin. In our calculations we assume for simplicity a uniform background charge of $-e$ per unit cell to achieve charge neutrality. We have checked for the case of free dye molecules whether the replacement of the counter ion by the uniform back-ground charge has a serious consequence for the steric arrangement of the atoms. At least for monomethin we observe practically no difference in conformation.

2.1 $[C_{19}H_{17}N_2O_2]^+$ *(trimethin) on the NaCl surface*

We discuss now results of total energy calculations for the flat dye molecule trimethin. This type of dye molecule has the tendency to form sandwich-like structures if many molecules are placed on the surfaces. Theoretical

Table 1. Change of energy with the distance between two trimethin molecules.

distance (Å)	3.6	3.7	3.8	3.9	4.0	4.1
energy (eV)	0.055	0.016	0.001	0.000	0.003	0.023

Table 2. Change of energy while sliding one of the trimethin molecules.

slided (Å)	0.0	1.0	2.0	3.0	3.5
energy (eV)	0.000	-0.079	-0.121	-0.160	-0.320

slided (Å)	4.0	5.0	6.0	7.0	7.56
energy (eV)	-0.124	-0.064	-0.060	-0.068	-0.088

investigations[4] have led to the proposal of different arrangements of the dye molecules. There are two possibilities for a perpendicular arrangement of the dye molecules on surfaces. One possibility is to put the methyl branches near the surface and the other is to put them far from the surface. Our calculations show that the first configuration has a lower energy. In addition it is energetically favorable to let the methyl group place itself on top of the Cl ion. The size of the supercell is $16.38 \times 10.92 \times 10.92$ Å^3. In the calculations we first place two dye molecules on top of the NaCl surface as shown in Fig. 3 and gradually change the distance between the two dye molecules, while the distance between the dye molecules and the surface is kept fixed at 2.5 Å. Table 1 shows that the optimal distance of the two dye molecules is about 3.9 Å. Keeping this distance, we move one of them. The associated variation of the energy is shown in Table 2. The configuration which has minimal energy, is a 3.5Å- slided structure as shown in Figs. 3 and 4. This configuration corresponds to a situation in which the positions of the nitrogen atoms (light blue) are as remote as possible from each other. For this type of dye molecules, the positive charge is mainly on the nitrogen atoms, and the structure of minimum energy is consistent with this. The total energy also depends on the inclination of the dye molecules, but the perpendicular arrangement with respect to the surface has the minimum energy. These results support the following adsorption processes: The flat dye molecules which hit the surface erect themselves and make a perpendicular aggregation on the surface.[5] It would be interesting to confirm the stability of this sandwich-like structure by MD calculations for more than two dye molecules but so far we have only simulated two molecules per supercell.

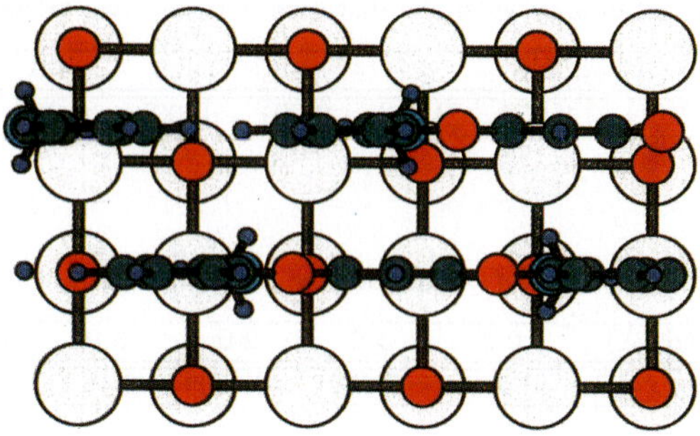

Figure 4. Configuration of two trimethin molecules per supercell on NaCl(100) which has minimal energy (top view).

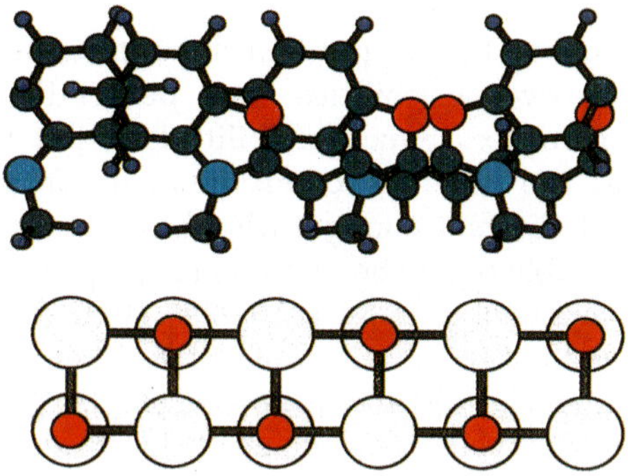

Figure 5. The same configuration as in Fig.4 (side view).

2.2 $[C_{21}H_{23}N_2]^+$ *(monomethin) on the NaCl(100) surface*

The cyanine molecule monomethin has a twisted ground state configuration due to the steric repulsion of the methyl groups at the nitrogen atoms. The conformation shown in Fig. 2 has been obtained by starting with a planar configuration and using the conjugate gradient method[1] for minimizing the total energy. The optimized structure has a central angle of about 150° between the two indolin groups which is much larger than the actual angle of 130° and almost planar with a dihedral angle of 1° instead of 45°.[6] The conjugate gradient method seems to be unable to yield the correct ground state. The reason may be that the system includes many atoms which must be optimized and, therefore, has many metastable structures. This planar configuration has

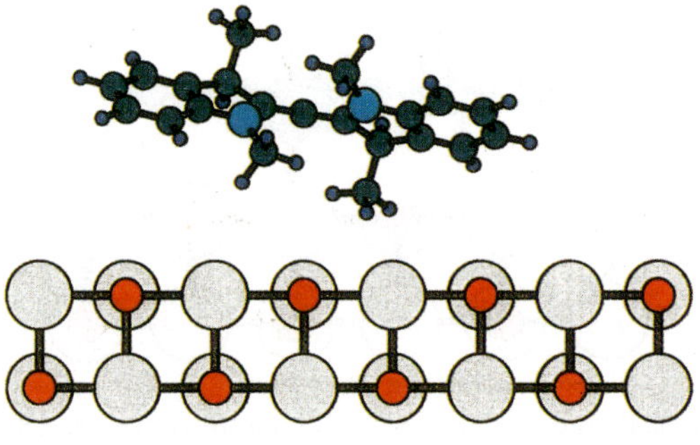

Figure 6. Initial configuration of monomethin on the NaCl(100) surface.

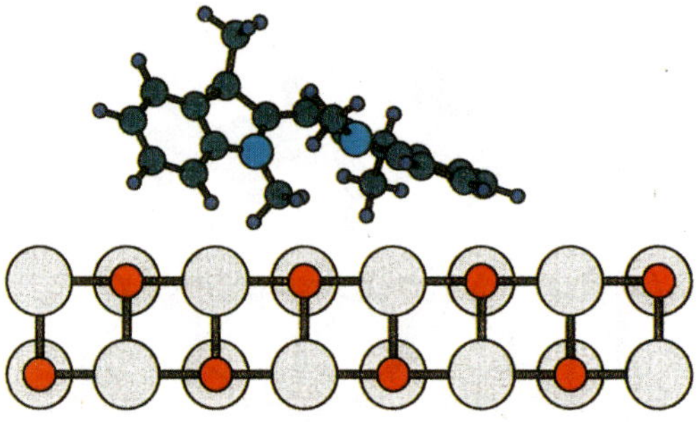

Figure 7. Configuration obtained after 600 fs of relaxation on the NaCl(100) surface.

been taken as the starting configuration of the MD runs. The MD simulations have been done for a supercell of 10 × 15 × 14 Å^3. All simulations have been performed for a microcanonical ensemble and 0.5 fs time steps. After several hundred steps, the two indolin groups of the molecule start to move out of the common plane and the molecule starts to vibrate around this conformation. Then we set the kinetic energy equal to zero and restart the MD simulation. The total energy decreases by 0.45 eV during a time of about 600 fs relaxation. During the MD runs the central angle decreases and the central dihedral angle increases. The final optimized structure has a central angle of about 130° and a dihedral angle of about 50°, which is in good agreement with the optimized structure which had been obtained by using the restricted Hartree-Fock method (Gaussian 95).[7] We have now place this optimized molecule on the 8 × 4 × 2 NaCl(100) surface and done further MD simulations. The size of the supercell is now 21.2 × 10.8 × 16.2 Å^3. The initial configuration of

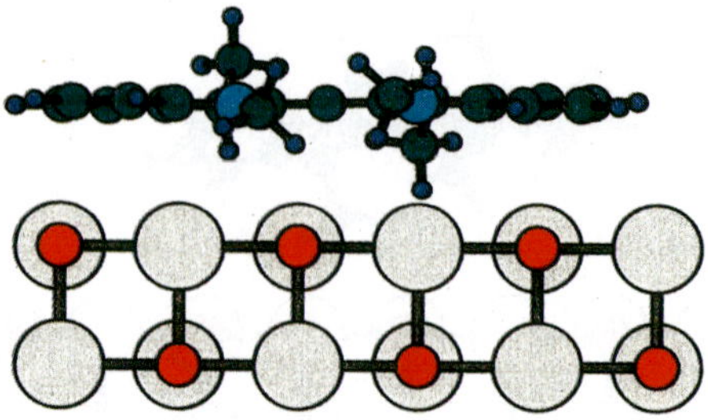

Figure 8. A second initial configuration of monomethin on the NaCl(100) surface (planar arrangement of the two indolin groups).

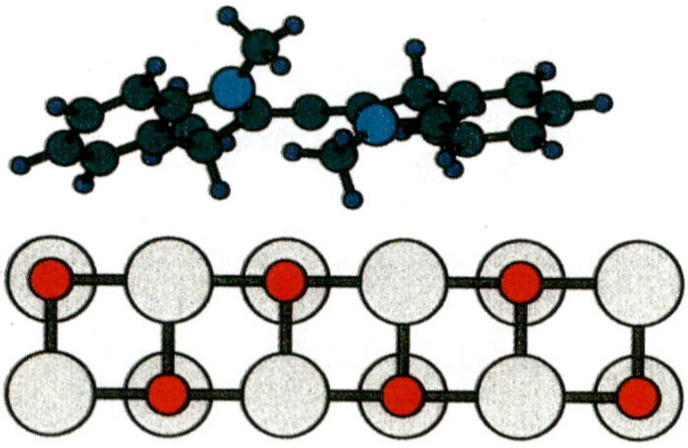

Figure 9. Configuration obtained after 1000 fs of relaxation of the NaCl(100) surface.

the molecule is shown in Fig. 6. During a relaxation time of about 400 fs the total energy decreases by 0.4 eV when the molecule approaches the surface with vibrating methyl groups. The configuration obtained after 400 fs is shown in Fig. 7. We have also done MD simulations for monomethin with a planar structure. In order to reduce the CPU time, a smaller surface of $6 \times 4 \times 2$ and a supercell of $16.2 \times 10.8 \times 16.2$ Å^3 have been chosen. The molecule approaches the surface with vibrating methyl groups as observed in the previous simulation. After 1000 fs of relaxation, the energy gain is about 0.4 eV compared to the initial flat structure of the molecule on the surface. The adsorption process in this simulation is shown in Figs. 8 and 9. Both simulations have shown that monomethin has a twisted ground structure,[6] which is consistent with the experimentally observed structure.[8]

3 Conclusion

We have placed two molecules of trimethin per supercell on a NaCl surface and done total energy calculations for different packing structures. Recently the aggregation process of dye molecules on a AgBr surface has been observed by using atomic force microscope (AFM).[9] According to the authors, the dye molecules adsorb on the AgBr surface and form a striped pattern with an epitaxial relation to the AgBr lattice. The actual packing structure of these dye molecules on the AgBr surface is still unclear. Our total energy calculation proposes an energetically favorable packed structure of trimethin at the surface. The other molecule that we have simulated is monomethin, which has a twisted structure. We have performed two MD simulations for one monomethin molecule on the NaCl(100) surface. The simulations show that the initially flat molecule changes its form, which is in good agreement with the experimentally observed structure. For simplicity, the calculations have been done by assuming a uniform background charge instead of realistic counter ions. The role of counter ions during the adsorption process is not yet clear[10] and it is left for further investigations.

4 Acknowledgements

This work has been supported by the Graduate College *Structure and Dynamics of Heterogeneours Systems* at the Gerhard-Mercator-University of Duisburg.

References

1. G. Kresse and J. Furthmüller, *Phys. Rev.* B **54**, 11, 169 (1996).
2. H. Meyer, P. Entel, and J. Hafner, submitted to *Surf. Sci.* (1999).
3. E. E. Jelley, *Nature* **138**, 1009 (1936).
4. V. Czikkely, H. D. Försterling, and H. Kuhn, *Chem. Phys. Lett.* **6**, 11 (1970).
5. O. Wörz and G. Scheibe, *Z. Naturforschg.* **381**, 246 (1969).
6. H. Meyer, Y. Sakamoto, and P. Entel, unpublished work.
7. V. Buss, S. Falzewski, and K. Kolster, *J. Org. Chem.* **64**, 1071 (1999).
8. R. Allmann and T. Debaerdemaeker, *Cryst. Struct. Comm.* **5**, 211 (1976).
9. H. Saijo and M. Siojiri, *J. Crystal Growth*, **166**, 930 (1996).
10. H. Saijo, T. Kitamura, and H. Ohtani, *Surf. Sci.* **177**, 431 (1986).

ELEMENTARY DIFFUSION PROCESSES IN AL-CU-ZN ALLOYS: AN AB INITIO STUDY

S. GRABOWSKI AND P. ENTEL
Theoretische Tieftemperaturphysik, Gerhard-Mercator-Universität Duisburg, 47048 Duisburg, Germany
E-mail: stefan@thp.uni-duisburg.de

The vacancy formation and migration energy of Al as well as the migration energies of Cu and Zn impurities were calculated at $T = 0\,\mathrm{K}$ using the Local-Density Approximation based on ultrasoft pseudopotentials. Diffusion processes were studied in supercells with 32 and 108 lattice sites. Furthermore the effect of Cu and Zn atoms on the activation energy for diffusion jumps of Al atoms in an ordered Al-Cu-Zn system was investigated. In each case relaxation of atoms around the defect was taken into account by relaxing all atoms in the supercell.

1 Introduction

Technologically important properties of alloys such as mechanical strength and toughness, creep and corrosion resistance are essentially controlled by the presence of precipitated particles of a second phase. This commonly results from decomposition of a solid after quenching. A fundamental understanding of the mechanism and kinetics of precipitation reactions and the arising microstructures in binary and ternary alloys is, therefore, of great interest. Phase separation in quenched systems is closely connected to diffusion processes via vacant or interstitial lattice sites. The aim of this work is to gain new microscopical insights into unusual fast diffusion processes in the early stages of spinodal ordering and decomposition in alloys. The Al-Cu-Zn system is particular interesting because, depending on composition, both spinodal ordering (Cu_3Zn) on the Cu-rich side of the phase diagram as well as spinodal decomposition (Al-Zn) on the Zn-rich side[1] are observed.

Vacancy migration is the most important diffusion mechanism in metals. Atoms diffuse by exchanging places with a vacancy at a nearest-neighbor lattice site. It is reasonable to assume that fast diffusion processes are a key to understand spinodal ordering and decomposition in Al-Cu-Zn alloys. Therefore, it is of interest to calculate the energy barriers for vacancy migration in Al-Cu-Zn systems from first principles. In our approach the interaction between valence electrons and the ionic core is described by ultrasoft pseudopotentials. Results of defect calculations in Al-Cu-Zn systems based on ultrasoft pseudopotentials have not yet been published. All calculations presented in this paper were performed using the Vienna ab initio simulation package.[2] In order to check the reliability of our calculations the results are compared with experimental results and theoretical calculations published by other groups.

In order to gain experience in handling Al-Cu-Zn systems, we replaced some atoms in the fcc Al lattice by Cu or Zn atoms and studied their influence on the diffusion of Al atoms. In fcc metals atoms diffuse by jumping via vacant nearest-neighbor lattice sites. The diffusing atom has to push apart four nearest neighbors when approaching the position halfway between the original lattice site and the vacancy site (the so called saddle point configuration, see Fig. 1). The energy a migrating atom has to take up to perform a diffusion jump arises mainly from the repulsive interaction with its four nearest neighbors in the saddle point configuration and is commonly interpreted as the migration energy. We exchanged two of these atoms by Cu or Zn atoms which are smaller than Al atoms. This allows us to study if and to what extent the Cu or Zn atoms lower the activation energy barrier for Al atoms while passing through the saddle point configuration.

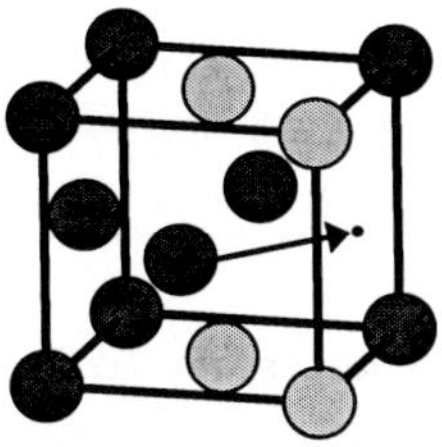

Figure 1. Nature of the constriction which inhibits motion of an adjacent atom into a vacancy in the fcc lattice. The four atoms indicated by light grey circles are the four nearest neighbors of the jumping atom in the saddle point configuration and must give way before the indicated jump can occur.

2 Method

In order to treat isolated defects and impurities in solids, one needs to consider large supercells because of the long range nature of the elastic interactions and because of loss of translational symmetry. A common approximation is to use supercells and artificially reimpose the translational symmetry by employing periodic boundary conditions. We start with the fcc lattice of Al and remove atoms in a periodic array so that each supercell of the lattice contains one vacancy. In the impurity calculations another Al atom in the supercell is substituted by an impurity atom so that each supercell contains one vacancy and one impurity. For a given atomic configuration we calculate the total energy of the supercell using a first-principles quantum mechanical method. Static relaxation of the atomic geometry was performed via a conjugate-gradient minimization of the total energy with respect to the atomic positions. After each update of the atomic coordinates the groundstate of the electronic system is self-consistently recalculated.

The total energy calculations presented in this paper are based on the self-consistent iterative solution of the Kohn-Sham equations[3] of Density Functional Theory.[4] We use the Local-Density Approximation (LDA) with the exchange correlation functional of Ceperley and Alder[5] which was parameterized by Perdew and Zunger.[6] The solutions of the Kohn-Sham equations are expanded in a plane-wave basis set and nonlocal ultrasoft pseudopotentials are used to describe the electron-ion interaction.[7,8]

Table 1. Valence configurations of the ultrasoft pseudopotentials used in our calculations.

	Al	Cu	Zn
Valence Configuration	$3s^2 3p^1$	$3d^{10} 4s^1$	$3d^{10} 4s^2$

Brillouin zone integrations necessary to calculate the band-structure energy were performed on Monkhorst-Pack[9] sets of special **k**-points. Partial occupancies of the Kohn-Sham one-electron states corresponding to a first order Methfessel-Paxton[10] function with $\sigma = 1.0\,\mathrm{eV}$, are introduced as a tool for reducing the number of **k**-points necessary to calculate an accurate band-structure energy. All these features of the computational technique are implemented in the Vienna ab initio simulation package which we used for our calculations. A more detailed description of the program can be found in the publications of its authors.[2]

2.1 Vacancy formation energy in Al

Denoting by $E(N, M, \Omega)$ the energy of the system of N atoms and M vacancies, which occupy $N + M$ lattice sites in a volume Ω, the vacancy formation energy at constant volume Ω is

$$\mathrm{E_{VF}} = E[N-1, 1, \Omega] - \frac{N-1}{N} E(N, 0, \Omega). \tag{1}$$

We took $\Omega = N\Omega_0$ with Ω_0 being the LDA equilibrium volume per atom at $T = 0\,\mathrm{K}$. The LDA lattice constant at $T = 0\,\mathrm{K}$ is $a_0 = 3.985\,\text{Å}$. The quantity of physical interest is obtained for $N \to \infty$, with $\Omega/N =$ constant. In oder to minimize the elastic interaction between vacancies, the supercell should be large. We present results calculated for supercells containing 32 and 108 lattice sites. Due to computational limitations it was not possible to perform calculations with the same supercell geometry for supercell sizes larger than 108 lattice sites. For both supercell sizes structural relaxation around the vacancy was taken into account by relaxing all atoms in the supercell.

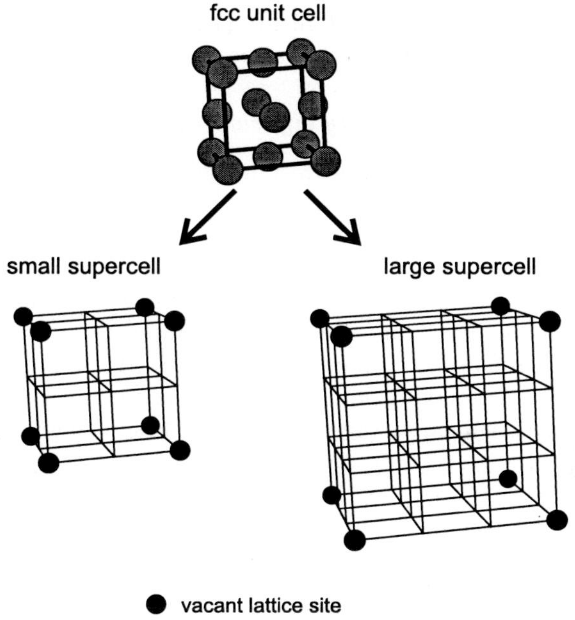

Figure 2. Geometry of the supercells used in the calculations. The small (large) supercell with 32 (108) lattice sites consists of 2^3 (3^3) cubic fcc elementary cells. Each supercell contains one vacancy so that the vacancies are distributed on a simple cubic lattice with lattice constant 2a (3a) with a=3.985 Å being the LDA lattice constant of the Al fcc lattice at $T = 0\,\mathrm{K}$.

2.2 Migration energy for self-diffusion in Al

In the framework of transition state theory[11,12] the mobility of vacancies is proportional to the Boltzmann factor $\exp(-E_{VM}/k_B T)$. The vacancy migration energy E_{VM} is given by the difference in potential energy for two relaxed static configurations, the first one with the considered atom in the saddle-point configuration between the initial and final state of the jump, and the second configuration with the atom at the initial potential minimum. For the calculation of the migration energy, the unrelaxed saddle-point configuration is obtained when a nearest-neighbor atom of the vacancy is displaced halfway in the [110] direction towards the vacancy. By fixing the displaced atom in the saddle point and relaxing the remaining atoms in the supercell the relaxed saddle-point configuration is reached. The vacancy migration energy E_{VM} is defined as the difference between the total energy of the saddle-point configuration and the energy $E[N-1,1,\Omega]$ in Eq. (1) which corresponds to the total energy of the initial minimum configuration.

2.3 Migration energy for impurity diffusion in Al

For the impurity calculations a nearest neighbor of the vacancy in the unrelaxed supercell was replaced by a Cu or Zn atom. After relaxation of all atoms in the supercell we obtained the total energy of the initial minimum configuration for the impurity jump. Starting with the unrelaxed initial minimum configuration the impurity was displaced halfway in the [110] direction towards the vacancy. By fixing the displaced impurity and relaxing the positions of all Al atoms in the supercell, we obtained the total energy of the relaxed saddle-point configuration. Just as in the case of self-diffusion the migration energy is defined as the difference of these two energies.

3 Results

3.1 Vacancy formation and migration energy of Al

A detailed investigation of the convergence with respect to the supercell size, Brillouin zone sampling and plane-wave cutoff is one of the major problems in order to obtain meaningful results for defect data from electronic structure calculations in metals. Vacancy formation and migration energies for Al were calculated for supercells with 32 and 108 lattice sites and various $n_{\mathbf{k}} \cdot n_{\mathbf{k}} \cdot n_{\mathbf{k}}$ Monkhorst-Pack sets of special **k**-points. The results are presented in Fig. 3.

From the data in Fig. 3 we conclude that the results for E_{VF} and E_{VM} are converged within 10 meV with respect to the total number of **k**-points $N_{\mathbf{k}} = n_{\mathbf{k}}^3$ for the small (large) supercell with $6 \leq n_{\mathbf{k}}$ ($4 \leq n_{\mathbf{k}}$). According to these calculations the vacancy formation and migration energy in Al is $E_{VF} = 0.7\,\mathrm{eV}$ and $E_{VM} = 0.6\,\mathrm{eV}$, respectively. The difference between the results for the small and the large supercell is smaller than 5 meV. Relaxation of the atomic coordinates reduces the calculated values of E_{VF} and E_{VM} by 0.07 eV and 0.3 eV, respectively, which indicates the importance of relaxation especially for E_{VM}.

Figure 4 shows the dependence of E_{VF} and E_{VM} on the plane-wave cutoff, E_{cut}, calculated for the small supercell. If E_{cut} is increased from 130.0 eV to 220.0 eV, the calculated value of E_{VF} and E_{VM} increases by 30 meV and 5 meV, respectively. Both E_{VF} and E_{VM} can be considered as converged values with respect to E_{cut} if $E_{cut} \geq 200$ eV.

From the data in Fig. 3 and 4 we conclude that the vacancy formation and migration energy for Al is 0.73 eV and 0.6 eV, respectively. This is in good agreement with the experimental results $E_{VF} = 0.67 \pm 0.03$ eV and $E_{VM} = 0.61 \pm 0.03$ eV.[13] Previous results of ab initio calculations published by other groups range from 0.6 eV to 0.84 eV for E_{VF} and from 0.5 eV to 0.9 eV for E_{VM} (see Table 2).

Table 2. Comparison of experimental and theoretical results for E_{VF} and E_{VM} calculated with different ab initio techniques.

Defect Energy	Expt. (eV)	Calc. (eV)	Method
E_{VF}	0.67 ± 0.03[13]	0.73	US-PP (this paper)
		0.84	FLAPW[17]
		0.66	NC-PP[18]
		0.6	KKR[19]
E_{VM}	0.61 ± 0.03[13]	0.6	US-PP (this paper)
		0.57	NC-PP[15]
		0.7 ± 0.2	NC-PP[16]

3.2 Migration energies for Cu and Zn impurities in Al

Our calculations of migration energies for Cu and Zn impurities in Al are restricted to a supercell with 32 lattice sites because of the much higher cutoff energy of the Cu and Zn pseudopotentials compared with the Al pseudopotential. For the Brillouin zone sampling we used $n_{\mathbf{k}} \cdot n_{\mathbf{k}} \cdot n_{\mathbf{k}}$ Monkhorst-Pack sets of special **k**-points with $3 \leq n_{\mathbf{k}} \leq 8$. The results are presented in Fig.

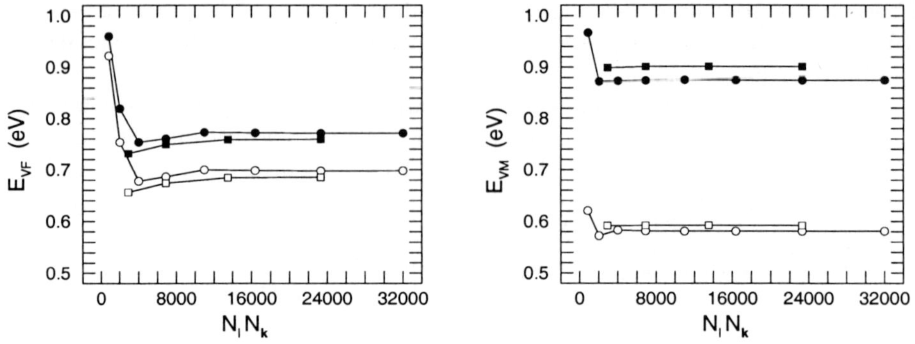

Figure 3. Vacancy formation and migration energy versus $N_l N_\mathbf{k}$. $N_\mathbf{k} = n_\mathbf{k}^3$ is the total number of **k**-points sampled in the Brillouin zone and N_l is the number of lattice sites in the supercell. Total energy calculations were performed using $n_\mathbf{k} \cdot n_\mathbf{k} \cdot n_\mathbf{k}$ Monkhorst-Pack sets of special **k**-points with $3 \leq n_\mathbf{k} \leq 10$ ($3 \leq n_\mathbf{k} \leq 6$) for the small (large) supercell. Circles (squares) represent data for the small (large) supercell. Open (filled) symbols represent data with (without) relaxation. For these calculations the plane-wave cutoff was chosen to be 130.0 eV.

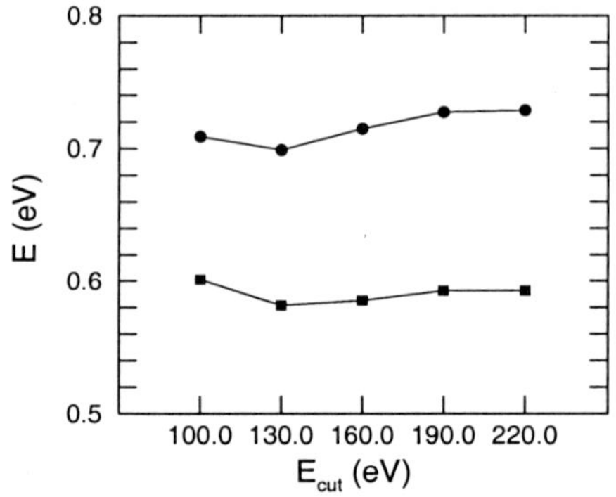

Figure 4. Vacancy formation energy (circles) and vacancy migration energy (squares) versus plane-wave cutoff E_{cut} calculated for the small supercell (32 sites) using an $8 \cdot 8 \cdot 8$ **k**-point mesh.

5. According to our calculations the relaxed value of the impurity migration energy of Cu (Zn) impurities is 0.56 eV (0.37 eV). Lattice relaxation reduces E_{VM} for Zn by 0.06 eV. For the Cu impurity E_{VM} increases by 0.1 eV if lattice relaxation is taken into account. With a Cu atom in the minimum configu-

ration the relaxation energy is approximately 0.15 eV larger than with a Zn or Al atom in the minimum configuration (see Table 3). Furthermore $\Delta d/d_0$ (see Table 3) for Cu is positive. This implies a repulsive interaction between the Cu atom and the vacancy at the nearest-neighbor lattice site.

Table 3. Relaxation energies E_{rel}^{sad}, E_{rel}^{min} with a Cu, Zn or Al atom in the saddle point or minimum configuration and the total relaxation energy $E_{rel} = E_{rel}^{sad} - E_{rel}^{min}$. Δd is the change of the distance between the Cu, Zn or Al atom in the minimum configuration and the vacancy if the lattice relaxation is taken into account. d_0 is the nearest-neighbor distance of the unrelaxed fcc lattice.

	E_{rel} (eV)	E_{rel}^{sad} (eV)	E_{rel}^{min} (eV)	$\Delta d/d_0$
Cu	+0.1	-0.12	-0.22	+2.4%
Zn	-0.06	-0.15	-0.9	-1.5%
Al	-0.23	-0.3	-0.07	-1.7%

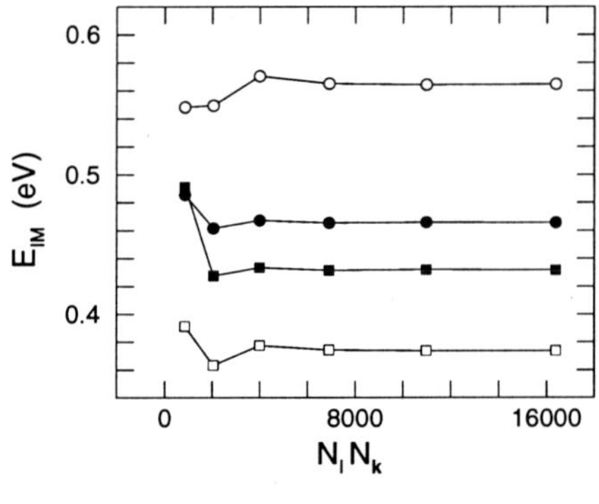

Figure 5. Migration energy of Cu and Zn impurities versus $N_l N_{\mathbf{k}}$ calculated for the small supercell containing $N_l = 32$ lattice sites. Total energy calculations were performed using $n_{\mathbf{k}} \cdot n_{\mathbf{k}} \cdot n_{\mathbf{k}}$ Monkhorst-Pack sets of special $\mathbf{k}$-points with $3 \leq n_{\mathbf{k}} \leq 8$. Circles (squares) represent data for the Cu (Zn) impurity. Open (filled) symbols represent data with (without) lattice relaxation. For these calculations the plane-wave cutoff was chosen to be 234.0 eV.

3.3 Comparison with experiment

For a cubic supercell with 32 lattice sites the calculated value for the activation energy $E_{VF} + E_{VM}$ of self-diffusion in Al is in good agreement with experimental values.[13] Our calculated values for E_{VF} and E_{VM} correspond to converged values with respect to the **k**-point density. Increasing the number of lattice sites from 32 to 108 does not change the result within 5 meV. The calculated activation energies for Cu and Zn impurities in Al are at least 0.1 eV lower than the experimental values.[14] One contribution to this error might be due to the pseudopotentials for Cu and Zn used in our calculations because it is difficult to describe perfectly the d-electrons within pseudopotential theory. Another source of error can be expected because of the use of the Local-Density Approximation. In general defect formation and migration energies in metals are temperature dependent so that systematic errors should be expected if zero-temperature theoretical results are compared with finite-temperature data obtained from experiments.

Table 4. Experimental activation enthalpies of self- and impurity diffusion in Al compared with the sum $E_{VF} + E_{VM}$ derived from the calculated values for E_{VF} and E_{VM}. In each case we took the calculated value $E_{VF} = 0.73\,\mathrm{eV}$ for bulk Al as theoretical value for E_{VF}. Experimental values for E_{VF} and E_{VM} for vacancies in Al are $0.67\,\mathrm{eV} \pm 0.03\,\mathrm{eV}$ and $0.61\,\mathrm{eV} \pm 0.03\,\mathrm{eV}$, respectively.

	Al in Al	Cu in Al	Zn in Al
Expt. (eV)	1.22-1.34	1.4	1.21-1.26
Theo. (eV)	1.33	1.29	1.1

3.4 Diffusion in Al-Cu-Zn alloys

In order to study the influence of Cu and Zn atoms on the activation energy barriers for Al diffusion, we investigated the Al-Cu-Zn system shown in Fig. 6. During the jump into a vacant nearest-neighbor site at least four atoms have to be pushed apart by the jumping atom according to Fig. 1. We simply exchanged two of these atoms by Cu or Zn atoms. Total energies were calculated starting from five different initial configurations which are explained in Fig. 7 (b). After relaxing all 31 atoms in the supercell, except for the Al atom at the intermediate position in configuration 2 and 4, we obtained the total energy of each configuration. The differences in total energy between configuration i and $i+1$ ($i = 1 \ldots 4$) are presented in Fig. 7 (a). If all atoms in the supercell were of the same type, configuration 2 and 4 would correspond to saddle point configurations. Although configuration 2 and 4 do not exactly correspond to saddle point configurations of the systems studied here, it is worth comparing energy differences between the minimum configurations 1, 3, 5 and the configurations 2, 4 with the vacancy migration energy of pure Al. For pure Al the difference $E_2 - E_1$ between the total energies E_1 and E_2 of configuration 1 and 2 is the vacancy migration energy. In the case of Al_6Cu_2, Al_6Zn_2 and pure Al, the configurations 1, 3, 5 as well as 2, 4 are equivalent because of symmetry reasons. Compared with pure Al the energy difference $E_2 - E_1$ in Al_6Cu_2 and Al_6Zn_2 is reduced by at least 0.17 eV. For Al_6Cu_2 the reduction in $E_2 - E_1$ is larger than for Al_6Zn_2. This can be explained if we assume that Cu is the smallest atom, Al the largest and the size of the Zn atom is in between the size of Cu and Al atoms. The results for Al_6CuZn are more complicated. In this case neither the configurations 1, 3, 5 nor the configurations 2, 4 are equivalent. The Al_6Cu_2 (Al_6Zn_2) system has two Cu (Zn) atoms and two Al atoms as the four nearest neighbors of the jumping Al atom in the unrelaxed configuration 2 (4). In configuration 2 (4) of the Al_6CuZn system also two Cu (Zn) atoms and two Al atoms are the nearest neighbors of the jumping Al atom. Surprisingly $E_2 - E_1$ is larger than for Al_6Cu_2 and $E_4 - E_3$ is smaller than for Al_6Zn_2.

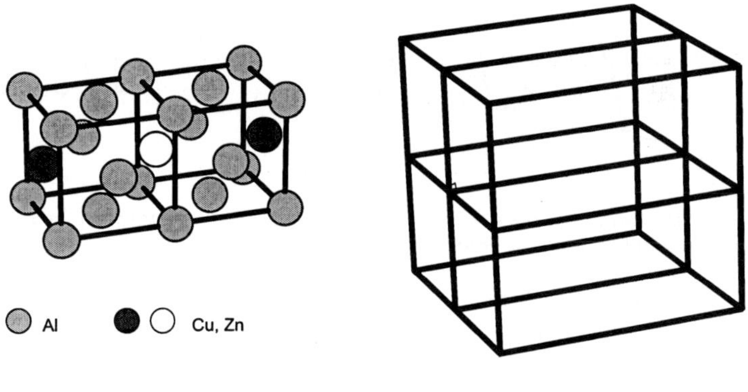

Figure 6. The elementary cell of the Al-Cu-Zn system investigated here is shown on the left hand side. This cell consists of two cubic fcc cells. Six lattice sites are filled with Al atoms (light grey circles) and the remaining two lattice sites (dark grey and white circle) are filled with Cu or Zn atoms depending on the system studied. The Figure on the right shows the geometry of the supercell for which we investigated activation energy barriers for Al diffusion. The supercell consists of four elementary cells shown on the left hand side and contains 32 lattice sites.

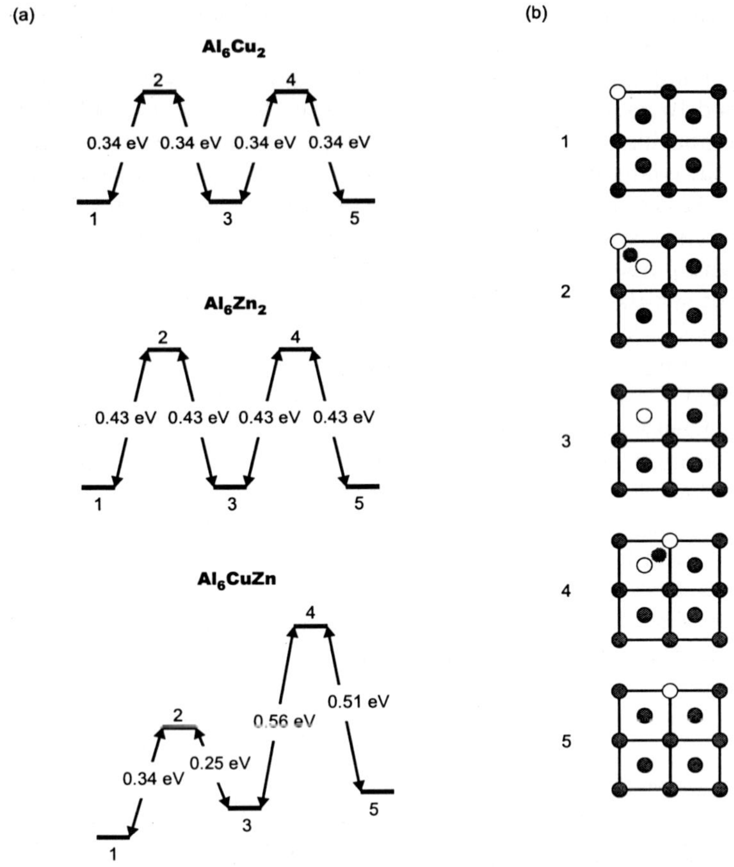

Figure 7. Fig. (a) shows the results of our calculations for the supercell in Fig. 6. Total energies were calculated for five configurations labeled by $1 \ldots 5$. The Figure shows the difference in total energies between the configuration i and $i+1$ ($i = 1 \ldots 4$) for three different compositions. For each configuration a plane cut through the supercell is shown in Fig. (b). White circles represent vacant lattice sites and grey circles represent lattice sites occupied by Al atoms. The configurations $1 \ldots 5$ only differ by the arrangement of atoms in this cut. This plane contains only Al atoms and a vacancy. In each case the supercell contains 31 atoms and one vacancy. Configurations 1, 3 and 5 are minimum configurations which are obtained by removing an Al atom from a regular lattice site and the remaining two configurations labeled by 2 and 4, are obtained by placing an Al atom in between the two vacant Al sites, see Fig. (b).

4 Conclusions

Using ultrasoft pseudopotentials we have calculated the vacancy formation and migration energy of Al and migration energies of Cu and Zn impurities in Al at $T = 0\,\mathrm{K}$. Our results are in good agreement with available experimental data even for the small supercell with 32 lattice sites. Furthermore we investigated the influence of Cu and Zn atoms on the migration energy of Al. Our calculations show that the migration energy of Al can be reduced by placing Cu or Zn atoms at the nearest-neighbor sites of a diffusing Al atom in the saddle point configuration.

In order to have fast diffusion processes in Al-Cu-Zn alloys, the energy barrier for the elementary diffusion process via nearest-neighbor vacancy sites must be much lower. For diffusion in bcc lattices it has been proposed that phonons can considerably help the diffusing atom.[20,21] For fcc lattices phonon-assisted diffusion has not yet been studied. Preliminary investigations show that for the elementary diffusion path for an atom at (0,0,0) to the vacancy position at $(\frac{1}{2}, \frac{1}{2}, 0)$ in the fcc lattice will not entirely occur in the (0,1,0) plane but will first follow a path along the $[1,1,1]$ direction and then going into the vacancy position. In this context it is of particular interest to note that all measured phonon spectra of Al-Cu-Zn systems alloys show very low lying TA $[1,1,1]$ branches[22] which help to lower the potential barrier of the diffusing atom. Work along this lines is in progress.

5 Acknowledgements

This work has been supported by the German Science Council (Deutsche Forschungsgemeinschaft) through the *Graduate College Structure and Dynamics of Heterogeneous Systems.*

References

1. H. Löffler, *Structure and Structure Development of* Al-Zn *alloys* (Akademie Verlag, Berlin, 1995).
2. G. Kresse and J. Furthmüller, *Phys. Rev.* B **54**, 11 169 (1996).
3. W. Kohn and L.J. Sham, *Phys. Rev.* **140**, A 1133 (1965).
4. P. Hohenberg and W. Kohn, *Phys. Rev.* **136**, B 864 (1964).
5. D.M. Ceperley and B.J. Alder, *Phys. Rev. Lett.* **55**, 566 (1980).
6. J.P. Perdew and A. Zunger, *Phys. Rev.* B **23**, 5048 (1981).
7. D. Vanderbilt, *Phys. Rev.* B **41**, 7892 (1990).
8. G. Kresse and J. Hafner, *J. Phys. : Condens. Matter* **6**, 8245 (1994).
9. H.J. Monckhorst and J.D. Pack, *Phys. Rev.* B **13**, 5188 (1976).
10. M. Methfessel and A.T. Paxton, *Phys. Rev.* B **40**, 3616 (1989).
11. C. Wert and C. Zener, *Phys. Rev.* **76**, 1169 (1949).
12. G.H. Vineyard, *J. Phys. Chem. Solids* **3**, 121 (1957).
13. *Atomic Defects in Metals*, ed. H. Ullmaier, Landolt-Börnstein, New Series Group III, Vol. 25 (Springer, Berlin, 1991).
14. *Diffusion in Solid Metals and Alloys*, ed. H. Mehrer, Landolt-Börnstein, New Series Group III, Vol. 26 (Springer, Berlin, 1990).
15. A. De Vita and M.J. Gillan, *J. Phys. Condens. Matter* **3**, 6225 (1991).
16. P.J.H. Denteneer and J.M. Soler, *Solid State Commun.* **78**, 857 (1991).
17. M.J. Mehl and B.M. Klein, *Physica* B **172**, 211 (1991).
18. N. Chetty, M. Weinert, T.S. Rahman and J.W. Davenport, *Phys. Rev.* B **52**, 6313 (1995).
19. T. Hoshino, N. Papanikolaou, R. Zeller, P.H. Dederichs, M. Asato, T. Asada and N. Stefanou, *Comput. Mat. Sci.* **14**, 56 (1999).
20. U. Köhler and C. Herzig, *Phil. Mag.* A **58**, 769 (1988).
21. G. Vogl, W. Petry, Th. Flottmann and A. Heiming, *Phys. Rev.* B **39**, 5025 (1989).
22. E.D. Hallman and B.N. Brockhouse, *Can. J. Phys.* **47**, 1117 (1969).

Nonmagnetic and magnetic clusters on surfaces

SCANNING TUNNELING SPECTROSCOPY ON SILVER CLUSTERS

K.-L. JONAS, A. BETTAC, V. RANK, AND K.-H. MEIWES-BROER
Fachbereich Physik, Universität Rostock
Universitätsplatz 3, 18051 Rostock, Germany

1 Introduction

The electron confinement in semiconductor nanostructures is known to induce a distinct electronic level structure. Thus it gives rise to a variety of interesting features which make those systems candidates for future technical applications. Whereas the scale length in such semiconductor structures is of the order of 100 *nm*, *metallic nanostructures* with sizes ranging from 1 *nm* to 5 *nm* should allow further minituarization with many novel properties. A wealth of information has been obtained on these *clusters* in gas phase[1,2] and on surfaces.[3] For example, the electron confinement in small metal clusters gives rise to photoelectron spectra which distinctly depend on the exact number of atoms inside the cluster. Fig.1 illustrates the situation of silver clusters with 12 atoms (Ag_N, N=12) which are prepared as anions in a cluster beam. By means of anion photoelectron spectroscopy (PES) richly structured spectra are obtained which show no similarity with those of the corresponding molecules or bulk surfaces.[4] Elaborate theoretical models have been developed to describe these many-particle systems. One result of effective medium calculations[5] is given by the vertical bars in Fig.1, demonstrating that the measured peaks do correspond to one-electron excitations in Ag_{12}.

By conserving the distinct energy spacings of a cluster when brought into contact with a surface, it would be possible to generate a *new surface* with adjustable electronic, and hence, also optical and chemical properties. Indeed, cluster size dependent surface reactivities have recently been observed, see[8,9]. Furthermore, many investigations dealt with N dependent optical absorption of clusters on surfaces or in matrices[10,11,12,13] which were well explained by collective electron excitations.

Whereas a wealth of theoretical information is available on the electronic properties of adsorbed atoms and molecules, the level structure of deposited clusters, on the other hand, has scarcely been considered. In the case of weakly interacting surfaces, e.g., sodium clusters on rock salt, it is found that the density of states (DOS) within the cluster is only weakly affected by the

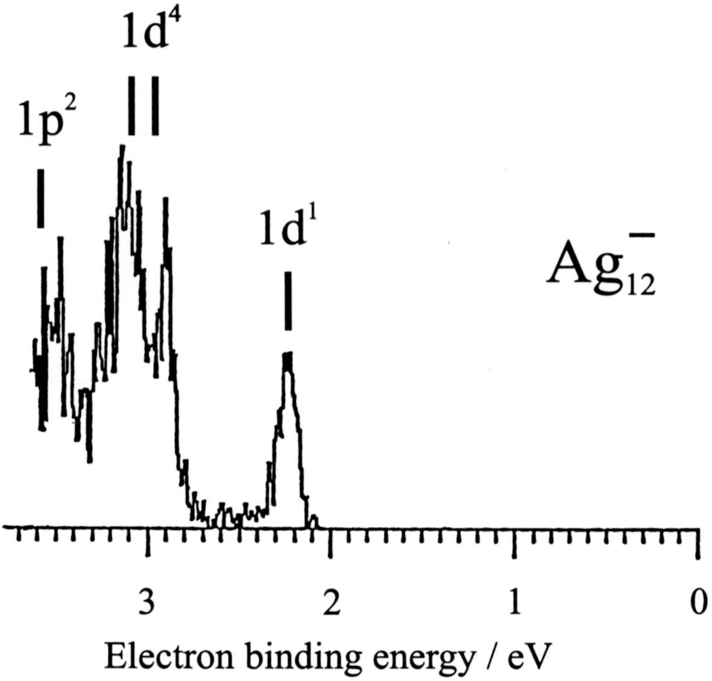

Figure 1. Photoelectron spectrum of mass-selected silver clusters with 12 atoms each in a beam. The energy zero-point is adjusted to match the vacuum energy. Note the low-intensity region at around 2.5 eV which can be interpreted as the HOMO-LUMO gap of the neutral Ag_{12}, but in the geometry of the anion. Excitation energy $h\nu = 4\ eV$. For the method see[6]. After Gausa et al.[7]

presence of the surface.[14]

On the experimental side, conventional photoelectron spectroscopy on deposited clusters did not give the expected insight. So far nothing similar to what was observed with PES on mass selected clusters in beams, see Fig.1, could be found with deposited metal clusters. One reason for this slightly discouraging situation is the fragility of the cluster on surface system. Possibly the irradiation by the vacuum UV radiation leads to a change or even a destruction/fragmentation process within the clusters. From gas phase experiments it is known that metal clusters can exhibit, at least in certain energy ranges, an enormous photo-fragmentation cross section with UV light.[15] At the same time, conventional photoelectron spectroscopy with scanning electron energy analyzers merely records a vanishing fraction of all electrons emitted. Thus the sample has to suffer a significant radiation dose before a spectrum is recorded with sufficient statistics (typical sampling time at a synchrotron lies around 500 s at a photon flux of about 10^{12} $s^{-1}cm^{-2}$).

Scanning tunneling microscopy (STM) is a suitable method for looking at nanometer-sized islands and supported clusters with a resolution on the atomic scale. It is important to note that no direct information can be obtained on the local chemical nature. One promising approach involves analysis of the electronic structure by carrying out Scanning Tunneling *Spectroscopy* (STS). Here, the current between tip and sample is measured as a function of the applied voltage. The current strongly depends on the density of states of the tip and sample and hence, in principle, extracts information about the local electronic structure and thus the chemical nature.

STS has been widely used for the spectroscopy of surfaces,[16,17] for a review see[18,19]. Indeed, recent analyses by STS[20] reveal even the vibrational levels of a molecule in surface contact. Most theoretical approaches use perturbation theory which treats both the tunneling tip, and the sample independently. In this representation the total current is proportional to the local charge density at the Fermi level evaluated at the tip end.[21,22] A simple picture evolves, whereby the tunneling current, or more precisely, its derivative dI/dU, maps the DOS of the sample whenever the DOS of the tip displays a smooth energy dependence.

As one goes to decreasing distances and higher lateral resolutions the tip-sample interaction plays an increasing role, and as a consequence, the theoretical perturbation breaks down. Now the tunneling must be treated as a scattering process where the convolution between the wave functions of the tip and the sample has to be considered. Up to now, this many particle problem has been tackled by first-principles local density functional calculations and the application of Green's functions. A theoretical review and applications to atomic adsorbates can be found in.[23] As there is yet no theoretical treatment of the conductivity of a cluster on a surface, we have to look into related fields, e.g., the physics of metallic atomic size contacts. Recent theoretical approaches deal with quantum electronic transport through microconstrictions with variable shapes. There it is shown that the conductance can be quantized due to the geometry of the constriction, i.e., even without consideration of the chemical nature of the contact. Along this line, Scherbakov et al.[24] showed that for three-dimensional free electron - like constrictions the charge transport is quantized. This is evident from peaks in the calculated dI/dU.

When considering additionally the orbital electronic structure and the local atomic environment around the region of constriction, the transmission eigenvalues are richly structured as a function of energy. This means that the conductance is strongly connected to the orbitals of the atoms involved, which gives hope that a measurement of the charge transport through a cluster might yield information about it's electronic structure. Finally, Coulomb charging has to be considered as another possible feature of a tunneling experiment on clusters. It is possible that under certain circumstances some substructure on the plateaus of the Coulomb staircase can originate from the electronic structure of the clusters.[25]

In order to summarize the current theoretical situation, it can be stated that there has been no study which would describe the charge transport through a deposited cluster on a molecular level. A complete picture has to include the electronic cluster levels in the given geometry in the presence of the surface, the influence of the electrical field under the tip (up to $10^8\ Vm^{-1}$) and possible charging effects.

Notwithstanding the lack of any concrete theory, several experimental investigations have nevertheless dealt with STS on clusters in order to explore electronic structure. The first result on mass selected clusters, i.e., Si_{10} on Au(001), shows a pronounced broad minimum in the dI/dU record, due to a band gap width of about 1 eV.[26] Iron clusters ($N \approx 13$) on GaAs(110) were found to exhibit a gap at the Fermi level which was attributed to the non-metallic character,[27] whereas for $N > 35$ the gap closes. Flat gold clusters on GaAs(110) give rise to pronounced band gap states.[28] The dI/dU spectra of 5 nm palladium clusters on $TiO_2(110)$ show an astonishingly broad low intensity range which is discussed in terms of a large gap in the cluster DOS.[29] Such a large gap, however, is unexpected and cannot be explained with current theories. Several further experiments have been reported, none of which show interpretable details of the electronic structure of the clusters.

Within our group we have investigated the electronic structure of metal clusters, mainly Pt_N, Na_N, and Ag_N by STS in ultrahigh vacuum at cryogenic temperatures.[30,31,32] The results on Ag_N presented below give evidence that these systems exhibit a pronounced energy level structure even when in contact with a surface. On the other hand, however, the charge transport itself gives rise to peaks in the STS which is due to the wave nature of the electrons.

2 Experimental Methods

The STM/STS measurements are carried out with a commercial low temperature scanning tunneling microscope (Omicron). Temperatures down to 5 K can be reached. At these conditions samples and tips can be exchanged. The vacuum system consists of two chambers, STM- and preparation chamber, both operating in the lower 10^{-11} $mbar$ range. The preparation chamber is equipped with up to four evaporators, a LEED and a flash station. Preparations are controlled step by step via STM and LEED. A cluster-on-surface system can be prepared for example by atomic diffusion or by depositing clusters from a beam under UHV conditions. In this contribution we use nanostructured highly orientated pyrolytic graphite (HOPG) as substrate for cluster aggregation on the surface. Following an old recipe[33,34] which has recently been rediscovered,[35,36] the HOPG is cleaved in air and subsequently sputtered by 1 keV ions, mostly Ar^+, but also C_{60}^+. Then, it is heated in air (750-800 K) for about 10 minutes. At this temperature oxidation takes place only at defects. Because of the gaseous oxitation products CO and CO_2, pits grow at point defects, most of them induced by sputtering. Depending on the ion energy and the oxidation time, depth and diameter of the pits can be adjusted. Final cleaning is done by heating the sample up to 670 K. Now, size, shape and the number of the produced pits is analyzed using the STM. In order to grow silver clusters, a small amount of silver is deposited onto the surface by thermal evaporation. At 500 K silver clusters are formed mainly within the pits. Depending on the amount of silver and the pit size, they can be filled partly or completely (Fig.2). Clusters also grow at atomic steps and remaining point defects.

For the STS measurements, either a single cluster is selected and investigated, or STS spectra are taken at regularly separated positions within a selected area. At each position, the tunneling current and the output of an analog lock-in amplifier are recorded while scanning the sample bias under open feedback conditions. The advantage of the usage of a look-in amplifier is the direct access of the derivative dI/dU and a better signal-to-noise ratio when compared to the numerical derivative of the tunneling current. Depending on measuring time, modulation frequency and amplitude, the noise can be suppressed by several magnitudes. The sinodial reference signal for the look-in amplifier has a modulation frequency of 9.6 kHz which is higher than other reported values. Using high modulation frequencies leads to an additional contribution to the measured current in the order of the tunneling current. However, this capacity-induced conductivity is constant while scanning the bias voltage in a STS-spectrum and therefore the information of the spectrum is not affected.

3 Results and discussion

The determination of the cluster size from STM measurements is often not straightforward. Due to the convolution of cluster and tip shapes a direct ob-

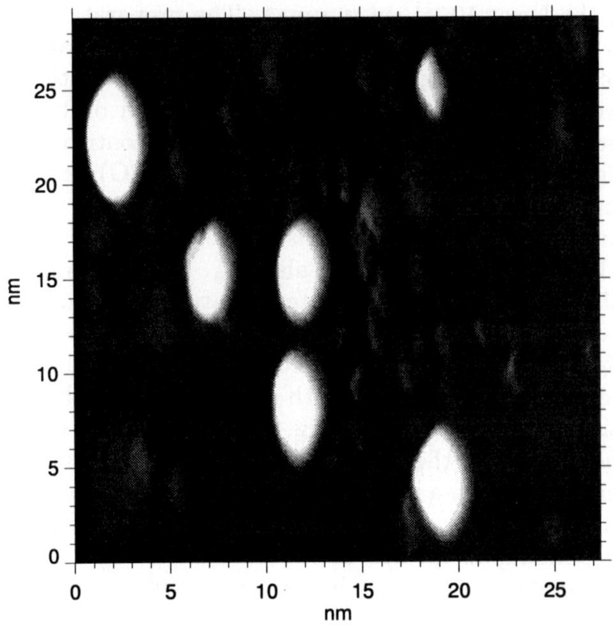

Figure 2. Silver clusters aggregated in pits and at defects on HOPG with heights of 1.2 *nm* and 0.9 *nm*.

servation of the lateral cluster size is difficult. Other authors report diameter-to-height ratios of 1.2 to 2.0, which appear to be typical values for coinage metal clusters on weakly interacting substrates.[37,38] In the studies presented here we choose a value of 1.5 in order to approximate the cluster size N from the height measurement. Additional uncertainties result from the unknown depth of the underlying pit in the case of completely filled pits.

All images and spectra shown are measured at 78 K. One typical example of an STS spectrum on Ag_N with a cluster height of 0.6 nm is depicted in Fig.3. Whereas the spectrum on the uncovered HOPG shows a clear parabolic shape, on top of the cluster a rich structure appears, consisting of two basically different features. First, strong peaked intensity at high negative and positive bias voltages shows up which is separated by a region of low intensity. Second, with higher sensitivity, a pronounced peaked structure emerges inside the apparent gap. Analogous behavior was observed for most clusters.

At first glance the gap could be interpreted as being due to the electronic structure of the silver cluster. On the other hand, effective medium calcu-

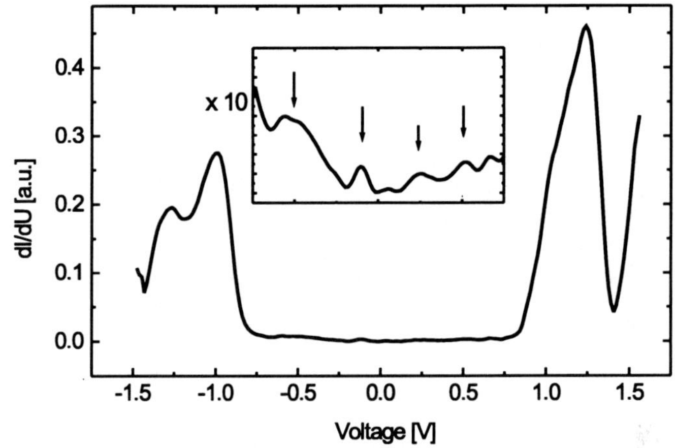

Figure 3. STS of a silver cluster containing about 15 atoms. Note the large apparent gap from -0.8V to +0.8V which shows only weak intensity.

lations and photoelectron spectra show that the gaps between the occupied and unoccupied levels, i.e., the HOMO-LUMO gaps, range around 1.2 eV for N=8 and 0.3 eV for N=25,[5,4] cf. also the photoelectron spectrum in Fig.1. There is no reason why a silver cluster should possess such a large energy gap as observed here in the STS. Thus, another mechanism must be responsible for the large apparent gap. With thin metal films[39] and on Pt clusters[30] it was demonstrated that electron transport can show interference effects. Considering the cluster as a square well potential vertical to the surface, a Fabry-Perot-like behavior of the scattered electrons will result because of the interference of waves reflected on both surfaces. With the cluster height d_Z the wave vector difference of the transmission maxima is $\Delta k = \pi/d_Z$. Using $E = \hbar\omega$, a simple equation relates the energy difference $E_n - E_{n-1} = \Delta E$ to the cluster height:

$$\Delta E = \frac{\pi \hbar v}{d_Z}, \tag{1}$$

with v the group velocity of the electron waves.

In order to check this model, we compile the measured apparent gap

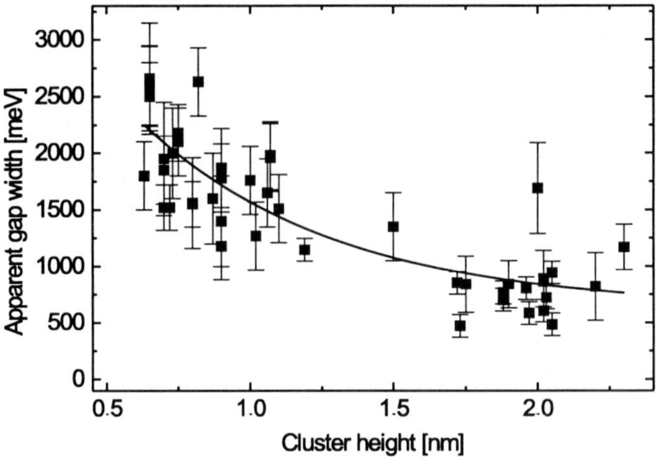

Figure 4. The apparent gap width decreases with increasing cluster size. The solid curve is a fit with eq.1 to the data using leaving the velocity as free parameter.

widths, see Fig.4. The measured height dependence clearly indicates the expected decrease for bigger clusters, changing in width from about 2 eV at N≈15 to about 0.7 eV for a big cluster with a height of 2 nm, i.e., about 500 atoms. For a comparison of the data with the model, the apparent gap widths are equated with $2\Delta E$. Then, a fit of the measured data leaving v as free parameter yields the curve in Fig.4. The resulting velocity of about 0.4 10^6 m/s is significantly lower than the Fermi velocity of bulk silver ($v_F = 1.39\ 10^6\ m/s$[40]), but it is in the order of those of heavy alkali metals (Cs bulk: $v_F = 0.7\ 10^6\ m/s$[40]). Many features of gas-phase silver clusters are known to exhibit properties of alkali clusters. Thus, a possibly reduced value of v_F might be an indication of the alkali-like behavior of silver clusters even when in contact with a HOPG surface.

The dependence of the apparent gap on the cluster height is not only observed for different clusters, but also at the same cluster. Fig.5 depicts an example of this behavior. The height variation of this 2 nm cluster is big enough to observe an increase of the gap when going to lower parts of the cluster. The existence of the height dependence at the same cluster supports

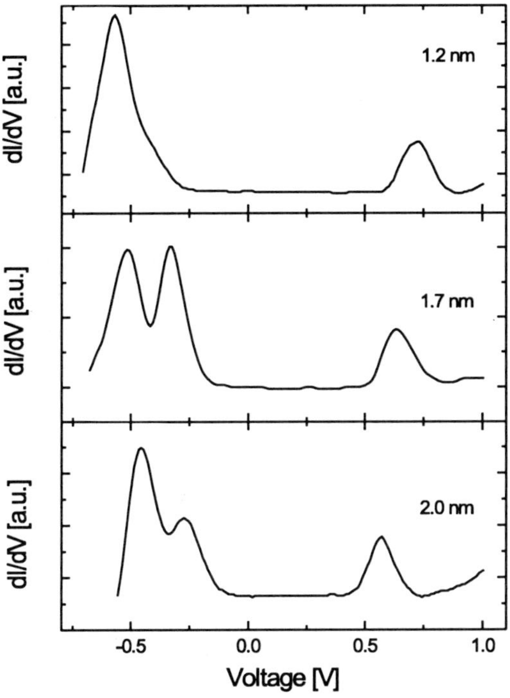

Figure 5. STS on a simple Ag cluster, but at different positions and corresponding cluster heights d_Z. The apparent gap width decreases with increasing cluster height.

the standing electron wave approach.

Thus, only the weak intensity in the apparent gap can be related to the cluster's electronic level structure which might lead to distinct conduction channels. The corresponding transmission eigenvalues depend on the orbital characters of cluster, which, so far, have not been investigated. With the method demonstrated by Cuevas et al.[41] it should be possible to calculate the microscopic origin of these STS spectra. Even the largest clusters with $N \approx 1000$ show no similarities to the DOS of bulk silver, clearly indicating that in this size range cluster-specific properties still prevail.

4 Conclusion

STS on supported metal clusters is a promising technique, because it provides high lateral resolution for electronic spectroscopy. With the scanning tunneling microscope, single clusters at a surface can be addressed. Silver clusters on HOPG surfaces clearly show discrete peaks in the tunneling spectra which can, in part, be related to the electronic level structure. Additionally, the wave nature of the electrons gives rise to pronounced peaks and gap-like features in the STS which are explained as interference effects. Thus, the charge transport through these nanostructures is determined by a convolution of a Fabry-Perot-like transmission and cluster-specific properties.

Financial help by the Deutsche Forschungsgemeinschaft and the EU (Brite/EuRam program) is gratefully acknowledged.

References

1. *Clusters of Atoms and Molecules*, volume I+II, ed. H. Haberland (Springer, Berlin, 1995).
2. S. Sugano and H. Koizumi, *Microcluster Physics* in *Springer Series in Material Sciences*, volume 20 (Springer, Berlin, 1996).
3. *Metall clusters at Surfaces*, ed. K.H. Meiwes-Broer, *Springer Series in Cluster Physics* (Springer, Berlin, Heidelberg, 2000).
4. K.H. Meiwes-Broer, *Appl. Phys.* A **55**, 430 (1992).
5. O.B. Christensen, K.W. Jacobsen, and K. Norskov, *Phys. Rev. Lett.* **66**, 2219 (1991).
6. G. Ganteför, M. Gausa, and K.H. Meiwes-Broer, *Z. Phys. D* **12**, 405 (1989).
7. M. Gausa and K.H. Meiwes-Broer, unpublished.
8. U. Heiz, F. Vanolli, A. Sanchez, and W.-D. Schneider, *J. Am. Chem. Soc.* **120**, 120 (1998).
9. M. Frank, S. Andersson, J. Libuda, S. Stempel, A. Sandell, B. Brena, A. Giertz, P.A. Brühweiler, and M. Bäumer, *Chem. Phys. Lett.* **279**, 92 (1997).
10. U. Kreibig and M. Vollmer in *Optical Properties of Metal Clusters*, *Springer Series in Materials Science*, volume 25 (Springer, Berlin, Heidelberg, 1995).
11. T. Wenzel, J. Bosbach, A. Goldmann, F. Stietz, and F. Träger, *Appl. Phys.* B **69**, 1 (1999).
12. W. Harbich, S. Fedrigo, and J. Buttet, *Chem. Phys. Lett.* **195**, 613 (1992).

13. T. Klar, M. Perner, S. Grosse, G. van Plessen, W. Spirkl, and J. Feldmann, *Phys. Rev. Lett.* **80**, 4249 (1998).
14. H. Häkkinen and M. Manninen, *Phys. Rev. Lett.* **76**, 1599 (1996).
15. J. Tiggesbäumker, L. Köller, K.H. Meiwes-Broer, and A. Liebsch, *Phys. Rev.* A **48**, R1749 (1993).
16. G. Binning and H. Rohrer, *Helv. Phys. Acta* **55**, 726 (1982).
17. R.M. Feenstra and P. Mårtensson, *Phys. Rev. Lett.* **61**, 447 (1988).
18. *Scanning Tunneling Microscopy and Related Methods*, ed. R.J. Behm, chapter R.M. Feenstra, page 211 (Kluwer Academic Publishers, Amsterdam, 1990).
19. T.A. Jung, F.J. Himpsel, R.R. Schlittler, and J.K. Gimzewski in *Scanning Probe Microscopy: Analytical Methods* (Springer Verlag, Berlin and Heidelberg, 1998), page 11.
20. B.C. Stipe, M.A. Rezaei, and W. Ho, *Science* **280**, 1732 (1998).
21. J. Tersoff and D.R. Hamann, *Phys. Rev.* B **31**, 805 (1985).
22. J. Bardeen, *Phys. Rev. Lett.* **6**, 57 (1961).
23. *Scanning Tunneling Microscopy III*, ed. R. Wiesendanger and H.-J. Güntherodt, *Springer Series in Surface Sciences*, volume 29 (Springer Verlag, Berlin and Heidelberg, 2nd edition, 1996).
24. A.G. Scherbakov, E.N. Bogachek, and U. Landman, *Phys. Rev.* B **53**, 4054 (1996).
25. T.P. Bigioni, L.E. Harrell, W.G. Cullen, D.K. Guthrie, R.L. Whetten, and P.N. First, *Eur. Phys. J.* D **6**, 355 (1999).
26. Y. Kuk, M.F. Jarrold, P.J. Silverman, J.E. Bower, and W.L. Brown, *Phys. Rev.* B **39**, 11168 (1989).
27. P.N. First, J.A. Stroscio, R.A. Dragoset, D.T. Pierce, and R.J. Celotta, *Phys. Rev. Lett.* **63**, 1416 (1989).
28. R.M. Feenstra, *Phys. Rev. Lett.* **63**, 1412 (1989).
29. C. Xu, X. Lai, G.W. Zajac, and D.W. Goodman, *Phys. Rev.* B **56**, 13464 (1997).
30. A. Bettac, L. Köller, V. Rank, and K.H. Meiwes-Broer, *Surf. Sci.* **402**, 475 (1998).
31. A. Bettac, V. Rank, and K.H. Meiwes-Broer, to be published.
32. A. Bettac, (thesis, Universität Rostock, 1999).
33. G.R. Hennig, *J. Chem. Phys.* **40**, 2877 (1964).
34. R.T. Yang and C. Wong, *J. Chem. Phys.* **75**, 4471 (1981).
35. H. Hövel, Th. Becker, A. Bettac, B. Reihl, M. Tschudy, and E.J. Williams, *Appl. Surf. Sci.* **81**, 154 (1997).
36. G. Bräuchle, S. Richard-Schneider, D. Illig, J. Rockenberger, R.D. Beck, and M.M. Kappes, *Appl. Phys. Lett.* **67**, 52 (1995).

37. H. Hövel, B. Grimm, M. Bödecker, K. Fieger, and B. Reihl, *Proceedings of the ECOSS-18*, 2000.
38. T. Götz, M. Vollmer, and F. Träger, *Appl. Phys.* A **57**, 101 (1993).
39. I.B. Altfeder, K.A. Matveev, and D.M. Chen, *Phys. Rev. Lett.* **78**, 2815 (1997).
40. N.W. Ashcroft and N.D. Mermin, *Solid State Physics* (Saunders College Publishing, New York, 1976).
41. J.C. Cuevas, A. Levy Yeyati, and A. Martin-Rodero, *Phys. Rev. Lett.* **80**, 1066 (1998).

SURFACE DIFFUSION AND MODELS FOR THE KINETICS OF EPITAXIAL GROWTH

K. SCHROEDER, A. ANTONS, R. BERGER, WI. KROMEN, AND S. BLÜGEL
Institut für Festkörperforschung, Forschungszentrum Jülich, D-52425 Jülich, Germany
E-mail: K.Schroeder@fz-juelich.de

We shortly review the surface processes relevant in epitaxy and discuss the conditions for the realization of different growth modes. We discuss the stability of 3-dimensional and 2-dimensional islands in terms of macroscopic thermodynamics and derive the critical island size. We introduce the kinetic equations governing epitaxy in the limit when island nucleation is the growth determining mechanism and discuss the important parameters. Finally, we discuss the action of surfactants, specifically of As, for homo-epitaxy and Ge hetero-epitaxy on Si(111). We present results of *ab initio* calculations for the rates of diffusion and incorporation of single Si, Ge, and Sn ad-atoms on As-terminated Si(111) and discuss growth models for these group-IV elements on Si(111):As.

1 Introduction

The mode of growth of crystals depends on many external parameters as well as materials properties. In Fig. (1) we have schematically sketched the possible processes an ad-atom can undergo when deposited on a low index flat surface. We shall discuss the processes in turn:

1.1 Adsorption and Desorption

The most important parameter governing the growth rate is the adsorption rate R. It is given by the flux of particles to the unit surface area (measured in $cm^{-2}s^{-1}$) and can be varied by gas pressure and temperature over many orders of magnitude. In most cases, epitaxy is performed for a large supersaturation of ad-atoms, which means high particle flux and such substrate temperatures that desorption is a very unlikely process. The desorption rate R_{des} depends on the substrate temperature T_S via an Arrhenius law containing the binding energy E_a of the ad-atoms to the surface and on the number of adsorbed atoms. Thus the rate in general is time dependent as long as the ad-atom density $n_1(t)$ is time dependent:

$$R_{des}(t) = n_1(t)\,\nu_a \exp\{-E_a/kT_S\} = n_1(t)/\tau_a \qquad (1)$$

τ_a is the mean residence time of the ad-atoms on the surface before it desorbs at a given temperature T_S. The pre-factor ν_a is determined by the curvature

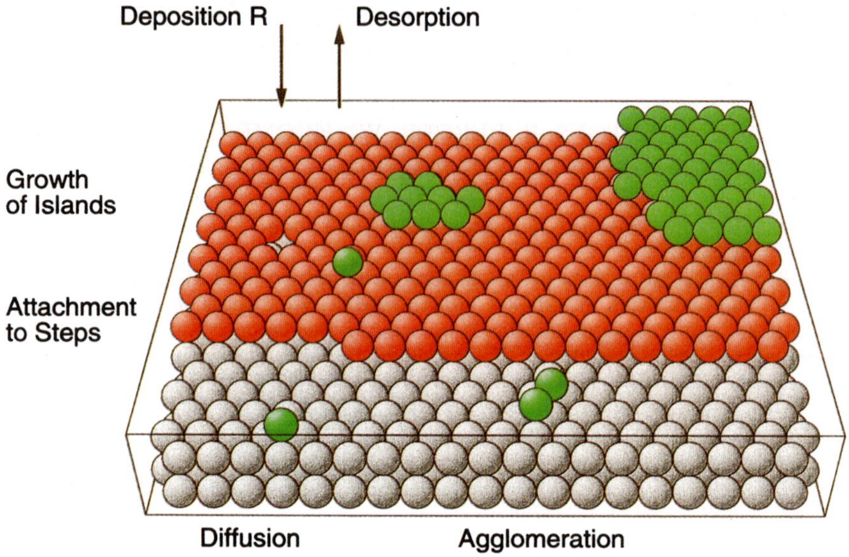

Figure 1. Processes for an ad-atom on the surface of a growing crystal: (a) Desorption is possible for high temperatures. (b) Diffusion on the surface leads to (c) growth of existing islands, (d) agglomeration into new clusters or (e) attachment to steps.

of the binding energy surface of an ad-atom as a function of vertical distance calculated at the equilibrium adsorption distance.

The ratio of the adsorption and desorption rates defines the so-called "sticking coefficient" β

$$\beta = 1 - R_{des}(t)/R = 1 - n_1(t)/(R\tau_a). \tag{2}$$

From this we can easily define the adsorption-desorption-equilibrium. This is reached when as many ad-atoms desorb as are deposited, i.e. $\beta = 0$:

$$\frac{dn_1(t)}{dt} = R - \frac{n_1}{\tau_a} = 0, \quad \text{i.e. } n_1 = R\tau_a : \quad \beta \to 0\,. \tag{3}$$

When other processes are possible this equilibrium usually is not reached, except at high temperature when the crystal is under its own vapor pressure. In microscopic terms, the binding energy E_a depends on the crystal face. This causes the faceted shape of crystals in equilibrium and of 3-dimensional clusters on surfaces in equilibrium conditions.[1] For epitaxial growth conditions the ad-atoms undergo reactions into states of higher binding energy, thus preventing desorption.

1.2 *Diffusion Reactions*

The ad-atoms have to move across the surface in order to meet each other or existing defects. On flat terraces the equilibrium sites of ad-atoms on the surface form a regular lattice. At ambient temperatures the ad-atoms perform a random walk by thermally excited jumps to any of the nearest neighbor sites with a rate

$$\Gamma_d = \nu_d \exp\{-E_d/kT_S\}, \tag{4}$$

where ν_d is an attempt frequency and E_d the activation energy for a diffusion jump.

The mean square displacement of an ad-atom depends linearly on time:

$$\langle r^2(t)\rangle = \langle x^2(t)\rangle + \langle y^2(t)\rangle = 4\,D\,t. \tag{5}$$

For a square lattice with lattice constant a the diffusion constant is given by

$$D = \nu_d\, a^2 \exp\{-E_d/kT_S\}. \tag{6}$$

The important quantity which governs the probability of reactions is the mean distance an ad-atom travels on the surface before it is desorbed, i.e. in the time interval τ_a:

$$x_a = (4D\tau_a)^{1/2} = 2a(\nu_d/v_a)^{1/2} \exp\{(E_a - E_d)/2kT_S\}. \tag{7}$$

If $x_a \geq d_{sink}$, the mean distance between sinks, then ad-atom reactions take place and a new adsorption equilibrium is reached.

A first reaction possibility is the attachment to steps and kinks, which is important for vicinal surfaces when the diffusion distance is larger than the terrace width: $x_a > L_T$. The binding energies E_{step} and E_{kink} determine the stability of the attached configurations. If this is the most probable reaction growth proceeds via step flow.

Another possibility is nucleation and growth of clusters (islands) on the flat terraces. For this reaction the density n_1 of single ad-atoms (called monomers in the following) has to be large, which will be the case for relatively low substrate temperatures T_S and large deposition rates R. The mean distance between monomers x_M can be estimated by $n_1 \pi x_M^2 = 1$, which means that one assigns a circle with radius x_M to each of the existing monomers which on the average contains no other monomer. If this distance is the smallest, i.e. $x_M < (L_T, x_a)$, agglomeration of ad-atoms into clusters (islands) of increasing size ($i \geq 2$) takes place. The binding energies E_i depend on the island size i, and for large islands they tend towards the binding energy E_{step}.

1.3 Scope of the Paper

In the following, we shall discuss the kinetics of the early stages of growth (less than 1 monolayer) via cluster nucleation and growth. The long time behavior (deposition of many monolayers) is usually studied by statistical methods[2] or by "Kinetic Monte Carlo" methods.[3] Depending on the balance between the rate for nucleation on top of existing islands and the growth rate of islands one expects more or less pronounced roughening of the surface[2] with growth of 3- or 2-dimensional islands.

The important parameters governing the growth mode for epitaxy include materials properties like lattice constants and surface energies, and activation energies for various kinetic processes. If substrate material **B** and deposited material **A** have a lattice mismatch Δa, strained layers result with the tendency towards growth of 3-dim islands.[4,5] Depending on the relation of the surface energies of the two materials two- or three-dimensional (2-dim or 3-dim) islands are favored (see section 2). The activation energies for desorption, diffusion and the binding energies of ad-atoms to islands determine the details of the growth kinetics, e.g. the critical island size and the temperature and deposition rate dependence of the growth rate (see section 3). The ratio of deposition rate to desorption rate defines a supersaturation ζ of ad-atoms on the surface which can be expressed as a difference in free-energy per particle compared to the gas phase:

$$\zeta = \frac{R}{R_{des}}; \qquad \Delta\mu = -kT_S \ln \zeta. \tag{8}$$

Supersaturations of the order of $\zeta \approx 10^{20}$ are not unusual for epitaxy. In this case desorption does not have to be considered.

As we will see in the last section 5, one can manipulate the thermodynamics and surface kinetics by adding small amounts of impurities to the surface. E.g., one monolayer of Sb or As on Si changes the growth mode of Ge from Stranski-Krastanov to layer-by-layer growth. Group-V impurities have a lower surface free-energy than Si or Ge and thus segregate to the surface. In addition, the diffusion behavior of ad-atoms is changed by the impurity layer. Materials which have such drastic influences on surface reactions are called "surface active" materials, in short "surfactants". Surfactant mediated growth is an active field of research, because it opens the possibility to grow flat layer hetero-structures also in very adverse circumstances. In order to understand the surfactant effect, a detailed microscopic study of the kinetic processes of ad-atoms on surfactant terminated surfaces is necessary. We will show in section 4, that *ab initio* calculations on the basis of density-functional-theory (DFT) can help to identify relevant reaction paths and products by calculating the energies and the atomic and electronic structures of ad-atoms and agglomerates as well as the activation energies for possible reactions. We present results of such calculations for group-IV ad-atoms on Si(111):As and discuss the difference between homo- and hetero-epitaxy.

2 Macroscopic Thermodynamics

For a general discussion of the concepts to determine the shape of growing islands and the critical nucleus for growth we employ macroscopic thermodynamics.

2.1 Three-dimensional Islands

In Fig. 2 we have sketched a 3-dim island of material **A** on a flat substrate of material **B**. Energetic considerations show that this situation is a stable configuration if the surface free energies γ_i ($i = \mathbf{A}, \mathbf{B}$; *=interface) have the following relation:

$$\Delta\gamma = \gamma_A + \gamma^* - \gamma_B > 0. \tag{9}$$

The new surface of deposited material **A** then costs energy. The 3-dim shape

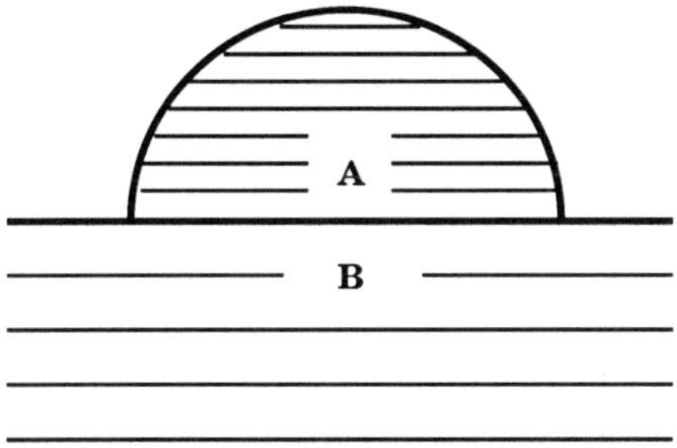

Figure 2. Three-dimensional islands are stable if the surface free-energy of deposited material **A** is larger than of the substrate material **B**.

minimizes the unfavorable surface for a given number of deposited atoms. Of course, the detailed shape of the islands depends on the relation of the surface energy of possibly exposed crystallographic faces and on the lattice mismatch between substrate and deposited material. The free-energy difference compared to the gas state of an island containing j atoms is:

$$\Delta G^{(3)}(j) = \underbrace{j\Delta\mu}_{volume} + \underbrace{j^{2/3}\overbrace{(F_A\gamma_A + F_B(\gamma^* - \gamma_B))}^{X>0}}_{surface}. \tag{10}$$

The first term $\propto j$ is the volume term with $\Delta\mu$ the difference of free-energy per **A**-particle in the island due to condensation to the **B**-substrate. Because of Eq. (9) this has to be negative if the **A**-island should grow. The second term $\propto j^{2/3}$ (the coefficient is abbreviated with X) is the surface term which is positive with the assumption Eq. (9).

The critical nucleus i is defined as the island size where the free-energy $\Delta G^{(3)}(j)$ is an extremum (for the parameters chosen this is a maximum):

$$\left.\frac{\partial G^{(3)}(j)}{\partial j}\right|_{j=i} = 0. \tag{11}$$

For $j > i$ the island grows because the free-energy decreases when an addi-

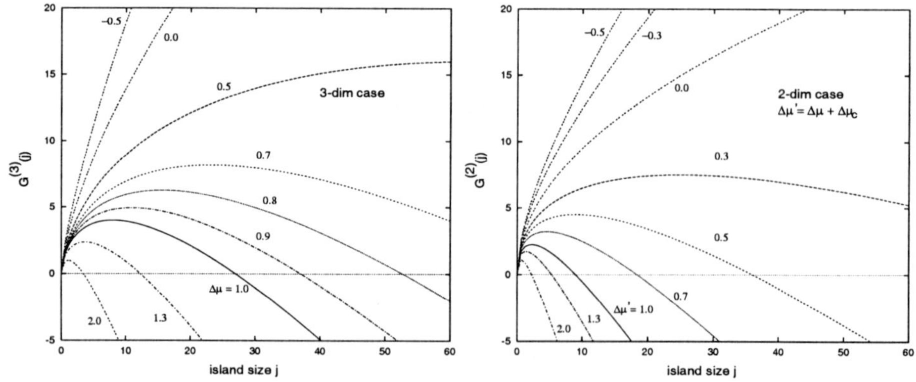

Figure 3. Change of the free-energy (relative to the gas phase) of deposited islands as a function of size j: (a) 3-dim case, (b) 2-dim case. The free-energy barrier $\Delta G(i)$ and the critical island size i (maximum of $\Delta G(i)$) decrease with increasing supersaturation. For given $\Delta\mu$ both quantities are smaller for the 2-dim case.

tional particle is attached, whereas for $j \leq i$ the islands can grow only due to kinetic and thermal fluctuations. The critical nucleus $i^{(3)}$ and the barrier height , i.e. the free-energy difference $\Delta G^{(3)}(i)$, for 3-dim islands are given by

$$i^{(3)} = \left(\frac{2X}{3\Delta\mu}\right)^3 ; \qquad \Delta G^{(3)}(i) = \frac{4}{27} \frac{X^3}{(\Delta\mu)^2} . \tag{12}$$

Fig. 3(a) shows the dependence on $\Delta\mu$ of both quantities. As can be seen, the critical island size and the barrier height decrease with increasing supersaturation $\Delta\mu$.

2.2 Two-dimensional Islands

For a 2-dim island which is sketched in Fig. 4 the situation is a little bit different. First of all, one expects 2-dim islands to be stable if

$$\Delta\gamma = \gamma_A + \gamma^* - \gamma_B < 0. \tag{13}$$

In this case the **A**-islands would like to maximize the surface which is achieved

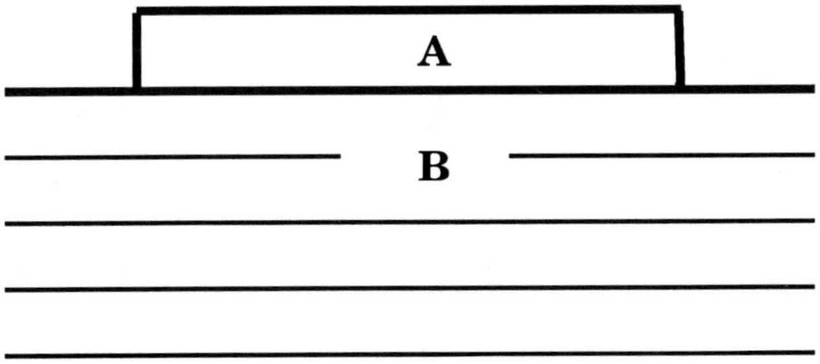

Figure 4. Two-dimensional islands are stable if the surface free-energy of deposited material **A** is smaller than of the substrate material **B**.

by 2-dim growth when each atom adds to the **A**-surface. Secondly, both the "volume term" ($\propto j$, the number of atoms) and the true surface term are $\propto j$. Thus, the island boundary term ($\propto j^{1/2}$) has to be taken into account in the free-energy. For two dimensions this reads :

$$\Delta G^{(2)}(j) = \underbrace{j\Delta\mu}_{volume} + \underbrace{j\,\overbrace{F_{at}\,\Delta\gamma}^{\Delta\mu_c<0}}_{surface} + \underbrace{j^{1/2}\overbrace{\sum_\ell c_\ell \gamma_\ell}^{Y>0}}_{boundary} . \tag{14}$$

Again, the critical nucleus is found where the free-energy has a maximum:

$$\left.\frac{\partial G^{(3)}(j)}{\partial j}\right|_{j=i} = 0. \tag{15}$$

This yields for 2-dim islands:

$$i^{(2)} = \left(\frac{Y}{2(\Delta\mu + \mid \Delta\mu_c \mid)}\right)^2; \qquad \Delta G^{(2)}(i) = \frac{1}{4}\frac{Y^2}{\Delta\mu + \mid \Delta\mu_c \mid}. \tag{16}$$

By comparing to the 3-dim case one sees that the negative $\Delta\mu_c$ helps the formation of islands. Thus, for the same supersaturation the critical size and the barrier height are smaller for 2-dim islands than in the 3-dim case. This can be seen in Fig. 3(b).

3 Growth Kinetics

In order to understand the time evolution of the growing crystal under a constant flux of deposited atoms one has to calculate the developing coverage of every newly created crystal plane due to agglomeration of the deposited atoms to islands and at terrace steps. We will only treat one case here by assuming that growth is due to agglomeration of ad-atoms to islands on the terrace. We thus have to study the time dependence of island densities and sizes which can be described by kinetic equations.

3.1 Kinetic Equations

We treat the case of homogeneous deposition and homogeneous nucleation only. Then we need to consider only the time dependence of spatially averaged densities of monomers $n_1(t)$ and clusters $n_j(t)$, $j > 1$. The derivation holds for 2-dim and 3-dim islands. We follow the formulation of the review paper of Venables,[6] where details can be found.

We assume that the monomers are the only diffusing species with diffusion coefficient D. Then we have to consider diffusion reactions of mobile monomers with existing clusters of any size and the decay of clusters by emitting monomers.

The diffusion reactions are bimolecular reactions of the general form:

$$U_j = D\sigma_j n_1 n_j, \tag{17}$$

where σ_j is a reaction cross-section determined by the strength and distance dependence of the interaction of a j-cluster with a monomer.

The decay reactions are mono-molecular reactions of the form:

$$K_j\, n_j \;=\; \exp\{-(E_j - E_{j-1})/kT_S\}\; n_j, \tag{18}$$

where the difference of the binding energies $E_j - E_{j-1}$ of two consecutive cluster sizes (binding energy of the detached ad-atom) determines the reactions rate.

The time dependence of the densities is then governed by the following equations:

$$\frac{dn_1(t)}{dt} = R - \frac{n_1(t)}{\tau_a} - 2\,U_1 - \sum_{j\geq 2} U_j + \sum_{j\geq 2} K_j\, n_j \tag{19}$$

$$\frac{dn_j(t)}{dt} = U_{j-1} - U_j - K_j\, n_j + K_{j+1}\, n_{j+1} \qquad (j > 1). \tag{20}$$

These equations have to be solved subject to the initial condition that all densities are zero at $t = 0$. For a (numerical) solution one has to know all activation energies (binding energies and diffusion barrier) and specify the deposition rate R.

3.2 Growth of Stable Islands

In view of the discussion in the previous section concerning the existence of stable clusters of size i it is reasonable to coarse grain the size distribution with the following arguments:

1. At long times the intermediate clusters with sizes $2 \leq j \leq i$ will reach a stationary density, the so-called Walton relation,[7] obtained by considering detailed balance of agglomeration and decay:

$$K_j\, n_j \;= D\,\sigma_{j-1}\, n_1 n_{j-1}, \tag{21}$$

$$\frac{n_j}{N_0} \;=\; \left(\frac{n_1}{N_0}\right)^j \; \exp\,\{-E_j/k\,T_S\}\,. \tag{22}$$

   Here N_0 is the planar density of atom sites.

2. We need not calculate in detail the time development of clusters with sizes $j > i + 1$. We make the reasonable assumption that such clusters do not decay anymore but rather absorb the attached ad-atoms and thus grow. The clusters with $j \geq i + 1$ are collected to one density n_x. Of course, the average size w_x of the stable x-clusters will also depend on time. The total number of atoms contained in these stable clusters then is $w_x\, n_x$.

3. The total coverage Z for 2-dim clusters can be expressed by

$$Z(t) = F_{at} \sum_j j\, n_j(t). \tag{23}$$

Since with the first assumption only the stable clusters grow, the time dependence of the coverage is entirely due to the change of the number of atoms in stable clusters:

$$\frac{dZ}{dt} = F_{at} \sum_{j \geq i+1} j \frac{dn_j}{dt} = F_{at} \frac{d(n_x w_x)}{dt}. \tag{24}$$

4. For larger coverages the direct deposition of atoms to stable clusters and the coalescence of clusters has to be taken into account.

With these assumptions the problem can be cast into the following kinetic equations:

$$\frac{dn_1(t)}{dt} = R - \frac{n_1(t)}{\tau_a} - \frac{d(w_x n_x)}{dt}; \qquad \text{monomers} \tag{25}$$

$$\frac{dn_j(t)}{dt} = U_{j-1} - K_j\, n_j - U_j + K_{j+1}\, n_{j+1} = 0 \qquad (2 \leq j \leq i)$$

$$\frac{dn_x(t)}{dt} = \sigma_x D\, n_1 n_x - 2\, n_x \frac{dZ}{dt}; \qquad \text{stable islands} \tag{26}$$

nucleation and coalescence

$$\frac{d(w_x n_x)}{dt} = (i+1)\, \sigma_i D\, n_1 n_i + \sigma_x D n_1 n_x + RZ; \quad \text{atoms in islands} \tag{27}$$

$$\frac{dZ}{dt} = F_{at} \frac{d(w_x n_x)}{dt}; \qquad \text{coverage for 2-dim. islands} \tag{28}$$

Two limiting cases can easily be distinguished[6] (see Fig.5):

1. Incomplete condensation when substrate temperature is high. Then most deposited atoms desorb again because $1/\tau_a \gg \sigma_x D\, n_x$. The monomer density saturates and the stable clusters and the coverage increase very slowly (Fig.5a).

2. Complete condensation when substrate temperature is low. Then all deposited atoms stick to the surface and the coverage increases linear in time. For large times all deposited atoms end up in stable clusters, because $1/\tau_a \ll \sigma_x\, D\, n_x$. This case is depicted in Fig.5b.

The general behavior of the maximal density of stable clusters n_x can be written as

$$n_x \propto R^p\, exp\, \{E/kT_S\}, \tag{29}$$

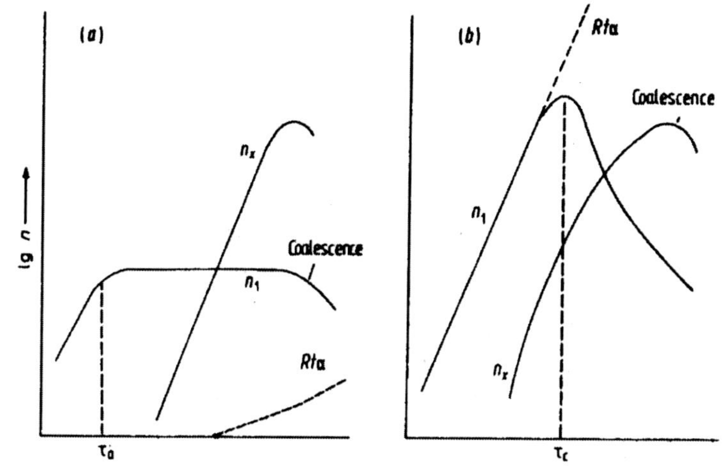

Figure 5. Time dependence of monomer concentration, $n_1(t)$, of growing islands, $n_x(t)$, and total number of atoms condensed, $R\,t\,\alpha$, for the two extreme cases: (a) incomplete condensation (T_S high, $1/\tau_a \gg \sigma_x\,D\,n_x$. Then desorption is important, the monomer density is limited and growth of stable islands is slow. (b) complete condensation (T_S low, $1/\tau_a \ll \sigma_x\,D\,n_x$). Then the monomer density reaches a maximum at $\tau_x = 1/(\sigma_x\,D\,n_x)$. The density of stable islands and the total number of deposited atoms increase quickly.

where the exponent p and the apparent activation energy E depend on the dimension of the islands, the substrate temperature T_S (i.e. on the relative importance of the desorption reaction), the size of the critical islands i and on the diffusion barrier E_d as well as the binding energy of atoms to the critical island E_i. The following table summarizes the important cases, the cases discussed above are underlined. More details can be found in the review paper by Venables.[6]

Table 1. Parameters determining the maximal island density

Regime	**3-dim. islands**	**2-dim. islands**
high T_S (with desorpt.)	$p = 2i/3$ $E = \frac{2}{3}\,[E_i + (i+1)E_a - E_d]$	$\underline{p = i}$ $\underline{E = E_i + (i+1)E_a - E_d}$
low T_S (no desorpt.)	$p = i/(i+2.5)$ $E = [E_i + iE_d]/(i+2.5)$	$\underline{p = i/(i+2)}$ $\underline{E = [E_i + iE_d]/(i+2)}$

From experiments on island nucleation for homo-epitaxy of Si on clean Si(111) Voigtländer *et al.* [8] deduced a critical island size of $i = 6$ and a diffusion activation energy of $E_d \approx 0.76eV$.

3.3 Important Microscopic Parameters

It is clear from the discussion in the previous section 3.2 and the description of the kinetic equations, section 3.1, that a number of microscopic parameters enter into the solutions. These include the diffusion barriers for mobile particles and the binding energies of ad-atoms to clusters of various sizes, and to steps and kinks. They have to be known before a microscopic path of epitaxial growth can be formulated. Some very basic quantities cannot directly be determined by experiments. E.g. to determine the diffusion energy of ad-atoms from the temperature dependence of the density of critical nuclei at low temperature T_S one has to know the binding energy E_i of ad-atoms to the critical island (see table 1). This quantity cannot be measured directly but has to be inferred from calculations. – Another example is the influence of surfactants on the ad-atom kinetics (see below). Experiments show that the density of nucleated islands is increased by a surfactant layer. These experiments cannot distinguish between an increase of the ad-atom diffusion barrier due to surfactants or a rapid incorporation of ad-atoms into the surfactant layer with the gain of a large binding energy. In both cases the *effective* diffusion barrier is larger, the diffusion length shorter, and the island density higher than for the pure surface, as is found. Theoretical calculation can help to clarify the picture.

In recent years *ab initio* calculations on the basis of the density functional theory (DFT)[9] have proven to give reliable results for total energies of molecules, solids, defects in crystals, and surfaces. The atomic and electronic structure can be determined and measurable quantities can be calculated. Thus, one can expect to identify relevant microscopic growth conditions such as the presence or absence of surface defects, impurities, steps or reconstructions and possible reactions of deposited atoms by such calculations. One can (i) analyze possible atomic structures (intrisic defects like vacancies, ad-atoms, clusters, steps on surfaces) and evaluate their relevance by comparing the total energies and (ii) calculate energy barriers (and in the future maybe also pre-factors and anharmonic dynamical effects) for reaction rates of relevant atomistic processes (deposition and desorption energies, barriers for diffusion on a terrace, diffusion across a step edge, diffusion along a step edge, diffusion around an island, diffusion of dimers, nucleation of islands, exchange of atoms).

4 Computational Method

We have developed `EStCoMPP`[10], a parallelized *ab initio* molecular dynamics program in the spirit of Car and Parrinello,[11] based on the local density approximation (LDA)[12] to the density functional theory (DFT), norm-conserving pseudo-potentials[13] of the Kleinman-Bylander form,[14] a plane-wave basis set for the wave functions, and iterative diagonalization schemes[15,16] for the eigenvalue problem. The convergence of the charge density to self-consistency is accelerated by quasi-Newton methods.[17] Force calculations combined with molecular statics are used to determine the minimum energy configurations and the reaction pathways.[18] The inhomogeneous system is simulated by a large repeated supercell. The computational procedure and parallelization strategy are described in some detail elsewhere.[19]

5 Surfactant Mediated Growth

It is hard to grow flat layer hetero-structures for systems with large lattice mismatch. An example is Ge/Si with a difference of lattice constants $\Delta a = (a_{Ge} - a_{Si})/a_{Ge} \approx 4\%$. The pure system shows the "Stranski-Krastanov" growth mode: 2 layers of Ge grow flat on Si, but for further deposition 3-dimensional islands develop. The reason is the reduction of the elastic energy: 3-dimensional Ge islands with free side-walls can expand much easier than full Ge layers on the Si substrate.

The use of so-called "surfactant layers" modifies the growth mode drastically. An overlayer of As or Sb changes the growth mode of Ge on Si(111) or Si(001) to layer-by-layer-growth.[20,21,22,23] Due to their extra electron group-V surfactant layers on Si or Ge have a reduced surface free-energy which makes them flow on the surface without being incorporated into the growing crystal. In addition, the kinetics of newly deposited ad-atoms is changed by the presence of the surfactant layer. On one hand, the surface diffusion is expected to be faster on top of the passivating overlayer. On the other hand, the ad-atoms eventually have to be incorporated underneath the overlayer. Thus, the competition between diffusion and incorporation of ad-atoms determines the growth mode in the presence of surfactants decisively.

If the diffusion length of ad-atoms (i.e. the distance the atoms move on average before they are incorporated) is large compared to the terrace width of the crystal surface, the atoms are likely to be incorporated at a step edge. Then, the crystal grows by step-flow. On the other hand, if the diffusion length is much smaller than the terrace width, e.g. due to trapping at surface impurities or incorporation into a surfactant layer, growth proceeds by nucleation and growth of islands. The type of islands (2-dim or 3-dim) is determined by surface energy and strain energy considerations.

5.1 Experiments for Si(111)

To get some insight into the *kinetic aspects of surfactant mediated growth* without the complication of strain, Voigtländer *et al.* [8] have performed experiments on homo-epitaxy of Si on clean and As(1 ML) covered Si(111). They find a drastic increase of the island density on the As (or Sb) covered surface and a decrease of the width of the island free (denuded) zone along step edges.

Two different scenarios of the effect of surfactants (and in particular of As) on the kinetics of ad-atoms have been discussed:

(i) Voigtländer *et al.* [8] interpreted their results as a decrease of the diffusion length of Si ad-atoms in the presence of surfactants due to a reduction of the *effective* diffusion coefficient. From the experiment one cannot decide whether this effect is due to an increase of the diffusion activation energy on the surfactant covered surface, or whether the termination of the diffusion path by incorporation of the Si ad-atoms under the surfactant layer plays the decisive role.

(ii) A different picture has been suggested by Kaxiras.[24,25] He assumes that deposited ad-atoms move rapidly across the passivated surface, and that incorporation is a very unlikely process. He argues further that surfactants not only passivate the flat surface but also the step edges,[26] which reduces the incorporation probability of ad-atoms. Thus, although in this scenario the diffusion length is much larger than the terrace width, homogeneous nucleation of islands on the terraces can take place, in spite of a high activation barrier for exchange. Kaxiras *et al.* [25] have substantiated these arguments by *ab initio* calculations for the exchange of *full monolayers* of Ge ad-atoms and Sb-surfactants on Si(111) for which they find a high energy barrier.

For hetero-epitaxy of Ge on Si(111) less quantitative experiments are available. It is found that on the As terminated Si(111) surface several dozens epitaxial Ge layers can be grown,[20] and on Sb terminated Si(111) the epitaxial Ge film can be made arbitrarily thick without substantial loss of Sb.[27] STM-pictures of sub-monolayer growth of Ge on Si(111):Sb show nucleation of Ge islands on the terraces.[8] The steps appear as widely fluctuating meanders. A detailed analysis has not been made and one can only speculate whether step motion due to the incorporation of Ge ad-atoms occurs.

5.2 Ab initio *Calculations for Si(111):As*

In order to clarify the kinetics of ad-atoms on surfactant passivated surfaces we have investigated the barriers for all kinetic processes for single Si (and Ge, Sn) ad-atoms on Si(111):As(1 ML) in a large surface supercell.[28] The results are summarized in Fig. 6 and in Table 2.

All calculations have been carried out with the `EStCoMPP`-code described above. The surface is simulated by a repeated slab model consisting of 6 atomic Si(111) layers in an inversion symmetric arrangement, 1 ML As on each Si surface, and 1 ad-atom on each As layer, separated by a vacuum equivalent to 6 atomic Si layers. We find that the p(3×3) periodic cell in the lateral directions, 4 $\mathbf{k}_{\parallel}$-points in the surface Brillouin-zone and a 13.69 Ry cut-off energy in the plane wave basis set are necessary to obtain well converged results. The symmetry-unrestricted geometry optimization includes all atoms except those of the 2 innermost atomic layers of the slab. To establish minimum energy configurations we converged the forces acting on the atoms to less than 0.1 mRy/a.u.

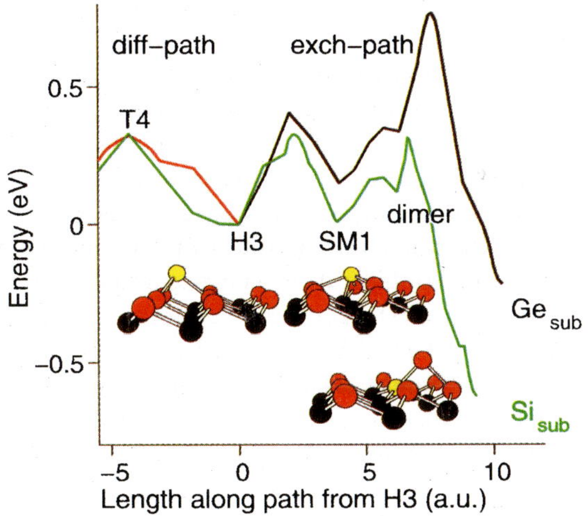

Figure 6. Energy (relative to H3 position of ad-atom) along the diffusion path (to the left of x = 0) and along the exchange path to the substitutional configuration (to the right). Compared are the energies for Si (dashed line) and Ge (full line). The atomic configurations for important minima are shown as insets: H3: ad-atom equilibrium ; SM1: asymmetric dimer; $Si(Ge)_{Sub}$: substitutional Si(Ge) and As ad-atom (light: Si ad-atom; gray: As; black: Si bulk).

First we performed calculations for Si ad-atoms to get some insight into homo-epitaxy.[28] We determined the equilibrium configuration of Si ad-atoms to be the H3 site (above the empty hexagon with 3 As neighbors in the top layer and 3 Si neighbors in the 2nd layer) by minimizing the total energy of the supercell with respect to all atomic coordinates, starting from an arbitrary ad-atom position on top of the As layer. Saddle points are determined by restricted relaxation similar to the "hyper-plane adaptive constraint (HAC)" method,[29] i.e. by minimizing the energy subject to additional constraints. In the case of the diffusion path we can use the symmetry of the surface unit cell to define the saddle plane. Since the H3-centered Wigner-Seitz-cell is bound by 6 equivalent lines *top – T4* a restriction of the ad-atom motion to the plane perpendicular to the surface and crossing the line $Top - T4 - H3$ contains all configurations relevant for the diffusion path of the group-IV atoms. We find that the ad-atom relaxes right into the H3 position, if started anywhere on this plane. This means that the T4 site (above the position of the Si-atoms of the 2nd layer) has the lowest energy of all boundary sites and must be the diffusion saddle. The energy difference to the H3 configuration is the activation energy E_d for surface diffusion, we find $E_d(Si) = 0.25$ eV.

We also determined the exchange path and calculated the activation energy E_{EX} for incorporation of a Si ad-atom by exchange with an As surfactant atom, and E_{REEX} for re-exchange. The search for the saddle point on the exchange reaction path Si_{Ad}+As_{Sub} $\rightarrow$ Si_{Sub}+As_{Ad} is exceedingly more complicated than for the diffusion saddle point. In principle one has to determine saddle points in a 6-dim configuration space spanned by the position vectors of the Si ad-atom and of the replaced As atom.[30] We used the following procedure in the spirit of the HAC-approach:[29] (i) we define new starting configurations *along the straight exchange path*, i.e. along the difference vector in 6 dimensional configuration space between the Si ad-atom configuration (Si_{Ad}+As_{Sub}) and the As ad-atom configuration (Si_{Sub}+As_{Ad}), at fractions of 1/32 of the total path length; (ii) the 6-dim force for the two exchanging atoms is projected to the plane perpendicular to the straight 6-dim path-vector; (iii) all other atoms are allowed to relax freely. In this way it is possible to determine energies at configurations which are not extrema on the energy surface. The application of the procedure just described is complicated by the existence of side minima (SM) along the exchange path which are to be expected in covalently bonded materials with directed bonds. Then, one first has to determine the side minima and apply the restricted minimization to all partial paths between neighboring minima. We are particularly inter-

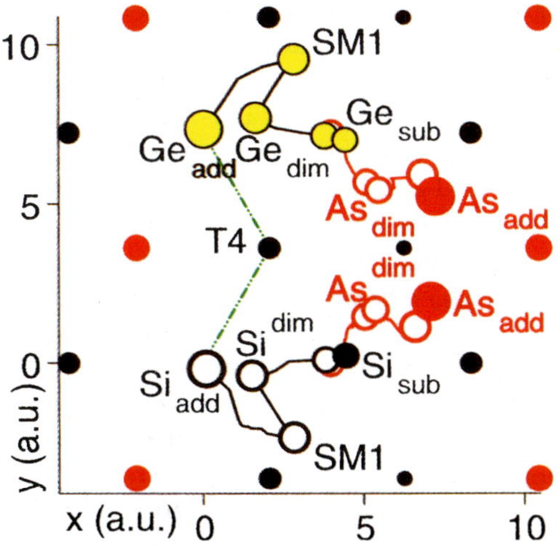

Figure 7. Top view of diffusion (dashed line $H3 - T4$) and exchange path of Ge (top part) and Si (bottom) on Si(111):As. The paths shown are virtually mirror images. The exchange process involves the concerted motion of the two exchanging partners and the relaxation of the neighboring atoms. The path consists of several sections between local equilibrium sites (SM1, dimer) which are indicated by open circles, whose size indicates the height above the As layer. The atoms of the Si(111):As surface are also indicated: top layer As atoms (big circles), second layer Si atoms (T4; medium circles) and 4th layer Si atoms (H3; small circles).

ested in paths with low barriers since they dominate the transition rate of a thermally activated system. This procedure with the movement of the 6-dim hyperplane along the direct path between minima should work well for systems in which all bonds are of the same strength as is the case in our system. For more complicated systems with very different types of atoms (e.g. dissociation of O_2 or H_2 on metal surfaces) more care has to be taken to ensure a continuous transition path between starting and end configuration. This can be achieved by the so-called "nudged band" method.[31] We found a complex exchange path which avoids to a large extent the simultaneous breaking of bonds. This leads to a surprisingly low E_{EX} which is comparable to E_d. The minimum energy exchange path is plotted in Fig. 7. It consists of several sections between local minima: First, the Si ad-atom moves partially towards the diffusion saddle point breaking the bond to only one of the 3 As neighbors. Then it dips down towards the As layer, while one of the As moves out of its substitutional position. There we find a first local minimum position SM1 where both exchanging partners bind to atoms in the layers below. Then the path proceeds to a dimer position where the Si and As atoms form a dimer molecule lying parallel to the surface with the center nearly vertically above the original substitutional As position. The final section brings the Si atom to the substitutional position with the formation of tetrahedral bonds while the As atom moves upward and finally ends up as an As ad-atom at a bridge position between the Si atom and a second As atom in the layer. We find saddle points between all local minima, the highest energy on the exchange path $Si_{add} + As_{sub} \rightarrow As_{add} + Si_{sub}$ is found on the first section, and $E_{EX}(Si)$ $= 0.27$ eV.

As a consequence of the relation between the activation energies, we suggested a new scenario, the re-exchange controlled diffusion of Si ad-atoms on surfactant covered Si(111):As, which can explain all features of the experiments of Voigtländer.[8]

Although the diffusion barrier of a Si ad-atom is reduced by the As layer (from $E_{d,pure} = 0.76$ eV (experimental value[8]) and 1.12 eV (calculated[32] on pure Si(111)) compared to $E_d(As) = 0.25$ eV on Si(111):As (our calculated value[28])) the diffusion path on the surfactant covered Si(111) surface is much shorter than on the pure surface. The reason is the very efficient exchange reaction $Si_{add} + As_{sub} \rightarrow As_{add} + Si_{sub}$ with a surprisingly low exchange barrier $E_{EX} = 0.27$ eV. The substitutional Si atom gains a binding energy of $E_B(Si) = 0.8$ eV. Thus, newly deposited Si ad-atoms quickly find their way from the ad-atom position on top of the As layer to the substitutional position under the As layer. One can estimate (see Table 3 below) that at $T = 600$ K a Si ad-atom on average performs 2 diffusion jumps before an exchange jump takes place.

Once the Si atom is in the substitutional position it can resume diffusion only after it is re-activated to the ad-atom position. For this process the re-exchange barrier $E_{REEX} = E_{EX} + E_B \approx 1.1$ eV has to be overcome. The effective diffusion of the Si ad-atom thus is a sequence of short paths on top of the As layer, then dropping to the substitutional position by an exchange process, and resuming diffusion when being re-activated by a re-exchange. The effective diffusion barrier is the large re-exchange energy: $E_{d,eff} = E_{REEX} =$ 1.1 eV.

5.4 Results for Hetero-epitaxy: Ge and Sn ad-atoms on Si(111):As

To check whether these results can be transferred to hetero-epitaxy, we recently performed calculations also for Ge and Sn ad-atoms on Si(111):As.[33]

Table 2. Energies of equilibrium and saddle point configurations of Si, Ge, Sn ad-atoms

Conf	$\mathbf{Si}_{ad}$	$\mathbf{Ge}_{ad}$	$\mathbf{Sn}_{ad}$
E_D	0.25 eV	0.25 eV	0.23 eV
E_{EX}	0.27 eV	0.70 eV	
E_{sub}	-0.80 eV	-0.20 eV	0.44 eV
E_{REEX}	1.06 eV	0.91 eV	

As can be seen from Fig. 6 and Table 2 the diffusion paths and barriers for the bigger group-IV atoms Ge and Sn are very similar to the one for Si. In fact, the equilibrium site (H3) and the diffusion saddle point configuration (T4) are also the same as for Si ad-atoms. Again this shows the very efficient passivation of the Si(111) surface by the As overlayer.

On the other hand, incorporated Ge and Sn atoms behave very differently from Si. The energy gain of a substitutional Ge atom (after an exchange reaction $Ge_{add} + As_{sub} \rightarrow As_{add} + Ge_{sub}$) is only $E_B(Ge) = 0.20$ eV, whereas substitutional Sn is *unfavorable.* The Sn binding energy is negative, $E_B(Sn)$ = -0.44 eV, i.e. substitutional Sn has a higher energy compared to the Sn ad-atom configuration. This can be understood by the trend in the strength of the binding energies for covalent bonds: The more extended the contributing p-states the smaller the energy gain of the bond. C bonds in diamond are the strongest, and Sn has a non-cubic metallic modification, $\beta - Sn$. In addition, there is some energy loss due to the strain which has to be built up when the bigger Ge or Sn atoms are squeezed into a lattice with the lattice constant of Si (the lattice constant of Si is smaller by 4% compared to Ge and more than 7% compared to α-Sn). We find that the energy differences of the ad-atom and substitutional configurations for Si and Ge without relaxation of the neighboring atoms are the same within 0.05 eV as with full relaxation. Thus, we conclude that the strain energy is a minor effect.

The bond energy difference apparently is already important when the Ge atom approaches the As layer on the exchange path. Although we found that the path coordinates (including the configurations of the relative minima) of the exchange path $Ge_{add}+As_{sub} \rightarrow As_{add}+Ge_{sub}$ are very similar to the path coordinates of the Si-As exchange, the energy along the path starts to deviate from the Si values when the Ge atom dips into the As layer. There is a nearly constant shift of the energy by ≈ 0.5 eV on the last portion of the path. Also, the saddle point for Ge exchange is found on this part of the path, and the saddle point energy for exchange $E_{EX}(Ge) = 0.7$ eV, i.e. by almost 0.5 eV higher than for Si exchange. Thus, Ge ad-atoms perform a large number of diffusion jumps before being incorporated under the As layer on Si(111).

Because of the negative binding energy Sn ad-atoms will not be incorporated into the As layer on Si(111). From other calculations[16] we know that Sn likes to sit underneath an As layer on InP(110) which has a much larger lattice constant than Si.

5.5 Growth-Mode Predictions

From the results of our calculation of the barrier heights for diffusion of ad-atoms on top of the As layer and for the exchange with an As atom we can quantitatively estimate the diffusion path length before incorporation.

From each H3 site an ad-atom has the possibility to jump over one of three T4 saddle points or to exchange with one of three As neighbors. The probability to make a diffusion jump and survive as an ad-atom is thus given by

$$q = \frac{\Gamma_d}{\Gamma_d + R_{EX}} \tag{30}$$

with the jump rate Γ_d (see Eq. 4) and a similarly defined exchange rate

$$R_{EX} = \nu_{EX} \exp\{-E_{EX}/kT_S\}\,. \tag{31}$$

If we assume that the pre-exponential factors are similar for the two processes (for bulk diffusion the attempt frequency is usually taken to be of the order of the Debye frequency, i.e. $\approx 10^{13} sec^{-1}$), the quotient of exchange and diffusion rate is determined by the difference of the respective activation energies only, and the survival probability becomes:

$$q = \frac{1}{1 + R_{EX}/\Gamma_d} = \frac{1}{1 + \exp\{(E_d - E_{EX})/kT_S\}} \tag{32}$$

The probability distribution $w(n)$ that an isolated ad-atom makes exactly n diffusion jumps before it is incorporated is given by

$$w(n) = (1-q)\, q^n, \tag{33}$$

and the average number of jumps is

$$\begin{aligned} \bar{n} &= \sum_{n=0}^{\infty} n\, w(n) = (1-q) \sum_{n=1}^{\infty} n\, q^n = \frac{1}{1-q} \\ &= 1 + \frac{\Gamma_d}{R_{EX}} = 1 + \exp\{(E_{EX} - E_d)/kT_S\} \end{aligned} \tag{34}$$

Table 3. Sugested growth mode of Si, Ge, Sn on Si(111):As at $T = 600K$.

stable conf	E_{EX}-E_d	$\bar{n}(600K)$	**growth mode**
Si: Si_{sub}	0.02 eV	$\bar{n} \approx 2$	2D islands
Ge: Ge_{sub}	0.50 eV	$\bar{n} \approx 10^4$	step flow (?)
Sn: Sn_{add}	–	$\bar{n} \approx \infty$	on top of As

The resulting number of jumps at $T_S = 600K$ are summarized in Table 3.

Si ad-atoms with a very small difference of the activation energies ($E_{EX} - E_d \approx 0.02$eV) make very few diffusion jumps, and a Ge ad-atoms with a large difference ($E_{EX} - E_d \approx 0.5$eV) make a large number of diffusion jumps before incorporation.

Thus, we expect that the main contribution to growth for Si homo-epitaxy on Si(111):As comes from nucleation of islands on terraces as is observed.[8]

On the other hand, because of their long survival Ge ad-atoms can reach step edges easily. Then growth would proceed mainly by step flow if the Ge ad-atoms can be incorporated at the steps. For growth on pure surfaces one would expect that the incorporation at steps is much easier than on terraces and leads to a larger binding energy. But it has been argued by Kaxiras[26] that As (and Sb) surfactants not only passivate the flat terraces but also the step edges. This would reduce the sticking of Ge atoms at steps which can thus be released to resume diffusion. In total, the probability of island nucleation on terraces will be increased. The crucial ingredients for a decision concerning the growth mode according to this model are the energies for exchange and binding of Ge atoms on flat terraces and at step edges in the presence of a surfactant layer.

Since Sn ad-atoms do not gain energy when incorporated under the As layer we would expect that Sn grows on top of As on Si(111).

6 Conclusions

All suggestions are still more or less speculative until the whole set of reaction rates and in particular activation and binding energies indicated in Sections 3.3 and 3 is known with some precision. This would have to include the whole sequence of structures from single ad-atoms via islands of various sizes and heights (single layer and double layer in the case of Si(111)) to steps and full double layers. In the case of Ge hetero-epitaxy also thicker layers have to be considered, and the stress as a function of thickness should be analyzed. Due to the presence of surfactants the kinetic models for island growth have to modified. There are some new approaches which take into account the competition of surface diffusion on top of the surfactant layer and incorporation into the layer by surfactant ad-atom exchange.[34,35,36]

Due to the progress in preparation techniques, experimental methods (particularly scanning tunneling microscopy (STM)), computational power and theoretical tools to simulate larger systems, we can be confident to gather enough information for a microscopic understanding of epitaxial growth in the near future.

Acknowledgements

We would like to thank the BMBF-Verbundprojekt 22, Halbleitergrenzflächen: Morphologie und elektronische Eigenschaften, for financial support.

References

1. A. Zangwill, *Physics at surfaces* (Cambridge University Press, 1988).
2. J. Krug, *Theoretische Aspekte der Molekularstrahlepitaxie* in *Dynamik und Strukturbildung in kondensierter Materie* (28. IFF-Ferienkurs 3. - 14. 1997, Forschungszentrum Jülich); J. Villain and A. Pimpinelli, *Physique de la croissance cristalline* (Éditions Eyrolles, Paris, 1995); A.-L. Barabási and H.E. Stanley, *Fractal Concepts in Surface Growth* (Cambridge University Press, 1995).
3. A. Papoulis, *Probability, Random Variables and Stochastic Processes* (Mc-Graw-Hill, New York, 1965); K. Binder, *Application of the Monte Carlo Method in Statistical Physics* in *Topics in Applied Physics* **36** (Springer, Heidelberg, 1987).
4. U. Köhler, O. Jusko, G. Pietsch, B. Müller and M. Henzler, *Surf. Sci.* **248**, 321 (1991).
5. B. Voigtländer and A. Zinner, *Appl. Phys. Lett.* **63**, 3055 (1993).
6. J.A. Venables, G.D.T. Spiller and M. Hanbücken, *Reports on Progress in Physics* **47**, 399 (1984).
7. D. Walton, *J. Chem. Phys.* **37**, 2182 (1962).
8. B. Voigtländer *el al.*, *Phys. Rev.* B **51**, 7583 (1995).
9. R.O. Jones and O. Gunnarson, *Rev. Mod. Phys.* **61**, 689 (1989).

10. *EStCoMPP: Electronic Structure Code for Materials Properties and Processes* (Forschungszentrum Jülich, 1999).
11. R. Car and M. Parrinello, *Phys. Rev. Lett.* **55**, 2471 (1985).
12. S.H. Vosko, L. Wilk, and M. Nusair, *Can. J. Phys.* **58**, 1200 (1980).
13. D. Vanderbilt, *Phys. Rev.* B **32**, 8412 (1985).
14. L. Kleinman, D.M. Bylander, *Phys. Rev. Lett.* **48**, 1425 (1982).
15. N. Kosugi, *J. Comput. Phys.* **55**, 426 (1984).
16. P. Richard (Ph.D. thesis, RWTH Aachen, Germany, 1996, published as Report of the Research Center Jülich, Jül-3331, December 1996).
17. S. Blügel (Ph.D. thesis, RWTH Aachen, Germany, 1988).
18. R.Fletcher, *Broyden-Fletcher-Goldfarb-Shannoo method* in *Practical Methods of Optimization* (John Wiley & Sons, 1987).
19. R. Berger, S. Blügel, A. Antons, Wi. Kromen, K. Schroeder, *Proceedings of the NIC-Workshop "'Molecular Dynamics on parallel Computers"'*, Jülich, 08.-10. Februar 1999 (World Scientific, 1999).
20. M. Copel, M.C. Reuter, E. Kaxiras, and R.M. Tromp, *Phys. Rev. Lett.* **63**, 632 (1989).
21. M. Horn-von Hoegen *et al.* , *Phys. Rev. Lett.* **67**, 1130 (1991).
22. G. Meyer, B. Voigtländer, and N.M. Amer, *Surf. Sci.* **272**, L541 (1992).
23. B. Voigtländer and A. Zinner, *J. Vac. Sci. Technol.* A **12**, 1932 (1994).
24. E. Kaxiras, *Mat. Sci. and Engin.* B **30**, 175 (1995).
25. D. Kandel and E. Kaxiras, *Phys. Rev. Lett.* **75**, 2742 (1995).
26. E. Kaxiras, *Thin Solid Films* **272**, 386 (1996).
27. D. Reinking, M. Kammler, M. Horn-von Hoegen and K.R. Hoffmann, *Appl. Phys. Lett.* **71**, 924 (1997).
28. K. Schroeder, B. Engels, P. Richard and S. Blügel, *Phys. Rev. Lett.* **80**, 2873 (1998).
29. J.J. Mortensen, Y. Morikawa, B. Hammer, and J.K. Norskov, *J. of Catalysis* **169**, 85 (1997).
30. This is already a simplification which assumes that the rest of the atoms are "spectator" atoms whose coordinates follow uniquely from the energy minimization for given coordinates of the two exchanging atoms. For semiconductors with directed covalent bonds this need not always be the case.
31. G. Mills, H. Jónsson and G.K. Schenter, *Surf. Sci.* **324**, 305 (1995).
32. K. Cho and E. Kaxiras, *Europhys. Lett.* **39**, 287 (1997).
33. K. Schroeder, A. Antons, R. Berger, Wi. Kromen, and S. Blügel, to be published.

34. J. Tersoff, A.W. Denier van der Gon, and R.M. Tromp, *Phys. Rev. Lett.* **72**, 266 (1994).
35. D. Kandel, *Phys. Rev. Lett.* **78**, 499 (1997).
36. I. Markov, *Surf. Sci.* **429**, 102 (1999).

HOMO- AND HETEROEPITAXY OF METAL ON METAL GROWTH

U. KÖHLER, C. JENSEN, K. RESHÖFT
Institut für Experimentalphysik / Oberflächenphysik,
Ruhr-Universität Bochum, D-44780 Bochum, Germany
E-mail: ulrich.k.koehler@ruhr-uni-bochum.de

A.C. SCHINDLER, D.E. WOLF
Theoretische Physik, FB 10
Gerhard Mercator Universität, D-47048 Duisburg, Germany

The atomistics of metal on metal growth is studied using time resolved STM, SPA-LEED and kinetic Monte-Carlo simulations (KMC). A comparison of the growth behavior seen with STM and KMC is used to show the influence of the fundamental diffusion steps on the morphology of the growing surface in the homoepitaxial system Fe on Fe(110). Especially the influence of a Schwoebel-Ehrlich barrier and an anisotropy in edge diffusion are shown. Details of the transition from a pseudomorphic layer to the relaxed bulk like structure is shown in the examples Fe on W(110). The transition proceeds stepwise and extends over several layers. It involves different metastable arrangements of the layer depending on the deposition temperature.

1 Introduction

In magnetic multilayers iron layers are often combined with non-magnetic materials with non-matching lattice parameters. This paper will give an overview over phenomena found for the system Fe on W(110). Even in heteroepitaxy eventually iron will relax to its own lattice structure and therefore adopt its own intrinsic growth behavior. The structure of the interface between the layers, on the other hand, is strongly influenced by the lattice mismatch. To optimize the quality of the multilayers, a detailed understanding is necessary of the processes controlling growth on an atomic level both for homoepitaxial growth and for heteroepitaxy. In the present study, scanning tunneling microscopy (STM) has been applied in situ directly during growth to follow the evolution of a specific location on the surface through different stages of growth. Whereas the dynamics of island formation can be followed on a time scale of seconds up to hours, the motion of individual atoms is too fast to be resolved at RT for most metal on metal systems. For the case if homoepitaxial growth of Fe(110) we therefore compare the results determined with STM with a Monte-Carlo simulation, which includes the exact crystal symmetry of the bcc(110) surface.

2 Results

2.1 Fe on Fe(110)

To avoid intermixing at the interface, often low deposition temperatures are used. Low growth temperatures, on the other hand, cause problems with the crystalline quality and surface roughness during growth even in homoepitaxy. As an example we like to show the influence of the local diffusion behavior on the morphology of Fe(110) layers. Fig. 1 shows selected frames from a sequence of STM micrographs of Fe/Fe(110) homoepitaxial growth at RT. The images have been high-pass filtered (as illuminated from the left).

Steps in the Fe(110)-substrate are visible in the upper left and the lower right. In fig. 1 a) islands have nucleated on the flat terrace and finger like half-islands extend from the step edges. These islands are strongly anisotropic, elongated in the [001] direction. In the submonolayer range the island edges are indented, showing a high density of kinks. The coalescence of islands sets in only in the [001] direction (see arrow in fig. 1 c), while in $[1\bar{1}0]$ islands stay separated even at a high coverage. The second layer nucleates long before the first one is closed, indicating the existence of a very effective step edge barrier (Schwoebel-Ehrlich-barrier[5]) preventing inter-layer mass transport at RT. A quantitative analysis shows that the layer distribution is roughly described by a Poisson-distribution.[1] The islands in higher layers become increasingly anisotropic (fig. 1 c)-f)) up to a ratio of length to width of ≈ 10. With increasing coverage more and more layers nucleate and the surface roughens.

For a better understanding of the observed surface morphologies in terms of basic processes on an atomic scale, we performed Kinetic Monte Carlo (KMC) simulations which includes the full symmetry of the bcc(110)-surface. In fig. 2 the local geometry and possible diffusion events are shown.

In a microscopic model of crystal growth, atoms are deposited randomly on the substrate with a constant rate F. In KMC surface diffusion is implemented as a thermally activated process, with an Arrhenius transition rate from lattice site i to j given by

$$R^{ij} = R_0 \, e^{E_A^{ij}/k_B T}. \tag{1}$$

R_0 is the *attempt frequency* of order $10^{12}s^{-1}$. T denotes the substrate temperature and k_B is Boltzmann's constant. E_A^{ij} is the activation energy for an individual diffusion process from i to j. Diffusion upwards across a step edge and desorption from the surface are neglected since the energy barriers for these events can be assumed to be too large at the simulated temperatures.

Two ways of choosing the activation energies were used in the simulations: First they were be parameterized in terms of bond numbers of the neighborhood of an adatom before and after a diffusion event:

$$E_A^{ij} = E_D + (n_N - \mu m_N)E_N + (n_{NN} - \mu m_{NN})E_{NN} + E_S \qquad (2)$$

In this scheme E_D is the diffusion barrier of single adatoms, E_N and E_{NN} the bonding energies for nearest and second nearest neighbors in plane and E_S an extra barrier at a step edge, which hinders an atom to cross a step

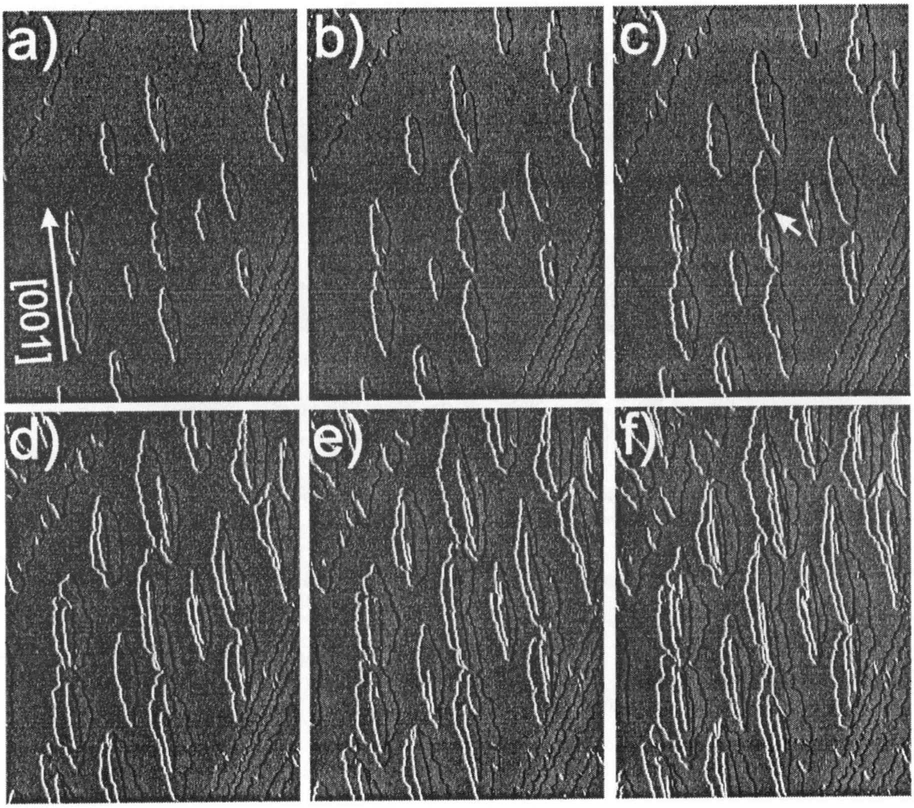

Figure 1. Selected frames of set of STM images showing the in situ growth of Fe on Fe(110) at RT. The frames are highpass filtered simulating an illumination from the left. (frame size 2700Åx 3600Å).

edge (Schwoebel-Ehrlich barrier). The numbers of adjacent neighbors before a jump are denoted n_N and n_{NN}. For lateral neighbors that would be present after a jump the numbers are given by m_N and m_{NN}. The factor $\mu \in [0,1]$ is used to set the influence of the neighbors after a jump. As μ was intended to increase the diffusion along the step edges it is set to zero in cases of attachments to lateral neighbors and inter-layer transport. The parameters were chosen such that desired features like island density and the shape of islands were obtained. Using such a bond counting model we study the general behavior of the growth on bcc(110)–surfaces.

The second approach used was to calculate a full set of activation energies – one for each possible configuration of neighboring atoms – using a Finnis–Sinclair potential.[3] Even though only adatom hops in the ⟨111⟩ and [001]

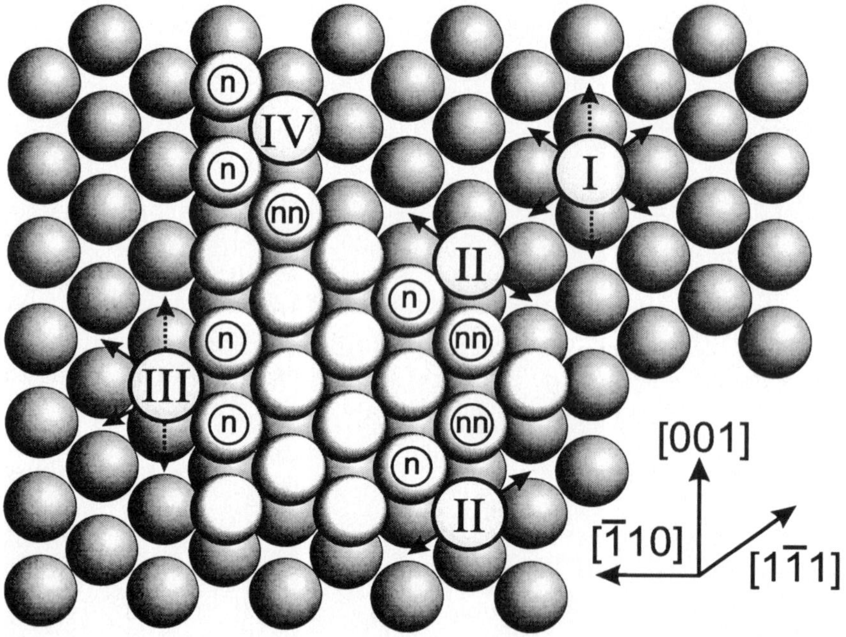

Figure 2. Possible jumps and configurations on the bcc(110) surface. Nearest- and next-nearest-neighbors are denoted by n and nn, respectively. Adatoms predominantly move along ⟨111⟩ (solid arrows) while a jump in [001] is nearly impossible. Atoms (denoted II) are able to diffuse along step edges along ⟨111⟩ and easily reach a better coordinated kink site (atom labeled IV) while atoms attached to a [001]–edge can not move along the step edge and detach before they are incorporated at kink sites.

directions are taken into account, there remains a set of $2 \cdot 2^8 = 512$ barriers to be calculated, plus three Schwoebel-barriers down the [001], ⟨110⟩ and ⟨111⟩ edges. This number can only slightly be reduced due to symmetry arguments. When calculating all those barriers, the system and configuration atoms are only relaxed in the initial step and additionally, configuration atoms are only relaxed in respect to their interaction with the (110) surface, not with each other. This is necessary to conserve a high symmetry. Suppressing the relaxation of surface and environment of course causes some numerical errors, but is necessary in order to avoid artifacts due to the fragility of the system. In agreement with Ref. [4], where barriers in a fully relaxed system were systematically lower than in a frozen system, our calculations for selected sample configurations gave barriers approximately 30% lower compared to the ones obtained with a frozen configuration. For further details on this see Ref. [1].

Possible reasons for the anisotropy of the islands as seen in fig. 1 were tested using KMC with the bond counting energetics. A very strong influence of an anisotropy in the diffusion along step edges was found. According to our calculations based on the FS–model and FIM-data[2] diffusion proceeds predominantly along ⟨111⟩– directions. For free adatoms this is already apparent by looking at the geometry of the bcc(110): An atom (labeled I in fig. 2 moving along a ⟨111⟩– direction (solid arrows) can easily cross a 2-fold bridge site, while in case of a jump along [001] (dotted arrows) it has to go over an on top position which should be much harder. The same situation can be found for the diffusion along the two main step edges ⟨111⟩ (atoms labeled II in fig. 2) and [001] (labeled III). The difference in mobility at step edges has a drastic effect on the growth of islands. An adatom attaching to a ⟨111⟩–edge (atom labeled II) has a high mobility and can easily meet another atom or reach a better coordinated kink site (IV). An atom attaching a [001]–edge (atom labeled III), on the other hand, has no chance to diffuse along the edge and to be incorporated in a kink site and will detach easily. This results in a faster growth at the ⟨111⟩–edges leading to the observed elongation of the islands along [001]. From the above discussion it is evident, that the island shape is mainly kinetically stabilized. On the other hand, there are also energetic effects supporting the anisotropic shape of the island. The simulations show that an interaction with next nearest neighbors (nn) is necessary, otherwise the islands develop a compact shape.

To test the set of material specific activation energies calculated using the Finnis–Sinclair potential we selected a set of 3 islands from fig. 1, for which the temporal behavior during growth is known, as input for the KMC-simulation. Fig. 3 a)-c) shows the growth as seen with STM, d)-f) the result of the MC-run. Two things had to be adjusted when comparing the simulated growth behavior to the real one: the temperature had to be set to 580 K instead RT to obtain the correct island density, and the step edge barriers were chosen slightly higher than those provided by the calculations based upon the Finnis–Sinclair potential. The theoretically correct barriers proved to be too low so that the islands of the lower level grew too wide before the next nucleated. As is obvious in the figure, many details of the real growth reproduce in the simulation, like the anisotropic shape of the islands with increasing aspect ratio at higher layers, the rough 3D–growth and the

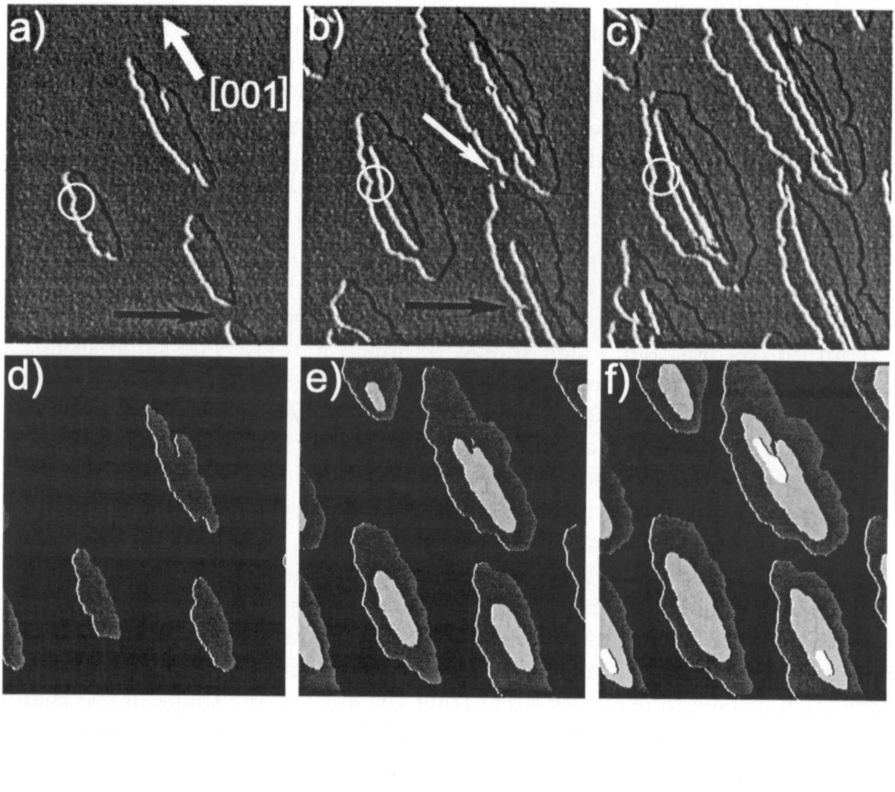

Figure 3. Comparison of the growth behavior (a-c) of three islands of the sequence of fig. 1 with Monte Carlo–Simulation based on the Finnis and Sinclair model (d-f) using a digitized version of the first STM-image as input.

reproduction of kinks in higher layers. The anisotropic shape of the islands in higher layers, on the other hand, is not as pronounced in the simulation than in reality. Nevertheless, it can be stated that, despite the simple model for the interaction potential used in the simulation, most of the features of the growth on the bcc(110)-surface are correctly represented.

With increasing total coverage the side walls of the islands along $[1\bar{1}0]$ convert into step regular trains. Further growth reduces the average step separation until characteristic facets are formed.[6] As a consequence of the 3d-growth mode the slopes of these edge facets increase with coverage. In addition to the in situ growth experiments, Fe–films of about 100ML grown at RT were analyzed by STM and SPA-LEED. At this coverage an equilibrium growth mode has developed, in which a stable set of facets is present. The preferred coalescence in [001]–direction leads to elongated 3d-islands in [001] which are more than 1000Å long. For a SPA-LEED analysis the surface morphology can be approximated as a formation of up and down stairs in $[1\bar{1}0]$–direction with infinite extent in [001], i.e. facets that can be constructed by sequences of {100} and {110} microfacets.

The SPA-LEED survey scan in fig. 4 a) shows a splitting of the normal spots characteristic for a complete facetting of the surface. When the energy of the electron beam is varied, the position of the extra spots periodically shifts in the (11)–direction.

A small stripe along (11) around the (00)–spot is shown in fig. 4 b) for beam energies between 158 eV and 216 eV. The vertical position of each stripe corresponds to the wave vector perpendicular to the surface q_z of each measurement. Therefore fig. 4 b) shows a view of the q_x/q_z–plane in reciprocal space. Additionally, lines showing the expected positions for different facets on the (110)–surface are sketched. Apparently, the intensity is spread over a wide area indication a variety of facets. On the other hand, there is obviously no intensity at the position of the {100}–facet, i.e. only facets with smaller angles are stable.

These findings concerning the stabile facets are consistent with results of Thürmer et al.[7,8] for the growth of Fe on Fe(100) at 400-450 K. On both surfaces, the (110) and the (001), the system doesn't adopt the steepest facet ({001} and {011} respectively) with an inclination of 45° to the surface. Instead, facets with smaller angles are favored. This behavior can be explained in terms of surface currents of impinging adatoms. Atoms landing on {001}–microfacets are assumed to funnel downwards to the next site with a fourfold coordination to the substrate. These atoms contribute to a downward current which causes the surface to flatten and a creation of {110}–facets. On the other hand, atoms landing on a {110}–facet cannot diffuse downwards due to an effective step edge barrier. They will move to a step edge of the next higher layer and therefore contribute to an upward current. Only if the

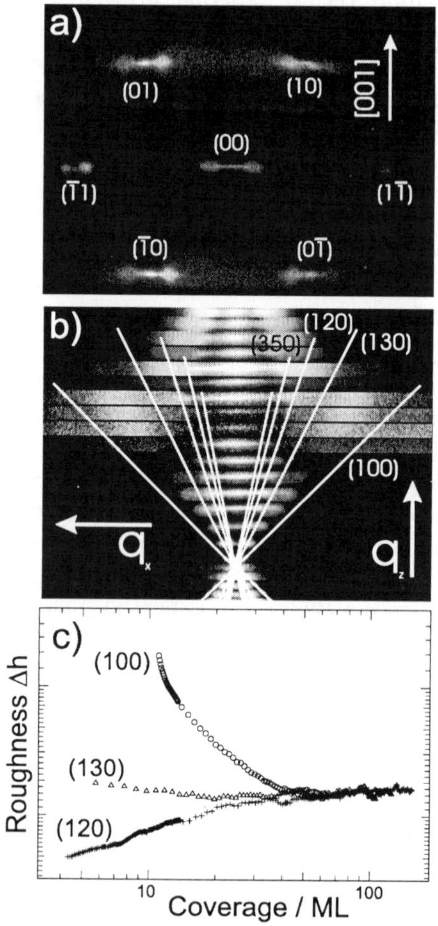

Figure 4. a) Kinetic facetting of Fe/Fe(110) observed with SPA-LEED as a splitting of the spots. b) Stripes around the (00)–spot arranged vertically to show a q_z/q_x–view of the reciprocal space. c) Development of the roughness of a KMC–simulation during growth starting with different ideal facets.

up– and downward currents are in equilibrium the slope of the facets stays constant during further growth. The equilibrium condition is fulfilled if the probability for an atom to land on a {001}–microfacet is as high as to land on a {110}–facet. In this case the respective areas of the facets projected onto the starting surface have to be equal. For the (001)–surface this is given by {120}–facets,[7,8] for the (110)–surface by {130}–facets.

KMC can be used to determine the stability of facets during growth and to prove the above model. Starting with ideal {100}, {130} and {240}–facets the trend of the roughness of the film as a measure for the average slope during growth was followed. Fig. 4 c) shows that the roughness of the {100}– and the {120}–facets converge towards the value of the {130}–facet after a coverage of about 50ML. Steeper facets flatten out, facets which initially have a lower inclination become steeper. This confirms the assumption of our model containing a downward funneling mechanism for atoms that land on sites that are not fourfold coordinated to the substrate.

2.2 Fe on W(110)

The misfit between the substrate and an epitaxial film often dominates the properties of thin heteroepitaxial layers. It may lead to several different structural defects depending on coverage and temperature. For the system Fe on W(110) we like to illustrate that STM applied in situ during growth can directly show the strong interrelation between the island morphology and the formation of dislocations. Fe and W belong to the same bcc-crystal class but have a relatively large misfit of 9.4%. For the first monolayer Fe grows pseudomorphically to the W-substrate which means that Fe has adopted the tungsten lattice constant.[9]

Fig. 5a) starts already with a coverage of 1.1 ML. The brightest areas represent the second layer while the dark channels are part of the substrate. The islands of the second layer increase in size and grow together in [001]-direction (fig. 5 b). In fig. 5 c), some small islands of the third layer and an enhanced corrugation are visible on a part of the second layer island (see also fig. 5 g)). These features are aligned along the [001]-direction and seem to spread initially only along this direction. In a second stage, from fig. 5 c) to e), the enhanced corrugation in the second layer and the nucleation of islands proceed also along the [1$\bar{1}$0]-direction. The average distance of nuclei is of the same order as the width of the unit cell of the two-dimensional distortion network which is clearly visible after the third layer islands have coalesced[10] (see fig. 5 f)). The periodicity in the third layer is consistent with an extra iron-row every 1/10 rows of strained iron which is inserted to compensate the misfit between Fe and W.

In fig. 5 g), the enhanced corrugation which precedes the formation of the third-layer nuclei forming zigzags and hexagonal lateral structures is shown in a perspective view. The structure which is only weakly ordered in this stage, is caused by the onset of strain relaxation with the creation of dislocations in the second layer. Thus the STM-sequence directly allows the conclusion that the relaxation starts in the second layer *before* the nucleation of the third layer begins. The beginning network of dislocations in the second layer,

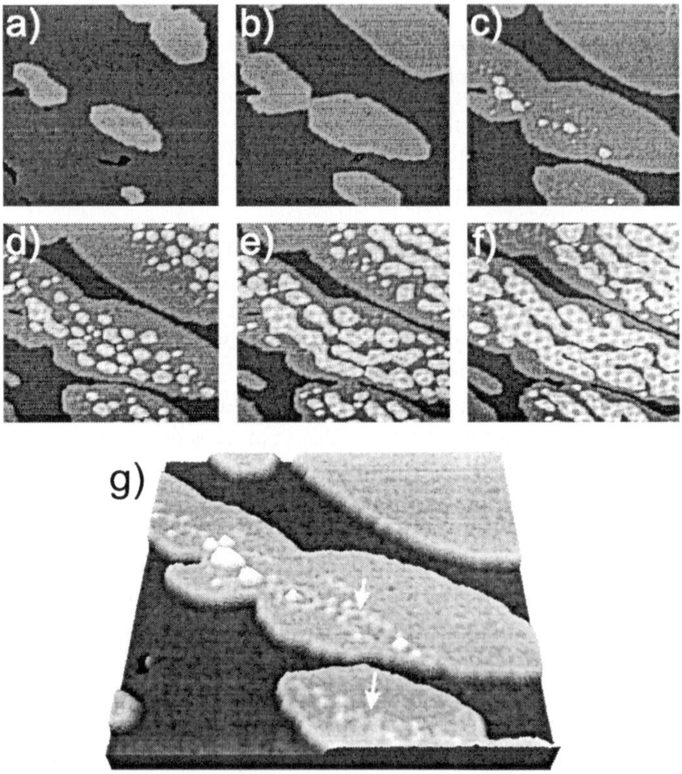

Figure 5. a)-f) Selection from a sequence of STM-images showing the RT-growth of the second layer of Fe on W(110) followed by the nucleation of the third layer and the formation of a dislocation network. g) Pseudo-3D-view of an STM-image chronologically between b) and c).

on the other hand, provides sites for an inhomogeneous nucleation of the next layer leading to a much higher density of island. There were no stable distortion-lines observed on the second layer without nuclei of the third layer. Further, a rearrangement of the distortion lines visible on top of third layer during the coalescence of third-layer islands can be seen by watching the whole sequence represented by fig. 5 as a movie. The sequence of fig. 5 shows that the appearance of the epitaxial layer on a larger scale is dominated by the relaxation to the own lattice constant and the formation of dislocations. The further growth behavior at RT is described in Ref. [10,11].

Whereas at RT only one type of dislocation network is present, the relaxation behavior is more complex at higher temperature. Fig. 6 shows an STM-sequence of Fe-growth on W(110) at 570 K. In this experiment the shading effect of the STM-tip leads to a inhomogeneous deposition with a higher rate in the lower part of the image. In fig. 6 a) the first ML of Fe is nearly completed. Because of the higher mobility at elevated temperature growth proceeds by step flow and the first ML is completely close before higher layers grow. In fig. 6 b) the terraces start to fill with a second layer. At the same time an island of the third layer nucleates in the middle of the lower half of the image which develops into a 3D-island in fig. 6 c)-f). In g) details of the appearance of the different layers are shown. On top of the 3D-island, which is five layers high in the end of the sequence, the two-dimensional dislocation pattern known from the RT-growth is visible. It shows that here the iron is completely relaxed to its own lattice structure. Additionally, also in the second and the third layer dislocation arrangements with different periodicities are visible. These arrange in one-dimensional line structures along [001] indicating that here the Fe-layer is relaxed only in one direction. Figs. 6 d)-f) show that as soon as there is completely relaxed iron present in the fifth layer (d) the coverage in lower levels in the neighborhood diminishes again (see arrows in d), e), f)). Fe-material flows from the second and third layer to the 3D-island. This indicates that these one-dimensional relaxation structures in the second and third layer are only metastable present during growth. In these layers there is still some residual strain left. Therefore, in the end there are only two-dimensionally relaxed 3D-islands stable together with one layer of pseudomorphic iron covering the W-substrate.

If the growth is slow and the temperature is high enough that the layer is close to thermodynamic equilibrium a stable Stranski-Krastanov growth mode is observed. At lower temperature, also larger areas with the metastable second and third layer can be grown and also non-equilibrium arrangements can be frozen in. This offers the possibility to study the magnetic and electronic behavior of these metastable structures and also use their properties for special multilayer structures.[12]

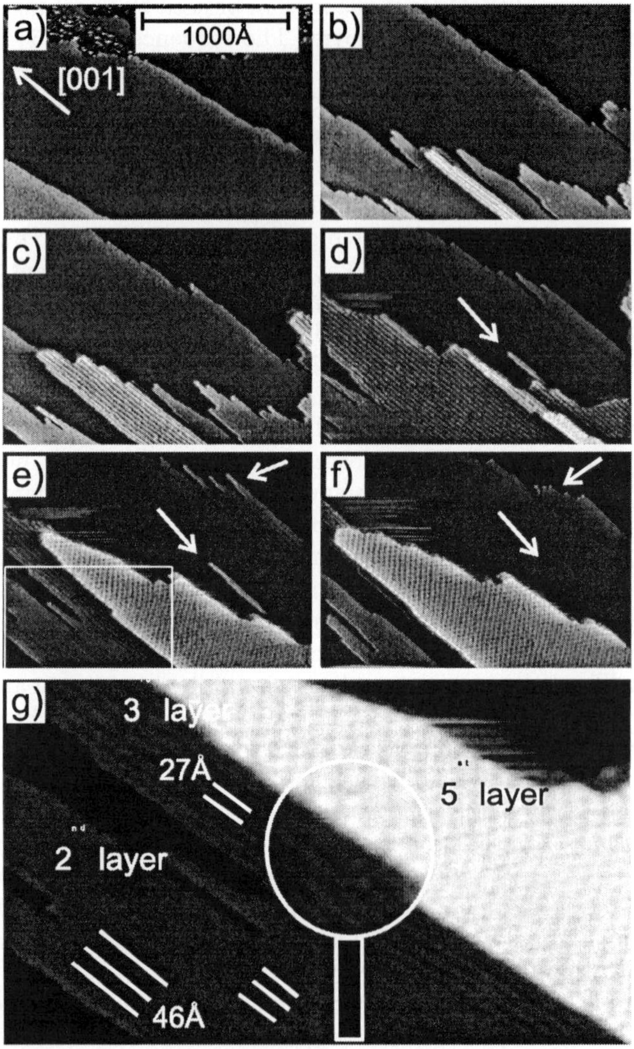

Figure 6. a)-f) Sequence of STM-images showing the growth of Fe on W(110) at 570 K. g) zoom-in on the region marked in e).

Acknowledgements

The authors would like to acknowledge the funding by the Deutsche Forschungsgemeinschaft.

References

1. U. Köhler, C. Jensen, A.C. Schindler, L. Brendel, D.E. Wolf, accepted by *Phil. Mag.* B.
2. G. Ehrlich and F.G. Hudda, *J. Chem. Phys.* **44**, 1039 (1966).
3. M.W. Finnis, and J.E. Sinclair, *Phil. Mag.* A **50**, 45 (1984).
4. M. Schroeder and D.E. Wolf, *Surf. Sci.* **375**, 129 (1997).
5. R.L. Schwoebel, *J. Appl. Phys.* **40**, 614 (1969).
6. M. Albrecht, H. Fritsche, and U. Gradmann, *Surf. Sci.* **294**, 1 (1993).
7. K. Thürmer,R. Koch, M. Weber, and K.H. Rieder, *Phys. Rev. Lett.* **75**, 1767 (1995).
8. K. Thürmer, R. Koch, P. Schilbe, and K.H. Rieder, *Surf. Sci.* **395**, 12 (1996).
9. U. Gradmann, G. Waller, *Surf. Sci.* **116**, 539 (1982).
10. H. Bethge, D. Heuer, C. Jensen, K. Reshöft, and U. Köhler, *Surf. Sci.* **331-333**, 878 (1995).
11. C. Jensen, K. Reshöft, and U. Köhler, *Appl. Phys.* A **62**, 217 (1996).
12. M. Bode, R. Pascal, M. Dreyer, and R. Wiesendanger , *Phys. Rev.* B **54**, 8385 (1996).

HIGH-TEMPERATURE-STM INVESTIGATION OF EPITAXIAL GROWTH OF SI AND GE ON SILICON

BERT VOIGTLÄNDER AND NILS THEUERKAUF

Institut für Grenzflächenforschung und Vakuumphysik, Forschungszentrum Jülich, 52425 Jülich, Germany

E-mail: B.Voigtlaender@fz-juelich.de

We use a scanning tunneling microscope (STM) capable of imaging the growing layer during MBE-growth at high temperatures. This method (MBSTM) opens the possibility to follow MBE growth processes dynamically on the atomic scale and gives access to the evolution of specific features during growth. The influence of surface reconstructions on growth kinetics can be studied directly. For the case of growth of Si islands on Si(111) we find lateral growth of rows of the width of the 7×7 reconstruction unit cell at the edges of two-dimensional islands. This leads to a kinetic stabilization of magic island sizes. The evolution of size and shape of individual {105} faceted Ge islands (hut cluster) on Si(001) is measured during growth. A slower growth rate is observed when an island grows to larger sizes. This behavior can be explained by kinetically self-limiting growth.

1 Introduction

Recently scanning tunneling microscopy (STM) has become a powerful method to study epitaxial growth on the atomic level. However, STM has been limited in the past to "snapshots" of certain growth stages. The growth was interrupted at a specific coverage and the sample was quenched to room temperature and transferred to the STM for imaging. The main disadvantage of this mode of STM operation is that local growth structures which are accessible by a real space microscopy cannot be studied in their dynamic evolution during growth. Recently some effort was made to overcome these disadvantages.[1,2,3]

In this letter we describe a mode of STM operation which opens the possibility to observe the dynamics during epitaxial growth "in vivo" (MBSTM). The experiments were performed in a standard ultra-high vacuum system. Due to the open design of the STM the molecular beam from an evaporator can be directed towards the sample while the STM is scanning the surface.[4] During imaging the sample is held at MBE growth temperatures of 600-900 K.

Using this method of simultaneous scanning and molecular beam epitaxy the area behind the tip is shaded. This area can be measured if we evaporate Antimony at room temperature (small diffusion length) while the tip is fixed

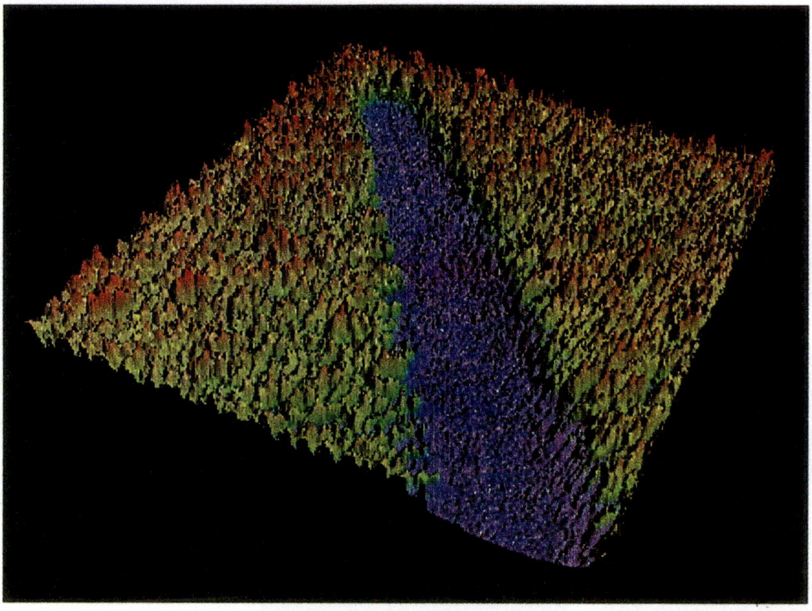

Figure 1. perspective view of the silhouette of the STM tip on the surface produced by the molecular beam. The antimony covered region is shown in yellow. The area shaded by the STM tip is blue. The tip diameter can be measured to 1000 Å(image size 5000×5000 $Å^2$.

(i.e. not scanning). Scanning this area afterwards shows the silhouette of the tip on the surface (fig. 1). The tip diameter can be measured to about 1000 Å so that for images of several thousand Å scan range the shaded area is sufficiently small.

2 Influence of the (7×7) reconstruction in Si/Si(111) epitaxy

During the epitaxial growth of 2D Si islands on Si(111) the growth of a selected island can is observed as a function of time. Figure 2(a-f) shows STM images from a growth sequence of such an island. In (a), the shape of the island is triangular. Images (b-d) show the same island at a later stage during growth. As shown by images (b-d), growth proceeds by advancement of a row of a certain width along the right island edge. The position of the kink at which the row is ending is shown by an arrow in (b-d). An analysis of the width of this row and further atomically resolved images show that the width of such

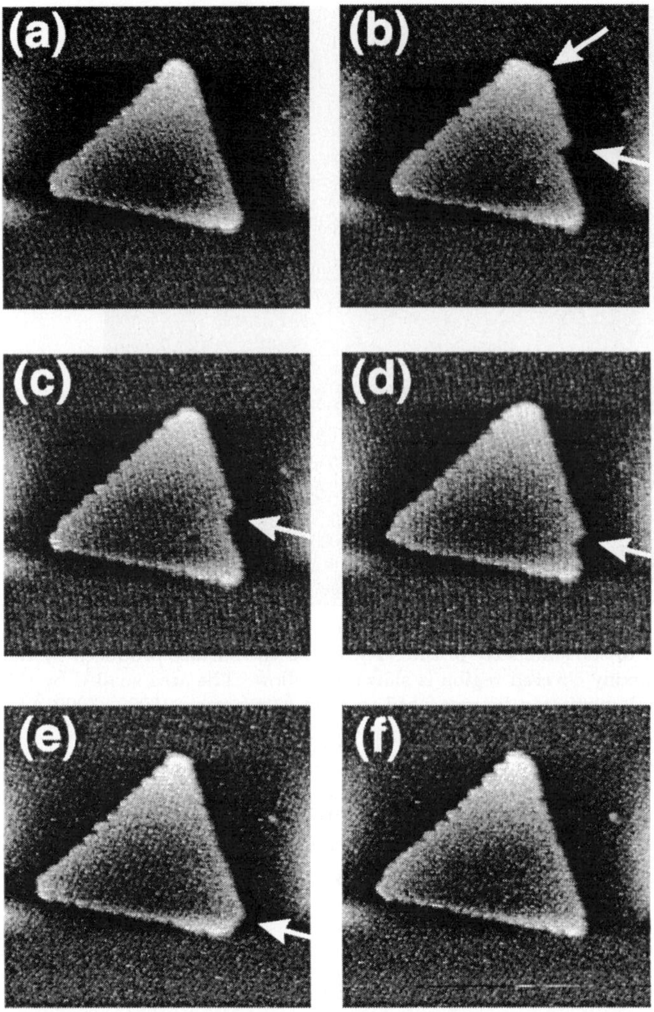

Figure 2. Sequence of images showing the lateral growth of a triangular Si(111) island. A row of the width of the 7×7 unit cell is growing along the right edge of the island (b-f). The image size is 500×500 Å^2, T=575 K. The complete growth sequence is available as a movie on the World Wide Web: http://www.fz-juelich.de/video/voigtlaender.

a row is 27 Å, which is just the width of one 7×7 reconstruction unit cell.[5]

In the following we discuss the influence of the 7×7 reconstruction on

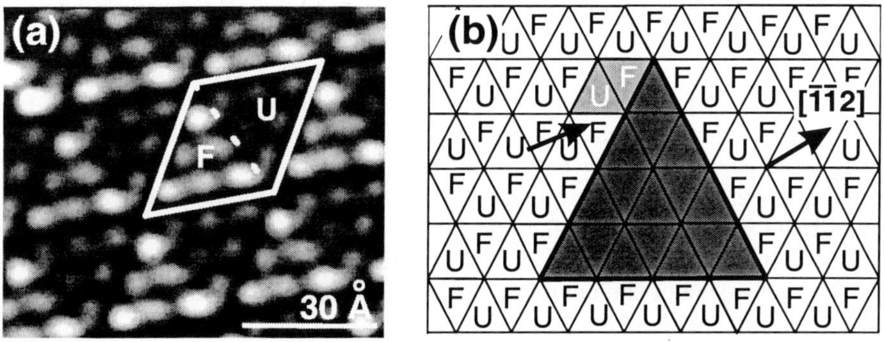

Figure 3. (a) STM image of the Si(111) - 7×7 reconstruction. The white protrusions are the Si adatoms. The unit cell is indicated by a white rhombus. A dashed line divides two triangular subunits. Since the image is taken at negative sample bias (-2 V), the adatoms in the faulted half of the unit cell (F) appear brighter than those in the unfaulted half (U). (b) Arrangement of the U and F parts of the 7×7 unit cell on the substrate around a triangular island (shown in gray).

the growth behavior. The rhombic unit cell of this reconstruction consists of two triangles. One of this triangles has a stacking fault in the surface layers, relative to the substrate stacking (F-HUC). The other triangle of the reconstruction unit cell is unfaulted relative to the substrate (U-HUC), fig. 3(a).

During lateral growth of an island, the surface reconstruction of the substrate has to be lifted and the substrate atoms have to rearrange to the bulk structure. This transformation of the reconstructed surface layer towards the bulk structure is a general phenomenon which has to occur in any epitaxial growth at reconstructed surfaces. It requires a higher energy to lift the reconstruction of the F-triangle as compared to the U-triangle.[6] To lift the reconstruction in the U-triangle, only atoms in the uppermost adatom layer have to rearrange. This is associated with a relatively low energy barrier. Lifting the reconstruction of the F-triangle requires the removal of the stacking fault in the layer below the adatoms. This rearrangement of atoms in deeper layers is associated with a larger energy barrier. This should lead to a high activation barrier for overgrowth of the F-triangle compared to overgrowth of the U-triangle.

In Fig. 3(b), a Si island (gray) and the U and F triangles of HUCs of the surrounding reconstructed substrate surface are shown. Due to the crystallographic orientation of the island, it is surrounded only by substrate F-HUCs. This means that further lateral growth (initiated by the overgrowth of an

F-triangle) is hindered by a high energy barrier. Once an F-triangle has nucleated, the neighboring U-triangles can be overgrown more easily (no stacking fault has to be removed). The overgrowth of the next F triangle is facilitated by the existence of a "macro kink" (arrow in Fig. 3(b)). Here the cost of the stacking fault energy is reduced by a gain in the island edge energy: The edge length is reduced after growth of an F triangle. Therefore, neighboring U and F units can be overgrown in quick succession, leading to the fast growth of a stripe of the width of the 7×7 unit cell.

Due to the quick growth of a row once it is nucleated and a longer time until the nucleation of the next row the island shape of a completed triangle is found often in snapshot images of the island morphology. Islands of complete triangular shape are kinetically stabilized. This higher stability for the closed shell triangular islands leads to pronounced peaks of islands of magic sizes in the measured island size distributions.[7]

3 Kinetically self-limiting growth of Ge islands on Si(001)

In Ge epitaxy on Si(001) three-dimensional Ge islands with {105} facets nucleate after the completion of the 2D wetting layer.[8] The island edges are oriented along $< 100 >$ directions, i.e. at 45° to the direction of the reconstruction dimer rows. The strain of these islands is partially relaxed elastically, but they are coherent with the substrate lattice (i.e., free of dislocations between the substrate and the hut clusters).[8] The evolution of the growth of these hut clusters is observed by STM during growth. A sequence of STM images for increasing coverage is shown in fig. 4. From the images for instance the volume of the hut clusters can be calculated.

Figure 5 shows the evolution of the volume of several individual islands as a function of total deposited coverage. The initially higher growth rate of individual islands just after the nucleation, indicated by initially large slopes in fig. 5, decreases when the islands grow to a larger size. In the following, we use a model which shows that kinetically self-limiting growth explains the observed slower growth for larger island sizes. This kinetic model for the growth of hut clusters relies on a barrier for the nucleation of each successive atomic layer on the {105} facets.[9,10] The barrier for the nucleation of a new facet increases with the facet size. This size dependent nucleation barrier for the repeated overgrowth of the {105} facets is the reason for the self-limiting growth behavior.[10] Results of the model calculations for the evolution of the hut cluster volume are shown as solid and dashed lines in fig. 5 and are in good agreement with the experimental data. The experimentally observed slower growth rate for larger islands is clearly reproduced. This indicates, that a

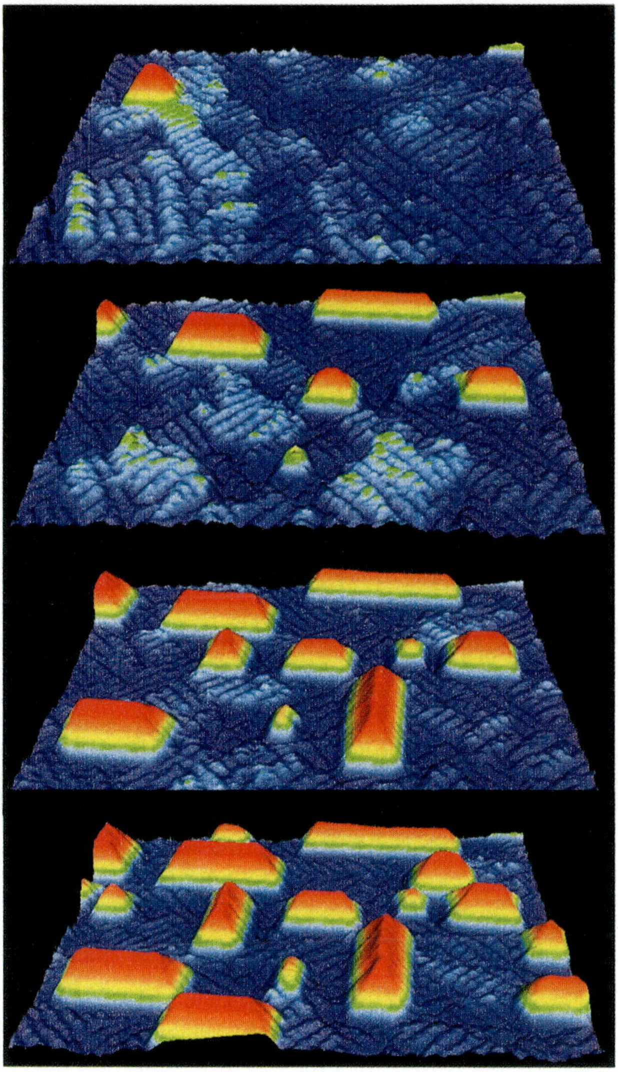

Figure 4. Perspective view of STM sample images of the hut cluster growth as function of coverage beyond the wetting layer. (0.07 ML, 0.46 ML, 1.33 ML and 1.81 ML from the top to the bottom; Image area: 1300×1000 Å^2, T=575 K. The complete growth sequence is available as a movie on the World Wide Web: http://www.fz-juelich.de/video/voigtlaender/.

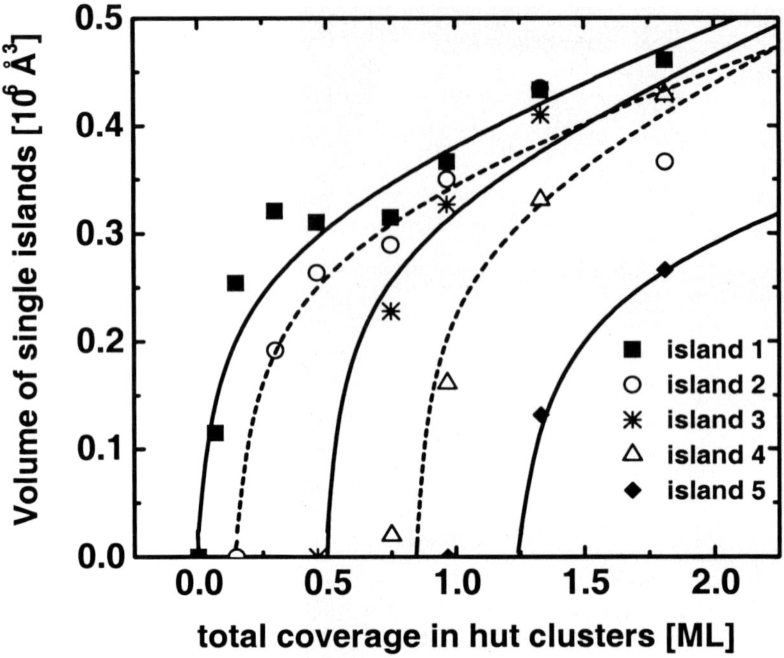

Figure 5. Evolution of the volume of individual hut clusters. The different symbols correspond to different individual islands. The size evolution shows self-limiting behavior: The initially larger growth rate (large slope) just after the nucleation decreases when the islands grow larger. Results of a model calculation of kinetically self-limiting growth including a kinetic energy barrier for the nucleation of new material on the facets are shown as solid and dashed lines.

kinetic self-limitation is effective during the growth of larger hut clusters. This kinetic self-limitation arises due to an energy barrier for the nucleation of new material on completely filled {105} facets.

4 Summary

In summary we have shown that STM can be operated in real time during molecular beam epitaxy at high temperatures. MBSTM opens the possibility to observe the growth history of single growth features in real time. In the case of the homoepitaxial growth of Si on Si(111) the influence of the surface reconstruction on the lateral growth of Si islands was demonstrated. For the case of heteroepitaxial growth of Ge on Si(001) a slower growth rate is observed when an island grows to larger sizes. This behavior can be explained by self-limiting growth. A kinetic growth model involving a nucleation barrier for each repeated growth of a new atomic layer on the {105} facets agrees with the experimental results for the evolution of the island volume.

References

1. A. K. Schmid, and J. Kirschner, *Ultramicroscopy* **42-44**, 483 (1992).
2. T. Hasegawa, M. Kohno, S. Hosaka, and S. Hosoki, *Phys. Rev.* B **48**, 1943 (1993).
3. B. Voigtländer and A. Zinner, *Appl. Phys. Lett.* **63**, 3055 (1993).
4. B. Voigtländer, A. Zinner, and Th. Weber, *Rev. Sci. Instrum.* **67**, 2568 (1996).
5. B. Voigtländer, and Th. Weber, *Phys. Rev. Lett.* **77**, 3861 (1996).
6. W. Shimada and H. Tochihara, *Surf. Sci.* **311**, 107 (1994).
7. B. Voigtländer, M. Kästner, and P. Smilauer, *Phys. Rev. Lett.* **81**, 858 (1998).
8. Y. W. Mo, D. E. Savage, B. S. Schwartzentruber, and M. G. Lagally, *Phys. Rev. Lett.* **65**, 1020 (1990).
9. K. M. Chen, D. E. Jesson, S. J. Pennycook, T. Thundat, and R. J. Warmack, *Mat. Res. Soc. Symp.* **399**, 271 (1996).
10. M. Kästner and B. Voigtländer, *Phys. Rev. Lett.* **82**, 2745 (1999).

LOW-TEMPERATURE SCANNING TUNNELLING SPECTROSCOPY

RICHARD BERNDT

2. Physikalisches Institut, RWTH Aachen, D-52056 Aachen, Germany

Using low-temperature scanning tunnelling microscopy we have investigated confinement of electronic surface states to hexagonal islands and other nanoscale structures on Ag(111). Local spectroscopy and spatial maps of the differential conductance are analyzed using simple models of the electronic structure. Moreover, we studied electron scattering from isolated Ce ad-atoms on Ag(111). dI/dV tunnelling spectra reveal a characteristic antiresonance around the Fermi energy which is interpreted a Fano interference between delocalized Ag states and the localized Kondo resonance.

1 Introduction

The development of low-temperature scanning tunnelling microscopes (STM) operating under ultra-high vacuum conditions has provided new opportunities for investigating electronic states at metal surfaces. At low temperatures, broadening of the Fermi levels is reduced enabling rather high energy resolution in scanning tunnelling spectroscopy. Further advantages are the absence of diffusion which, together with the spatial resolution of the STM, is a prerequisite for detailed studies of electronic states on and near single adsorbed atoms and other nanoscale structures. In this short note, we present two examples of spectroscopic studies: imaging and spectroscopy of surface state confinement and spectroscopy of a single Kondo impurity.

2 Methods

The experiments were performed in a custom-built ultra-high vacuum (UHV) STM operating at $T = 5$ K.[1] Hexagonal Ag islands on Ag(111) were prepared by Ar^+ bombardment and Ag deposition at room temperature. Subsequently, the sample was cooled in order to freeze the island structure. Isolated Ce and Ag atoms were deposited by evaporation from a tungsten filament onto the Ag substrate at $T = 5$ K.

For spectroscopy of the tunnelling current I and the differential conductance dI/dV versus the sample voltage V, which in simple models is directly related to the local density of states (LDOS) above the surface, the STM feedback loop is opened. A voltage modulation (amplitude $V_m = 2 - 4\ \mathrm{mV_{rms}}$) is added to V and a lock-in amplifier is used to measure dI/dV.

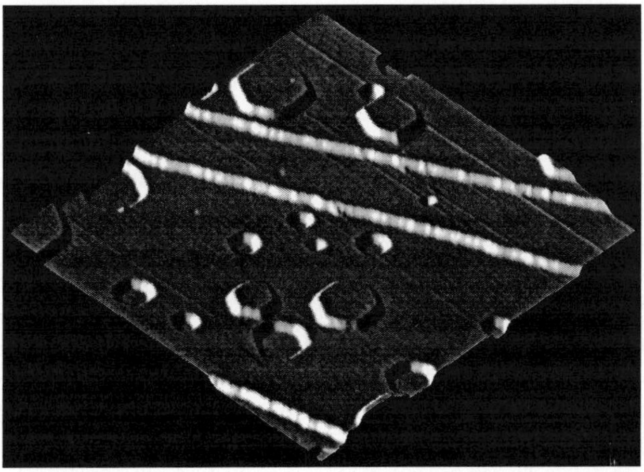

Figure 1. Pseudo-three dimensional STM image of monolayer-high Ag adatom islands and vacancy islands on Ag(111). The crystalline symmetry leads both to adopt approximately hexagonal shapes. Also present are several monatomic steps; image size approximately 160×160 nm^2. From Ref. [5]

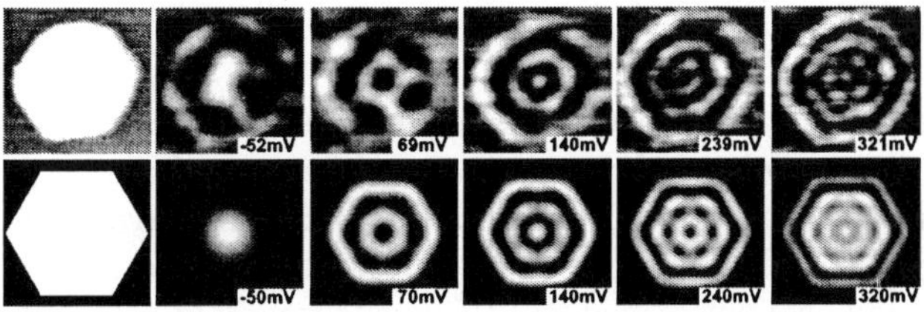

Figure 2. Upper row: constant-current topograph of an approximately hexagonal Ag island on Ag(111) (inner diameter $\Phi \sim 10.4$ nm), and a series of dI/dV maps recorded at various bias voltages (at $T = 50$ K). Lower: geometry of a hexagonal box confining a two-dimensional electron gas, and the resulting local density of states. Experimental values have been used for the surface state onset E_0, electron effective mass m^*, and the island size. The calculations include broadening via a self-energy. From Ref. [5]

3 Spectroscopy of hexagonal islands

Taking advantage of the improved energy resolution at low temperatures we investigated surface state confinement in hexagonal Ag islands on Ag(111) (Fig. 1) in analogy to a room-temperature experiment by Avouris and Lyo.[2] Figure 2 shows a series of experimental dI/dV scans from an island.

Strongly voltage-dependent features are observed which correspond to standing waves patterns following the confinement of surface state electrons by the edges of the island. For sample voltages around −65 mV and lower the interior of the island is featureless in the images. Around −50 mV we find a single peak near the center of the island. It gradually evolves with increasing voltage to multiple rings with six-fold azimuthal modulations, which are partially distorted by deviations of the island shape from hexagonal symmetry.

The experiments are complemented by electronic structure calculations (Fig. 2). At the simplest level we analyse the energy levels in terms of two-dimensional "particle-in-a-box" eigenstates, corresponding to electrons with effective mass m^* (which we take to be 0.42, the value determined from both STM[3] and photoelectron spectroscopy[4]), confined within a two-dimensional domain by infinitely high barriers, the domain corresponding to a regular hexagonal region for a perfect island or some closely related shape when modelling imperfect islands. The use of this simple two-dimensional model is justified by the successful modelling of the experimental data, below, but also by comparison with the results of more realistic electronic structure calculations.[5]

Owing to the small separation in energy of the eigenstates, the patterns in Fig. 2 cannot be identified as the state densities of individual confined states, but instead represent superpositions of a number of such states from a range of energies which is determined by intrinsic (lifetime) and thermal broadening.[2,6,7]

A clearer picture of the electronic structure emerges when we utilize the symmetry of the islands: Only eigenstates which transform according the A_1 representation of the group C_6 (6mm) of the hexagon[8] have a non-zero amplitude at the center. Therefore, in order to obtain resolved spectral structures in the experiment,, we performed dI/dV spectroscopy at the centers of islands where A_1 states play a dominant role.
Figure 3 shows data for island sizes from a single Ag adatom to an extended Ag(111) terrace. On the terrace (Fig. 3g) we find the familiar spectrum of defect free Ag(111).[3] In addition to a background conductivity from bulk states (suppressed in Fig. 3) a sharp rise of dI/dV occurs at the surface state edge $E_0 \approx -70$ meV. At higher energies within the surface state band, a gradual decrease of the conductivity is observed although the detailed spectral shape in this voltage region is found to be tip-dependent.

Spectra recorded above the center of hexagonal islands with diameters $\Phi = 32, 50, 68, 98$ Å (Figs. 3c–f) show considerably more structure, with a series of resonant levels identifiable above E_0. The spectra from these islands are strongly island-size dependent. The lowest level, identified as E_1 in the figures, rises rapidly in energy as the island size decreases. An increase in the

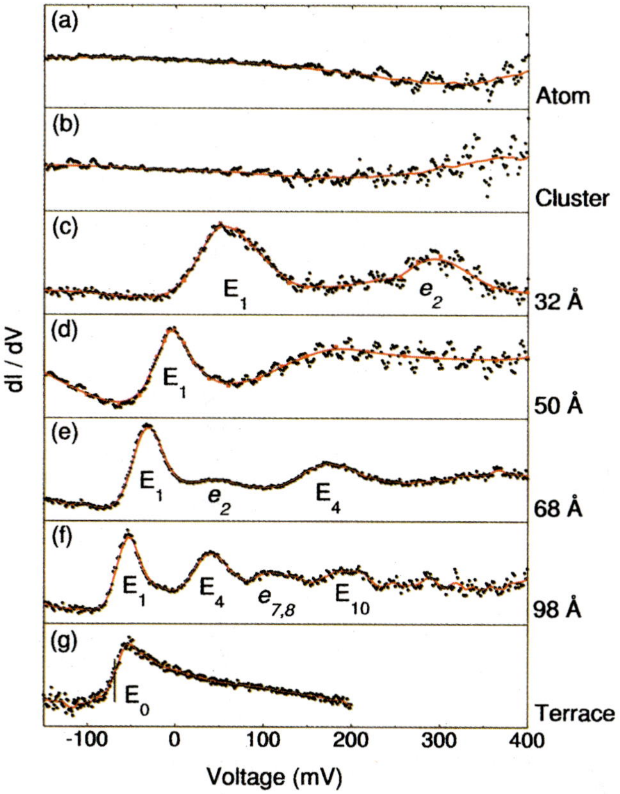

Figure 3. dI/dV-spectra recorded above (a) a single Ag atom, (b)–(f) the center of nearly hexagonal islands, and (g) a clean Ag(111) terrace. The island diameters Φ are indicated. A constant background has been subtracted from the spectra. Moreover, the spectra have been scaled arbitrarily. E_0 is the energy of the surface state band edge. E_n denotes the energy of state n. Italics are used to indicate states that are symmetry forbidden at the center of a perfect hexagon. From Ref. [9]

spacing between the resonances, and in their widths, is also apparent as the island size reduces. Even smaller structures such as a cluster of few atoms (Fig. 3b) or from a single Ag adatom (Fig. 3a) do not exhibit any discernible surface state related feature in the energy range investigated. In these smallest structures the confinement raises the energies of the surface state electrons above the energy window under consideration. A semi-quantitative understanding of these spectra through comparison with the predictions of the "2d box" model can be achieved.[9]

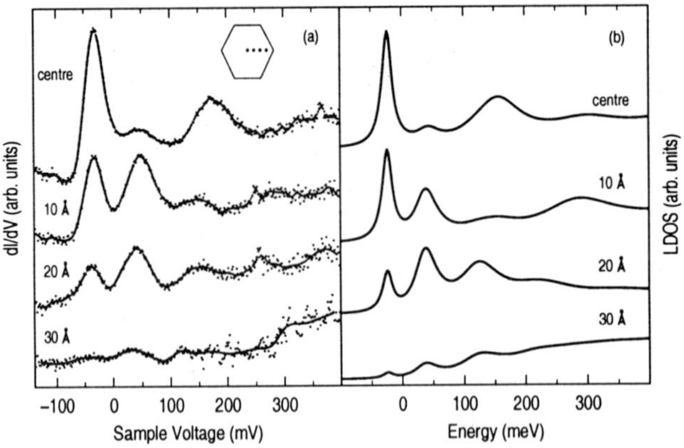

Figure 4. (a) Experimental dI/dV and (b) calculated LDOS from a slightly distorted hexagonal island. Spectra are shown for tip positions at the island center and at distances of 10 Å, 20 Å and 30 Å from the center towards a corner of the island. From Ref. [9]

This model can also describe dI/dV scans in specific directions. In Fig. 4 we compare experimental dI/dV spectra with the calculated LDOS, measured at the center and at three off center points for a slightly distorted island with $\Phi \sim 68$ Å. For this island the topological scans indicated a smaller upper half than lower which was taken into account in the calculation. As a result of this broken symmetry, there are contributions to the LDOS at the island center from states with symmetries other than A_1, visible as a weak peak near +50 mV, which increases in amplitude at points moving out radially. The good overall correspondence between the features seen experimentally and theoretically further illustrates how the model, although very simple, may be used to describe both spectral and spatial features of the electronic structure.

4 Spectroscopy of a single Ce impurity

A particularly interesting subject for single-atom spectroscopy are magnetic impurities. Recently, Yazdani et al. succeeded in observing modifications of the density of states caused by a single magnetic impurity on a superconductor.[10] Here, we are concerned with Kondo systems of magnetic impurities in nonmagnetic, non-superconducting metals. These systems exhibit an unconventional ground state and unusual effects in their transport

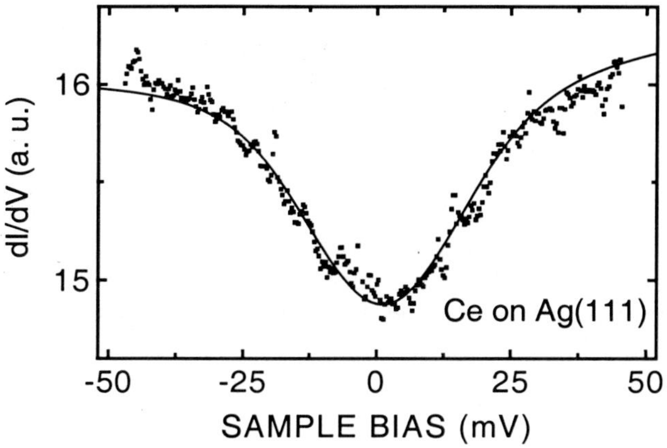

Figure 5. Comparison of an experimental dI/dV spectrum (dots) from a single Ce impurity on Ag(111) with a calculated result (line). Lifetime and instrumental broadening where taken into account by applying a 16 meV Gaussian broadening. From Ref. [15]

properties and the specific heat: Below a characteristic Kondo temperature T_K, the local d or f-moment is screened to yield a many-body non-magnetic singlet ground state.[11] This screening gives rise to low-energy excitations which cause the so-called Kondo resonance near the Fermi energy. Such a resonance has been observed for a number of homogeneous dense Kondo compounds with photoelectron spectroscopy.[12,13] So far, however, no direct observation of the Kondo effect from a single magnetic impurity on an atomic length scale had been reported

Figure 5 displays a dI/dV tunnelling spectrum (dots) measured on an isolated Ce adatom on Ag(111). There is a clear dip in dI/dV around the Fermi energy which disappears when the tip is moved laterally by a few Angstroms. Tunnelling spectra of a prototype nonmagnetic adatom, namely Ag, are featureless throughout the range from -300 mV to 300 mV (Fig. 3a). These data suggest that the dip near E_F is indeed characteristic of the magnetic Ce adatom. Our observation appears to contradict the recent theoretical prediction of a sharp resonance in dI/dV near the Fermi energy.[14]

To interpret our result we performed model calculations of f-spectra within the Anderson single impurity model for a single magnetic $4f$ impurity embedded in a bulk conductor.[15] For reasonable parameters "Kondo" peaks appear on an energy scale which matches that of the observed antiresonance as predicted in Ref.[14] and in apparent contradiction to Fig. 5.

Therefore, we took a fresh look at the tunnelling process: The localized Kondo resonance is degenerate with the delocalized Ag band states. This scenario resembles the spectroscopy of a discrete autoionized state, which has been elucidated by Fano.[16] Fano found that, in the limit where the transition matrix element for the localized state approaches zero ($q \approx 0$ case in Ref. [16]), the autoionization spectrum shows a dip or antiresonance, as a consequence of the interference effects at the site with the localized state. In our case of tunneling at a large tip-to-surface distance, the STM is known to probe *sp* wave functions rather than the confined $4f$-derived Kondo ground state.[17] This is exactly the $q \approx 0$ limit! Thus, instead of a f-spectral peak, the dI/dV spectrum should show an antiresonance as observed experimentally. As discussed in detail in Ref.[15], using this Fano-picture, we obtain the line-shape of Fig. 5 (line) which is in encouraging agreement with the experimental observation (dots). We conclude that a Fano interference effect transforms the Kondo resonance known from photoelectron spectroscopy into an antiresonance in the tunnelling spectra. A similar interpretation has recently been proposed to explain similar observations of a typical Fano-line-shape from Co atoms on Au(111).[18]

It is a pleasure to thank J. Li, W.-D. Schneider, S. Crampin and B. Delley for their contributions to this work.

References

1. R. Gaisch, J.K. Gimzewski, B. Reihl, R.R. Schlittler, M. Tschudy, W.-D. Schneider, *Ultramicroscopy* **42–44**, 1621 (1992).
2. Ph. Avouris and I.-W. Lyo, *Science* **264**, 942 (1994).
3. J.T. Li, W.-D. Schneider, R. Berndt, *Phs. Rev.* B **56**, 7656 (1997).
4. R. Paniago, R. Matzdorf, G. Meister, A. Goldmann, *Surf. Sci.* **336**, 113 (1995).
5. J.T. Li, W.-D. Schneider, R. Berndt, S. Crampin, *Phys. Rev. Let.* **80**, 3332 (1998).
6. M.F. Crommie, C.P. Lutz, D.M. Eigler, *Science* **262**, 218 (1993).
7. Ph. Avouris, I.-W. Lyo, R.E. Walkup, *J. Vac. Sci. Technol.* B **12**, 1447 (1994).
8. M. Tinkham, *Group Theory and Quantum Mechanics* (McGraw-Hill, New York, 1964).
9. J.T. Li, W.-D. Schneider, S. Crampin, R. Berndt, *Surf. Sci.* **422**, 95 (1999).

10. A. Yazdani, B.A. Jones, C.P. Lutz, M.F. Crommie, D.M. Eigler, *Science* **275**, 1767 (1997).
11. O. Gunnarsson and K. Schönhammer in *Handbook on the Physics and Chemistry of Rare Earths*, Vol. **10**, pp. 103, eds. K.A. Gschneidner, L. Eyring, S. Hüfner (Elsevier Publisher, Amsterdam, 1987) and references therein.
12. F. Patthey, J.-M. Imer, W.-D. Schneider, H. Beck, Y. Baer, B. Delley, *Phys. Rev.* B **42**, 8864 (1990).
13. D. Malterre, M. Grioni, Y. Baer, *Adv. Phys.* **45**, 299 (1996).
14. S. Hershfield, J.H. Davies, J.W. Wilkins, *Phys. Rev. Let.* **67**, 3720 (1991).
15. J.T. Li, W.-D. Schneider, R. Berndt, B. Delley, *Phys. Rev. Let.* **80**, 2893 (1998).
16. U. Fano, *Phys. Rev.* **124**, 1866 (1961).
17. N.D. Lang, *Phys. Rev. Let.* **58**, 45 (1987).
18. V. Madhavan, W. Chen, T. Jamneala, M.F. Crommie, N.S. Wingreen, *Science* **280**, 567 (1998).

DIFFUSION AND GROWTH ON STRAINED SURFACES

A.C. SCHINDLER AND D.E. WOLF
Theoretische Physik, FB 10, Gerhard Mercator Universität Duisburg, 47048 Duisburg, Germany
E-mail: A.Schindler@Uni-Duisburg.de

Using an empirical tight binding potential, we study the strain dependence of hopping and exchange diffusion barriers for Platinum on the unreconstructed Pt(100) surface. Also, adatom binding energies and Ehrlich-Schwoebel barriers are calculated as a function of substrate strain ϵ. By analyzing the above quantities, we predict that for the growth of Pt on Pt(100) Layer by Layer growth should be improved under compressive strain.

1 Introduction

To improve the efficiency of electronic devices, the fabrication of highly ordered metal and semiconductor nanostructures is an important goal in technological applications.[1,2] Molecular Beam Epitaxy (MBE) seems to be a promising method for the construction of such devices. Besides nanostructures, another important aim is the growth of smooth films and multilayer systems, which also play an important role in the fabrication of electronic devices.

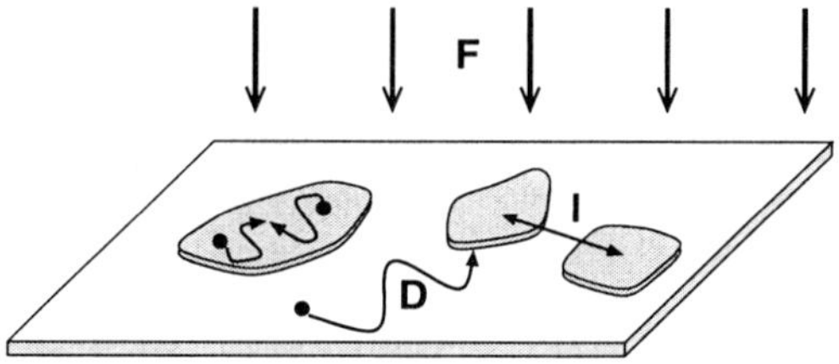

Figure 1. Important processes in MBE.

In a simplified model the process of growth in MBE consists of adatom deposition with a flux F and surface diffusion with a diffusion constant $D = \nu_0 \exp\{-(E_S - E_B)/k_BT\}$ (with $\nu_0 \approx 10^{12}/s$ being the so called attempt frequency, E_S the saddle-point or transition state energy, E_B the binding energy, k_B Boltzmann's constant and T the substrate temperature). The nucleation of islands will occur with an average distance,[3,4,5,6] $l \propto (D/F)^\gamma$. As the islands are growing, nucleations will also happen on top of existing islands, and island coalescence starts. Further on, the growth behavior will fall in one out of three categories: Layer by Layer (Frank van der Merwe) growth, 3D-islanding (Volmer-Weber) and a mixture of both, known as Stranski-Krastanov mode, where after the growth of a few smooth layers 3D islands start to form.

If now the substrate is under compressive or tensile strain, saddle-point as well as binding energies are affected by strain. In the case of a homogeneously strained surface (e.g. in the middle of a large perfectly smooth terrace), this will simply result in an increased or decreased diffusion constant D. On small islands or near island edges, the situation is more complicated. Islands try the relief stress at the edges, and therefore inhomogeneously strained areas occur, resulting in a gradient in binding as well as in saddle-point energies. This leads to an adatom current

$$\underline{j} = D\left(\nabla\rho + \frac{\rho}{k_B T}\nabla E_B\right). \tag{1}$$

Beside the first term, Fick's law, there is in general a drift towards positions of lower energy[7] (ρ is the adatom density).

Also, in systems which obey an additional barrier at step-edges, known as Ehrlich-Schwoebel barrier, this quantity will depend on the strain and will play an important role.

The aim of our work is to calculate the strain dependence of those three quantities important for surface diffusion : Binding and saddle-point energies and Ehrlich-Schwoebel barriers. By analyzing the strain dependence, we will try to make up a prediction for the growth of Platinum on Pt(100), whether Layer by Layer growth will be improved or hindered by a compressed or tensile strained substrate.

2 The model

Our calculations are based on an empirical tight-binding potential, which can be used to simulate various fcc metals[8,9] and is given by

$$E_i = A\sum_j \exp\left[-p\left(\frac{r_{ij}}{a_{NN}} - 1\right)\right] - \xi_0\left\{\sum_j \exp\left[-2q\left(\frac{r_{ij}}{a_{NN}} - 1\right)\right]\right\}^{1/2}. \tag{2}$$

a_{NN} is the equilibrium nearest neighbor distance in the fcc lattice and r_{ij} is the distance of two atoms i and j. The quantities A, p, q and ξ_0 are parameters which are used by the authors to fit the potential to the bulk properties of 10 different fcc metals.

We use a slab geometry with one free and one frozen (100) surface perpendicular to the z direction, and periodic boundary conditions in the x and y directions. By adjusting the lateral size of the simulation box, the system can be put under compressive or tensile strain. Diffusion barriers and binding energies are calculated using the molecular static method:[10] We move a "test" adatom across the surface in small, discrete steps. In each of these steps we relax the whole system (with the constraints given above), with the exception that the adatom is fixed laterally and can relax only in the z direction. Relaxation is done using a conjugate gradient method, and is stopped when both an energy ($|\Delta E| < 10^{-10}\, eV$) and a force ($|F_i| < 10^{-3} N$) criterium are fulfilled. In each of these steps we record the minimal energy, and by analyzing the resulting energy landscape we can calculate the binding and saddle-point energies, and thus the activation energy $E_A = E_B - E_S$.

To calculate the Ehrlich-Schwoebel barrier an additional island can be placed on the substrate before the calculation starts.

3 Pt/Pt(100) self-diffusion

In this paper we want to discuss growth and diffusion of Platinum on Pt(100). As observed for other fcc metals, like e.g. Iridium or Gold, under certain conditions the Pt surface obeys a (5x1) surface reconstruction, which shall be neglected for our calculations. This is valid if we restrict ourselves to small samples, like they are in use in field ion microscopy (FIM).[11]

In general, two diffusion modes are possible on the fcc (100) surface : hopping and exchange diffusion (see Fig. 2).

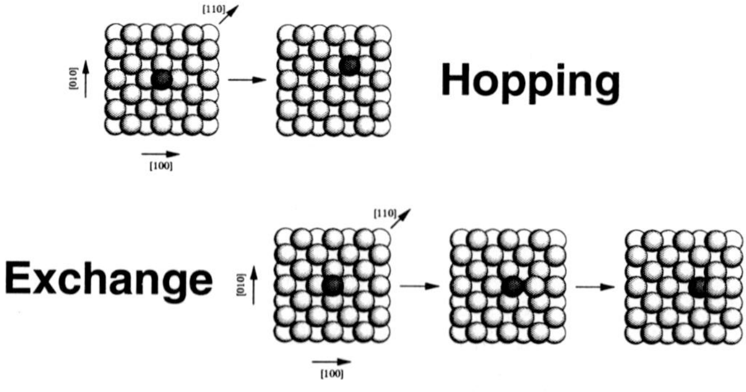

Figure 2. Possible modes for surface diffusion : hopping and exchange.

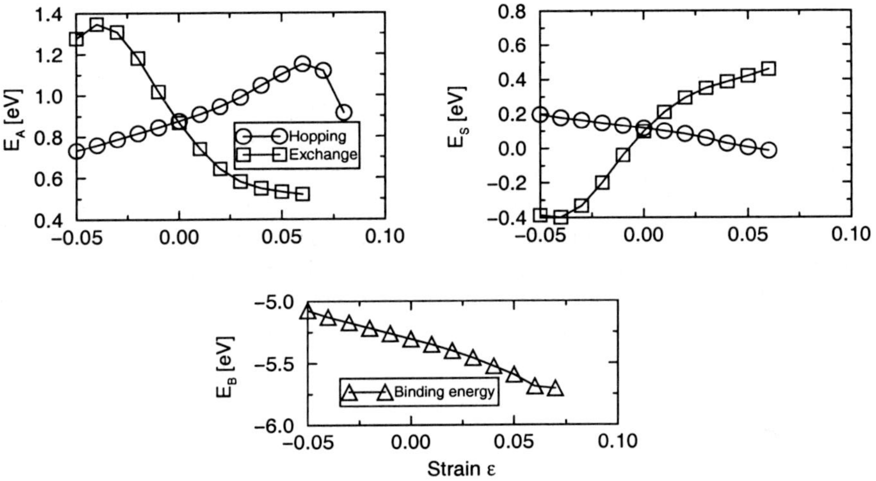

Figure 3. Results for the activation energy E_A, Ehrlich-Schwoebel barrier E_S, and adatom binding energy E_B as a function of surface strain. A negative ϵ denotes compressive isotropic strain parallel to the surface. In the reference state $\epsilon = 0$ the lattice constant along the surface is the same as in the unstrained bulk at zero pressure.

We calculated the important quantities for both diffusion modes. The results are shown in Fig. 3. Let us first look at the activation energy E_A. As observed before for Ag on Ag(100)[12] hopping and exchange diffusion behave totally different. While hopping diffusion is faster under compressive ($\epsilon < 0$) and slowed down under tensile ($\epsilon > 0$) strain, for exchange diffusion it is exactly the other way round. For vanishing strain the activation energy for hopping as well as for exchange is approximately $E_A \approx 0.87\,eV$. In the case of hopping this is almost in agreement with recent[13] density functional calculations, which give a value of $E_A = 1.04\,eV$.

To decide whether in the unstrained case hopping or exchange diffusion is the primary diffusion mode, it is not sufficient to look only at the diffusion barrier, but one also has to take the prefactor ν_0 into account. Studies[14] show that this prefactor is about a factor 10 - 50 larger for exchange than for hopping. This, together with the fact that our barrier for hopping diffusion seems to be to small, can lead to the conclusion that exchange diffusion is the preferred diffusion mode is exchange diffusion, in accordance to what was found in FIM studies.[11]

As observed before for hopping diffusion in various systems,[15,7,16] for small ϵ the activation energy of Pt on Pt(100) is an almost linear function of strain. This also holds in the case of exchange diffusion, however in this case we observe a negative slope.

Like the activation energy also the Ehrlich-Schwoebel barrier behaves completely different for hopping and exchange at an island edge. For exchange it is lowered and even gets negative for compressive strain, while under tensile strain it is increased. For hopping the strain dependence is the opposite.

Finally, the adatom binding energy is found to be lowered for tensile strain and increased for compressive strain. As we have shown before, this leads to a particle current. E.g. adatoms on strained islands on a compressed surface will drift to the island edge, as due to relaxation effects the island edges are weaker strained than the island center.

4 Prediction for Pt/Pt(100)

Now we summarize the results of the previous section and try to make a prediction under which type of strain Layer by Layer growth is preferred. Again, surface reconstruction will be neglected, and we stick to exchange diffusion, as this diffusion mode seems to be the relevant one from the experimental and theoretical point of view. It is important to note that we only need the fundamental trends we obtained in the previous section. While we do not claim that our values are close to the exact numbers for the diffusion barriers and binding energies, we trust the sign of the slope of their strain dependence. Our results are schematically summarized in Fig. 4

The first point to note is the island shape. Islands on a compressed surface will relax the edges outward, while on a tensile strained substrate they will relax inward. In the case of compressive strain the island relaxation leads to a decreased diffusion barrier on islands compared to the substrate, resulting in an increased interlayer transport of adatoms and better Layer by Layer growth.

Due to the reduced diffusion constant D in the case of compressive strain, the island density is increased. Empirically one finds that an increased island density often favors Layer by Layer growth, which is the reason why techniques like a short sputtering or temperature drop at the beginning of each monolayer,[17] or pulsed laser deposition[18] are applied. One possible explanation is an effective reduction of the Ehrlich-Schwoebel barrier for small islands. On the other hand, simulations and analytical considerations show[19] that in the absence of Ehrlich-Schwoebel barriers a reduced adatom mobility actually hinders Layer by Layer growth.

Analyzing the behavior of the binding energy, one can conclude that in

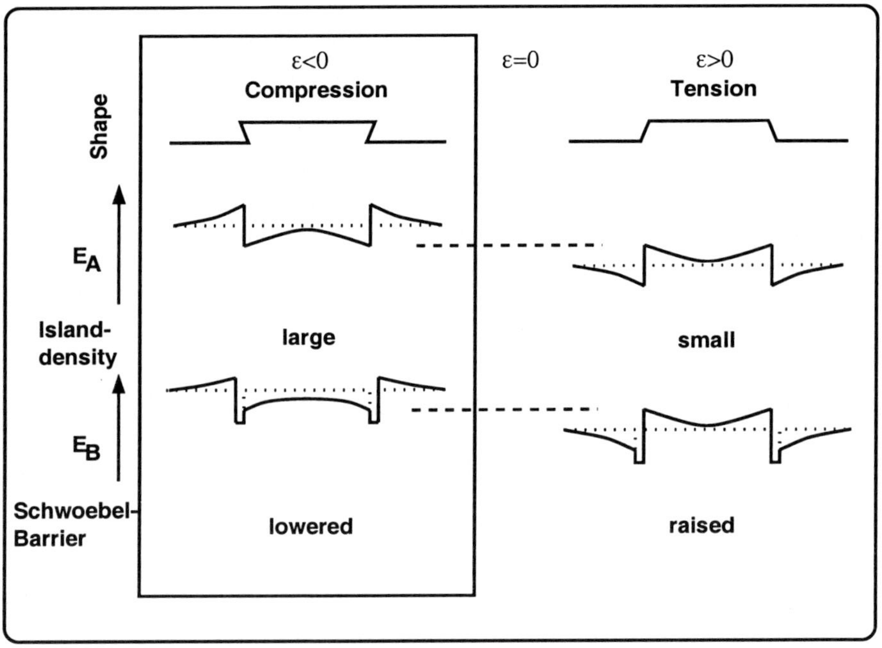

Figure 4. Summary of the results for the diffusion of Pt on Pt(100).

the case of compressive strain the relaxation of the island edges leads to a net adatom current to the less compressed island edges, while under tensile strain adatoms drift to the island center. From this point of view compressive strain should improve the interlayer transport and therefore the Layer by Layer growth.

Finally we look at the additional Ehrlich-Schwoebel barrier at the step-edges. Assuming that adatom exchange processes also prevail at the step edges, we see that the activation energy for interlayer transport is reduced under compressive strain and raised for tensile strain, again leading to improved Layer by Layer growth in the case of compressive strain.

Putting this altogether, 3 out of 4 arguments predict improved Layer by Layer growth under compressive strain. Additionally, it has been shown[20] in an example that the combined effects of particle drift and enhanced mobility on top of islands override the reduced average mobility of adatoms in favoring Layer by Layer growth. Both particle drift and inhomogeneous adatom mobility are due to strain relaxation at step edges. We conclude that compressive strain should lead to improved Layer by Layer growth for Pt on Pt(100), assuming exchange diffusion being the primary diffusion mode and neglecting the (5x1) surface reconstruction.

Acknowledgements

This work was supported by the DFG within SFB-166 and DFG grant No. WO 577/2-1.

References

1. M. Brune, M. Giovannini, K. Bromann and K. Kern, *Nature* **394**, 451 (1998).
2. S. C. Jain, A. H. Harker and R. A. Cowley, *Phil. Mag.* A **75**, 1461 (1997).
3. G. Zinsmeister, *Vacuum* **60**, 529 (1966).
4. D. R. Frankl and J. A. Venables, *Adv. Phys.* **19**, 409 (1970).
5. S. Stoyanov and D. Kashchiev in *Current topics in material science*, Vol.7, Ed. by E. Kaldis (North Holland, Amsterdam, 1981).
6. J. Villain, A. Pimpinelli, L. -H. Tang and D. E. Wolf, *J. Phys. I France* **2**, 2107 (1992).
7. M. Schroeder and D. E. Wolf, *Surf. Sci.* **375**, 129 (1997).
8. V. Rosato M. Guillope and B. Legrand, *Phil. Mag.* A **59**, 321 (1989).
9. F. Cleri and V. Rosato, *Phys. Rev.* B **48**, 22 (1993).
10. H. Spjut and D. A. Faux, *Surf. Sci.* **306**, 233 (1994).
11. G. L. Kellog, *Surf. Sci. Rep.* **21**, 1 (1994).
12. B. D. Yu and M. Scheffler, *Phys. Rev.* B **56**, R15569 (1997).
13. P. Feibelmann, preprint.
14. R. Stumpf and M. Scheffler, *Phys. Rev. Lett.* **72**, 254 (1994).
15. A. C. Schindler, Diploma-Thesis, Gerhard-Mercator-Universität Duisburg 1996.
16. C. Ratsch, A. Seitsonen and M. Scheffler, *Phys. Rev.* B **55**, 6750 (1997).
17. G. Rosenfeld, N. N. Lipkin, W. Wulfhekel, J. Kliewer, K. Morgenstern, B. Poelsema and G. Comsa, *Appl. Phys.* A **61**, 455 (1995).
18. H. Jenniches, M. Klaua, H. Höchle and J. Kirschner, *Appl. Phys. Lett.* **69**, 3339 (1996).
19. H. Kallabis, L. Brendel, J. Krug and D. E. Wolf, *Int. J. Mod. Phy.* B **11**, 3621 (1997).
20. A. C. Schindler and D. E. Wolf, in preparation.

EFFECTIVE REDUCTION OF COULOMB REPULSION IN CHARGED GRANULAR MATTER

T. SCHEFFLER, J. WERTH AND D.E. WOLF

FB 10, Theoretische Physik, Gerhard–Mercator–Universität Duisburg, 47048 Duisburg, Germany

This paper is an extension to a previous article by Scheffler and Wolf.[6] We study the rate of energy dissipation due to inelastic collisions in a charged granular gas. One observes that the electrostatic repulsion of two particles is effectively reduced by nearest neighbor interactions in a dense granular gas. We study the radial distribution function for dense systems, which leads to a better expression for the reduced energy barrier.

1 Introduction

Granular materials like sand can be observed in different states:[1] If an avalanche goes down a sand pile it flows like a fluid. But on the other hand the pile keeps a fixed shape like a solid and does not dissolve to the state of lowest potential energy. In this paper we study a strongly agitated granular "gas" at low density. One can think of grains falling through a vertical pipe or of planetary rings.[2] In this limit the dynamics is dominated by pair correlations.

When particles collide they usually become electrically charged. Much work has been done in the investigation of granular problems, but the particle charges had been neglected although they can become quite important in the sense that the electrical force will dominate the dynamics. Recently[6] we showed how electrical charges change the rate of energy dissipation in the basic phenomenon of collisional cooling in a granular gas. In the following section 2 we give a short introduction to this problem and recall results of previous work.

Section 3 extends the investigation and gives new results for the radial distribution function of charged granular gases. Taking this into account we give an improved expression for the reduction of the Coulomb barrier and explain deviations.

2 Collisional Cooling

If one understands the granular temperature as a measure for the mean energy of relative motion of the particles inside a granular gas

$$T := \frac{1}{3}\left(\langle v^2\rangle - \langle v\rangle^2\right) \tag{1}$$

the collision of two particles "cools" the system in that sense, that kinetic energy is lost due to the inelasticity of the collision. The amount of dissipated energy per collision is determined by the coefficient of (normal) restitution

$$e_n := |v'_n/v_n| \in [0,1] \tag{2}$$

where v'_n denotes the normal component of the relative velocity after the collision and v_n before the collision. This coefficient of restitution is kept fixed as a material constant in this model for a granular gas.

The amount of kinetic energy per unit volume, which is dissipated per time step is called dissipation rate γ. It can be calculated by using elementary kinetic gas theory:[3,4]

$$\gamma_0 = 2\sqrt{\pi} n^2 d^2 m \left(1 - e_{\mathrm{n}}^2\right) T^{3/2} \tag{3}$$

where $n = N/V$ is the number density, d the diameter of the identical particles and m the mass of one particle. This equation can be understood in the following way: $\sqrt{T}$ is a measure for the typical relative velocity in the system. So the number of collisions per time step is proportional to $n^2 d^2 \sqrt{T}$ where d^2 is proportional to the collisional cross section of hard spheres and the square of the number density n gives the probability to find two colliding spheres inside a given volume. In Eq. (3) the number of collisions per time step is multiplied with the typical energy loss of one collision, which is proportional to $m\left(1 - e_{\mathrm{n}}^2\right) T$ with Eq. (2).

Eq. (3) is correct only for the limit of dilute systems ($\nu \to 0$, where $\nu = \pi n d^3/6$ is the solid fraction). If the system gets denser, pair correlations because of the excluded hard-sphere volume become important. In that case the dissipation rate can be calculated by using the Enskog expansion of the kinetic theory for dense gases and one gets the dissipation rate:[4]

$$\gamma = \gamma_0 \cdot g_{\mathrm{hs}}(\nu) \tag{4}$$

where g_{hs} is the equilibrium pair distribution function of the non-dissipative hard-sphere fluid at contact. It only depends on the solid fraction ν:

$$g_{\mathrm{hs}}(\nu) = \frac{2-\nu}{2\left(1-\nu\right)^3} \tag{5}$$

(approximation by Carnahan and Starling.[5])

In order to study in which way the collisional cooling of granular gas changes if the particles are charged, we introduced a simple model for monopolar charged granular matter:[6] All particles carry the same electrical charge q fixed in the middle of the sphere. No charge transfer takes place during collisions. Between collisions two particles i and j interact via the Coulomb potential

$$\Phi_{ij} = \frac{q^2}{r_{ij}} \tag{6}$$

This makes the particles repel each other as they approach for collision. In that way the collisions become more unlikely compared to the uncharged granular gas and the dissipation rate gets smaller. We calculated the dissipation rate for the dilute limit ($\nu \to 0$):

$$\gamma = \gamma_0 \cdot \exp\left(-E_{\mathrm{q}}/mT\right) \tag{7}$$

where γ_0 is the dissipation rate in the dilute uncharged granular gas, Eq. (3). The influence of the particle charges is included by a Boltzmann factor and $E_{\mathrm{q}} = q^2/d$ is the electrostatic energy at contact.

3 Radial correlation in dense systems

In this paper we take a closer look at the radial distribution g_{chs} of the charged hard-sphere (CHS) fluid. It is obtained from molecular dynamics simulations[7] with $e_{\mathrm{n}} = 1$. Fig. 1 shows radial distribution functions for three different solid fractions. Each plot contains the data for three different granular temperatures. The distances $x = r/a$ are scaled by the ion-sphere radius $a = (4\pi n/3)^{-1/3}$, which is the natural length scale of the system. The peak at the hard-sphere distance ($x = d/a$) gets larger with increasing density as expected by Eq. (5). For medium densities the shape of the radial distribution function clearly changes if the particles are charged, whereas for the highest density (graph at the bottom), the Coulomb interactions do not influence the radial distribution function, at least for granular temperatures $T > \frac{1}{6}E_{\mathrm{q}}/m$.

For dense charged granular systems the excluded volume effect of the hard-sphere fluid becomes important, as before in the Enskog correction Eq. (4). Also the electrical interaction with the nearest neighbors of a colliding pair must be included in the description. It could be neglected in the dilute limit Eq. (7). In the aforementioned article[6] we show that the dissipation rate in a dense system can be described by the ansatz to a very good

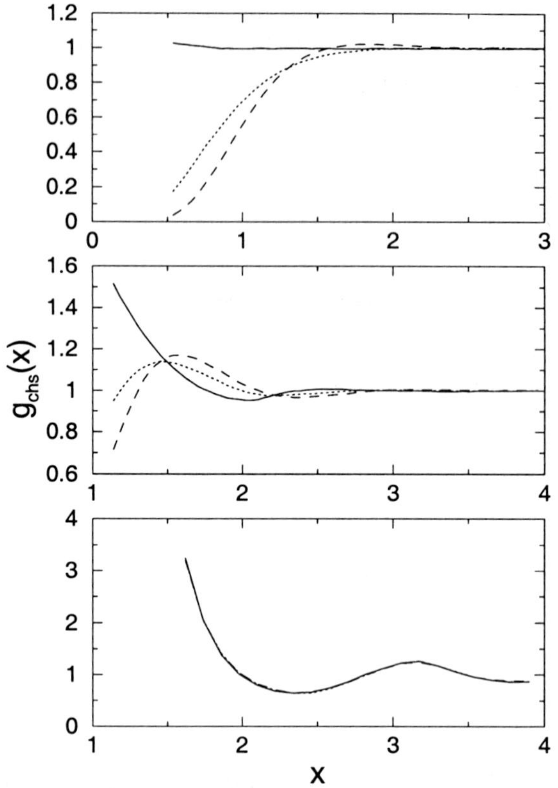

Figure 1. Radial distribution function $g_{\mathrm{chs}}(x = r/a)$. The uppermost graph shows systems with solid fraction $\nu = 1.5625 \cdot 10^{-2}$ or $d/a = 0.5$. Solid curve stands for $E_{\mathrm{q}}/mT = 0$ (no charge), dotted curve $E_{\mathrm{q}}/mT = 3.9$ and the dashed line $E_{\mathrm{q}}/mT = 7.8$. The middle graph shows systems with $\nu = 1.6637 \cdot 10^{-1}$ or $d/a = 1.1$. Solid curve is $E_{\mathrm{q}}/mT = 0$, dotted curve $E_{\mathrm{q}}/mT = 4.3$ and dashed curve $E_{\mathrm{q}}/mT = 7.8$. The lowest graph shows systems with $\nu = 0.45$ or $d/a = 1.53$ (close to the fluid-solid transition of a hard sphere system, $\nu \approx 0.5$). Here one cannot separate the curves for $E_{\mathrm{q}}/mT = 0$, $E_{\mathrm{q}}/mT = 3.9$ and $E_{\mathrm{q}}/mT = 6.2$ any more.

approximation:

$$\gamma = \gamma_0 \cdot g_{\mathrm{hs}}(\nu) \cdot \exp\left(-E_{\mathrm{eff}}\left(\nu\right)/mT\right) \tag{8}$$

where g_{hs} once again is the radial distribution function of the hard-sphere fluid at contact, Eq. (5), and the Coulomb barrier E_{q} of Eq. (7) is replaced by an effective energy barrier $E_{\text{eff}}(\nu)$, that depends on ν. This effective energy barrier includes the Coulomb interaction with the neighbors of the colliding particle pair, which can be neglected in the dilute limit:

$$E_{\text{eff}} \xrightarrow[\nu \to 0]{} E_{\text{q}}. \tag{9}$$

The idea behind Eq. (8) is, that in analogy to Eq. (4) the dissipation rate can be written as $\gamma = \gamma_0 \cdot g_{\text{chs}}$. Unfortunately the literature is lacking a satisfying analytic expression for g_{chs}. In 1972 Palmer and Weeks[8] did a mean spherical approximation for the CHS model and derived an analytic expression for g_{chs}, but this approximation is poor for low densities. Many methods[9] give g_{chs} as a result of integral equations, that can be solved numerically.

For Eq. (8) we made the ansatz, that g_{chs} can be approximated by a separation of the Coulomb interaction and the hard-sphere fluid in the following way:

$$g_{\text{chs}} = g_{\text{hs}} \cdot \exp\left(-E_{\text{eff}}/mT\right) \tag{10}$$

This ansatz covers with Eq. (9) the dilute limit, Eq. (7), and the uncharged limit, Eq. (4), as well.

As E_{eff} is a reduced Coulomb barrier compared to the dilute barrier E_{q}, we defined a dimensionless reduction function $f(d/a)$:

$$E_{\text{eff}} = E_{\text{q}} - \frac{q^2}{a} f(d/a) \tag{11}$$

It is instructive to extend the definition of the reduction function $f(d/a)$ for distances larger than the contact distance $x = d/a$. With Eq. (10) and (11) the natural definition is:

$$f(x) = \frac{1}{x} + \frac{mT}{q^2/a} \ln\left(\frac{g_{\text{chs}}(x)}{g_{\text{hs}}(x)}\right) \tag{12}$$

Fig. 2 shows $f(x)$ for the three systems from Fig. 1. The data collapse on a single curve with some slight deviation and the behavior can be approximated by a linear fit for $1 < x < 1.5$

$$f(x) = c_0 - c_1 x \tag{13}$$

The data fit to $c_0 = 1.20$ and $c_1 = 0.36$ (dashed line). Taking $x = d/a$ we get E_{eff} according to Eq. (11):

$$E_{\text{eff}} = E_{\text{q}} - \frac{q^2}{a}\left(c_0 - c_1 \frac{d}{a}\right) \tag{14}$$

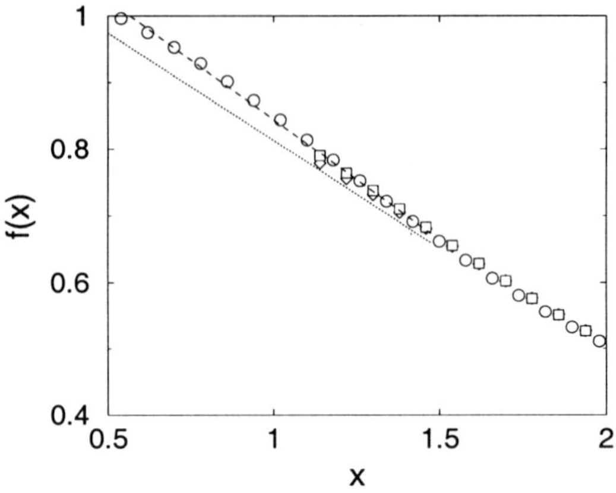

Figure 2. The circles denote the system $\nu = 1.5625 \cdot 10^{-2}$ and $E_q/mT = 7.8$, the diamonds $\nu = 1.6637 \cdot 10^{-1}$ and $E_q/mT = 4.3$ and the squares $\nu = 1.6637 \cdot 10^{-1}$ and $E_q/mT = 7.8$. The dashed line is a linear fit for $1 < x < 1.5$. It corresponds to $1.20 - 0.36x$. The dotted line assumes a BCC structure, it is $1.136 - 0.323x$.

or the ratio

$$\frac{E_{\text{eff}}}{E_{\text{q}}} = 1 - 2\, c_0\, \nu^{1/3} + 4\, c_1\, \nu^{2/3} \tag{15}$$

with $2\nu^{1/3} = d/a$.

In the previous article[6] we measured E_{eff} for systems of varying densities, Fig. 3 shows the results of the simulations.

The dashed curve corresponds to the fit from Fig. 2, while the dotted curve corresponds to the values $c_0 = 1.136$ and $c_1 = 0.323$. These values can be easily computed, if one assumes a BCC short range ordering in the charged granular gas.[6] This assumption was made in analogy[10] to the One Component Plasma (OCP) model, which is the charged granular gas in the limit of vanishing particle diameters $d \to 0$. The OCP crystallizes in a BCC structure at low temperature.[11] However, as Fig. 2 demonstrates, the assumption of a BCC structure does not fit our reduction function well.

In Fig. 3 the dashed curve approximates the simulation data quite well, but for lower densities the data deviate slightly. This is because the reducing function $f(x)$ becomes nonlinear for small distances $x < 1$. Fig. 4 compares one of the data sets from Fig. 2 with an approximate expression $f_{\text{ocp}}(x)$ given by Rosenfeld[12] for the One Component Plasma. It is obvious that a system

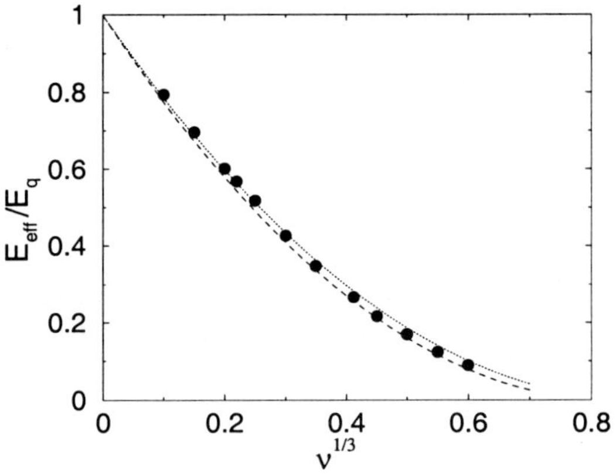

Figure 3. The dependence of the effective energy barrier on the solid fraction. Filled circles correspond to computer simulations. The dashed curve corresponds to the fit from Fig. 2, the dotted curve assumes a BCC short-range structure.

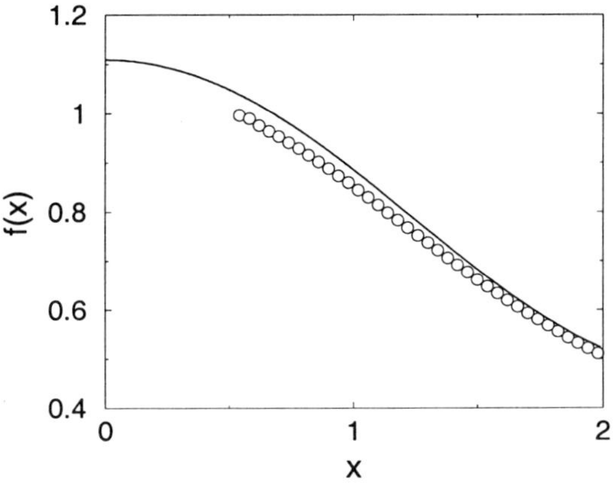

Figure 4. The circles denoted the system $\nu = 1.5625 \cdot 10^{-2}$ and $E_q/mT = 7.8$. The solid curve is a fit for $f(x)$ for the One Component Plasma (see text).

of charged hard spheres behaves differently from a One Component Plasma, if the ion sphere radius a becomes smaller than the hard core radius d, i.e. at high enough densities. This might also explain the difference between the curves in Fig. 4, although there $d = a/2$. Still, the qualitative agreement of the nonlinearities of both curves indicates, that they are of similar origin in both models. The linear fit Eq. (13) underestimates E_{eff} for dilute systems ($x = d/a < 1$ or $\nu^{1/3} < 0.5$). Therefore the data points for $\nu^{1/3} < 0.5$ in Fig. 3 lie above the dashed curve.

4 Results

We studied the radial distribution function of a charged granular gas. The distribution function can be approximated by a product ansatz separating the hard-sphere fluid contribution from the electrostatic interaction. The correlations coming from the electrostatic interaction show some qualitative similarities to the One Component Plasma, although quantitatively they differ. The charged granular gas also shows an effective reduction of the electrostatic repulsion for colliding pairs. Analyzing the radial distribution function, we made a linear approximation for the reducing function that fits previously published data[6] better.

Acknowledgements

This work has been funded by the Deutsche Forschungsgemeinschaft (DFG) through grant No. Wo 577/1-3. J. Werth gratefully acknowledges support by the DFG Graduiertenkolleg "Struktur und Dynamik heterogener Systeme".

References

1. H. M. Jaeger, S. R. Nagel, R. P. Behringer, *Rev. Mod. Phys.* **68**, 1259 (1996).
2. F. Spahn, O. Petzschmann, K.-U. Thiessenhusen, J. Schmidt; in *Physics of Dry Granular Media*, eds. H. J. Herrmann, J.-P. Hovi and S. Ludingn (Kluwer Acad. Publ., Dordrecht, 1998), pp. 401 - 406.
3. P. K. Haff; *J. Fluid Mech.* **134**, 401 (1983).
4. C. K. L. Lun, S. B. Savage, D. J. Jeffrey, N. Chepurniy, *J. Fluid Mech.* **140**, 223 (1984).
5. N. F. Carnahan, K. E. Starling; *J. Chem. Phys.* **51**, 635 (1969).
6. T. Scheffler, D. E. Wolf; *Phys. Rev.* E, submitted.
7. T. Scheffler, D. E. Wolf; in *Molecular Dynamics on Parallel Computers*, eds. R. Esser, P. Grassberger, J. Grotendorst, M. Lewerenz (World Scientific, Singapore, 1999).
8. R. G. Palmer, J. D. Weeks; *J. Chem. Phys.* **58**, 4171 (1973).

9. J. P. Hansen, J. J. Weis; *Mol. Phys.* **5**, 1379 (1977);
J. P. Rino, O. Hipólito; *Phys. Rev.* E **48**, 1375 (1993);
H. S. Kang, F. H. Ree; *J. Chem. Phys.* **103**, 9370 (1995).
10. E. E. Salpeter, H. M. Van Horn; *Astrophys. J.* **155**, 183 (1969).
11. W. L. Slattery, G. D. Doolen, H. E. De Witt, *Phys. Rev.* A **21**, 2087 (1980).
12. Y. Rosenfeld, *Phys. Rev.* A **46**, 1059 (1992).

LAYER-BY-LAYER PATTERN PROPAGATION AND PULSED LASER DEPOSITION

F. WESTERHOFF, L. BRENDEL, D.E. WOLF
Gerhard-Mercator-Universität Duisburg, 47048 Duisburg

In this article kinetic Monte Carlo simulations for molecular beam epitaxy (MBE) and pulsed laser depositon (PLD) are compared. It will be shown that an optimal pattern conservation during MBE is achieved for a specific ratio of diffusion to deposition rate. Further on pulsed laser deposition is presented as an alternative way to control layer by layer growth. First results concerning the island density in the sub-monolayer regime are shown.

1 Introduction

Multi-layered structures with sharp interfaces are of high interest, due to the new physical properties they exhibit resulting from the reduced dimensionality. In order to obtain such sharp interfaces the deposited material has to grow in layers instead of forming 3D islands. Two different methods of growing thin films will be compared here.

Molecular beam epitaxy (MBE) allows to produce mono-crystalline layer structures with high sharpness at the interfaces. Growth takes place in an ultra high vacuum chamber in which the material is thermally evaporated and thus is deposited on the substrate with a continuous flux F.

The fabrication of thin films with pulsed laser deposition (PLD) is a quite new technique which can improve layer by layer growth.[1] In addition PLD provides the possibility to deposit any desired material with the stoichiometry of the ablated material being conserved. In contrast to thermal evaporation in MBE many particles are ablated from the target simultaneously, resulting in a high adatom density directly after the pulse. This causes an increased islands density in the PLD as will be described.

2 Molecular Beam Epitaxy

For an initially flat surface layer by layer growth usually improves, the higher the temperature of the substrate is.[2] But a pattern (e.g. a structure produced by lithographic techniques) will in general not be conserved better during layer by layer growth, if the substrate temperature is raised. The reason is, that for high temperatures the adatoms do not nucleate on top of the islands any more because the diffusion length becomes too large. Instead they are incorporated at the edges of the islands and therefore flatten the surface which means that information about the pattern gets lost. As the diffusion length depends on the ratio of the diffusion rate to the deposition rate,[3] this value has to be optimized if a given pattern should be conserved during growth. One can also consider the opposite optimization problem: How to choose the growth parameters in order to heal any unwanted structures as efficiently as possible? As the two optimization problems are obviously two sides of the same medal, we consider only the first one in the following.

2.1 Model

In the simulation we use a simple cubic crystal with periodic boundary conditions and assume that growth takes place without overhangs and defects. The size of the substrate is $(L \cdot a)^2$, L^2 denoting the number of lattice sites and a the lattice constant. The atoms are deposited with a flux F and diffuse with the diffusion constant D until they are irreversibly bound to another adatom or an island edge.

By *pattern propagation* we mean the deterministic height increase by one lattice constant during the deposition of one monolayer. The probability for pattern propagation[4] is then the fraction of sites which were conserved in every monolayer until time t:

$$p(t) = \left\langle \prod_{s=1}^{t} \delta_{h(\vec{x},s),h(\vec{x},s-1)+1} \right\rangle \tag{1}$$

2.2 Results

Figure 1 (a) shows $p(t)$ for an initially flat surface. The exponential decay $p(t) = \exp(-t/t_c)$ is what one would expect, if the propagation of the surface sites from one layer to the next would be temporally uncorrelated with constant transition probability $p(1)$. Surprisingly, the exponential decay of $p(t)$ is found although there is clear evidence that transition probabilities are temporally correlated.[5] The *life time* t_c (Fig. 1 b) of the initial surface configuration is determined from the slope of $\ln(p(t))$. The data give a power law for t_c extending over 6 decades:

$$t_c \propto (D/F)^{0.20\pm0.01} \tag{2}$$

In the limit of $D/F \to 0$ (random deposition) we get $t_c = 1$.[6] That the last data point in Fig.1 (b) deviates from the power law is a finite size effect.

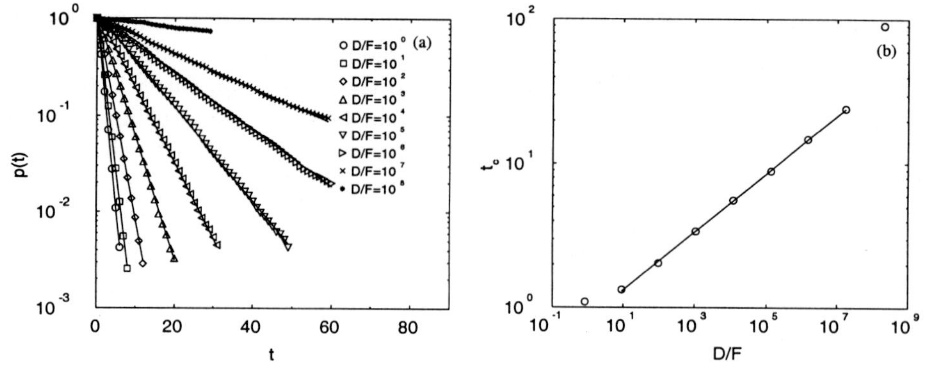

Figure 1. (a) Propagation for an initially flat surface. (b) Life time $t_c(D/F)$.

In a study of the one dimensional case[4,6] a logarithmic dependence ($t_c \sim \log(D/F)$) was found. Most recent calculations seem to show a cross over from power law to logarithmic dependence for larger D/F in two dimensions, as well.[7]

2.3 Structured surfaces

As long as the pattern is large compared to the diffusion length for an initially flat surface, the propagation probability should not change since the adatoms will not notice the pattern during their life time. Therefore we also expect an exponential decay for small ratios of D/F. Here we consider larger values for D/F, though, where we observe a different behavior. We find a maximum in $t_c(D/F)$ which is shifted to higher D/F with growing pattern size r (Fig. 2 a). This can be explained by the existence of two different mechanisms responsible for the pattern's decay.

For small D/F the life time t_c is independent of the island size. This is the regime where the pattern vanishes through kinetic roughening. To the right of the maximum the diffusion length l_D is large compared to the pattern size, hence there are almost no nucleations on top of the "artificial" islands. Instead, they take on the form of natural growing and coalescing islands. In this way the pattern is also lost.

We find (see Fig.2 b) that the maximum life time $t_{c,\max}$ for a given pattern length scale r is reached for

$$(D/F)_{\max} \propto r^{2.4\pm0.1} \quad \text{and scales like} \quad t_{c,\max} \propto r^{0.50\pm0.05}. \tag{3}$$

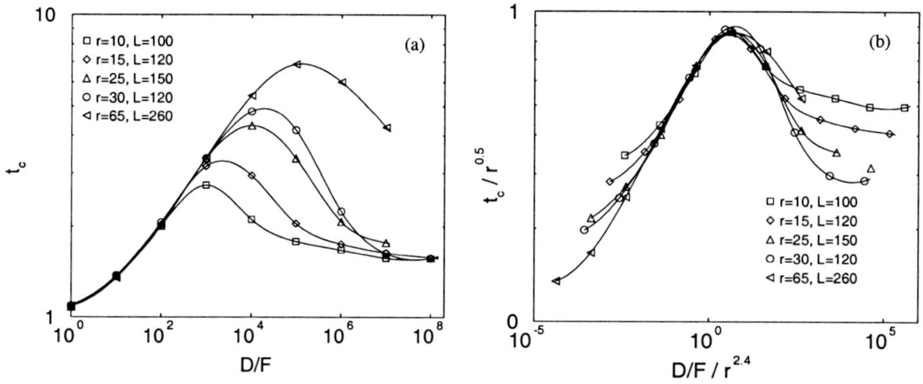

Figure 2. (a) Life time t_c for patterned surfaces. (b) Scaled data.

This is in agreement with (2): $t_{c,\max} \propto (D/F)_{\max}^{0.5/2.4} = (D/F)_{\max}^{0.21}$. As (2) crosses over to a much weaker dependence for very large D/F, we expect also that (3) will be modified for larger r.

3 Pulsed laser deposition

In pulsed laser deposition the atoms are deposited with a pulse intensity I (number of atoms per unit area) and frequency ω forming an effective (average) flux

$$F = I \cdot \omega \,. \tag{4}$$

If we reduce the intensity I to the limiting case of one atom per pulse and at the same time increase the frequency so that the effective flux stays constant, we will reach conditions equivalent to thermal deposition (MBE).[a] One should therefore expect that both methods show similar characteristics in the submonolayer regime.

If the intensity is increased, one should expect a higher island density, because the adatom density and therefore the probability for nucleation is also larger. Hence we predict a *critical pulse intensity* I_c which separates both regimes.

In the simulation the kinetic energy of the arriving atoms is neglected. This should be a valid approximation for experiments, in which the laser intensity

[a]Nevertheless in the simulation there is still a small difference since in PLD atoms are deposited at fixed time intervals, whereas in MBE fluctuations may occur.

is just above the ablation threshold[1] or the deposition takes place, e.g. in an oxygen atmosphere which slows down the arriving particles.

3.1 Critical pulse intensity

During thermal deposition the time evolution of the adatom density ρ is given by a gain term, the flux F and a loss term ρ/τ, where τ is the life time of adatoms on the surface:

$$\dot{\rho} = F - \frac{\rho}{\tau}\,. \tag{5}$$

In the quasi-stationary regime (after island nucleation and before island coalescence) the life time τ is determined by the typical distance l_D an adatom diffuses before reaching an island edge:[8] $\tau = {l_D}^2/D$. Then $\dot{\rho}$ can be neglected and one gets:

$$\rho \approx F \cdot \tau = F \cdot \frac{{l_D}^2}{D}. \tag{6}$$

With $l_D \propto \left(\frac{D}{F}\right)^\gamma$ and $\gamma = 1/6$ for two dimensions and irreversible aggregation[3] this leads to

$$\rho \propto {l_D}^{-4}. \tag{7}$$

Equation (7) describes the adatom density for thermal deposition. In PLD with a pulse intensity I much smaller than this density, the adatom density will oscillate around this value with the frequency ω. Then one expects the same scaling of the island density as for thermal deposition. By contrast a much larger pulse intensity would prevent the adatom density in PLD to become on average similar to the one in MBE. Hence the critical pulse intensity is expected to be of the order

$$I_c \propto {l_D}^{-4} \propto \left(\frac{D}{F}\right)^{-2/3}. \tag{8}$$

In the case of a large pulse intensity, the island density increases to such a degree that the life time of the adatoms is shorter than the time between two pulses. Hence the adatom density and therefore the island density becomes independent of the pulse frequency and only depends on I.

3.2 Results

Fig. 3 (a) depicts the island density as a function of the deposited monolayers. We expect three different regimes:[9] The regime for the nucleation of islands,

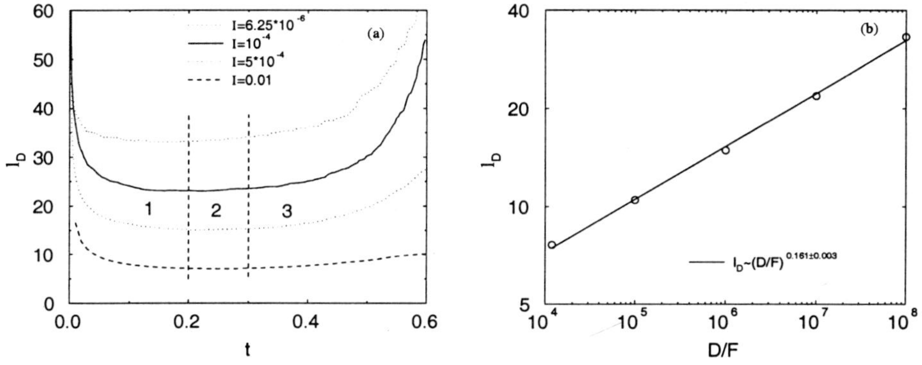

Figure 3. (a) Island distance l_D as a function of time t (in units of monolayers). (b) $l_D(D/F)$ for the limiting case of one atom per pulse.

the saturation of the island density and finally the coalescence of islands with an increasing island distance. An increase of the pulse intensity leads to an overall decrease of the island distance.

3.3 l_D *for the limit of small pulse intensities*

According to a classical result, confirmed by experiments[10] and simulations[9,11] the island distance scales like:

$$l_D \propto \left(\frac{D}{F}\right)^{1/6} \tag{9}$$

This behavior is confirmed in figure 3 (b) for the limit, that only one atom per pulse ($I \cdot L^2 \to 1$) is deposited.

3.4 Scaling behavior of the island distance

Figure 4 (a) depicts the island distance l_D as a function of the pulse intensity I. For small values of I the diffusion length approaches a constant value, which is the island distance for the MBE case. For larger pulse intensities l_D is described by a power law:

$$l_D \propto I^{-\nu} \, . \tag{10}$$

It turns out that l_D can be written in a scaling form:

$$l_D \propto \left(\frac{D}{F}\right)^{\gamma} \cdot f\left(\frac{I}{I_c}\right) \tag{11}$$

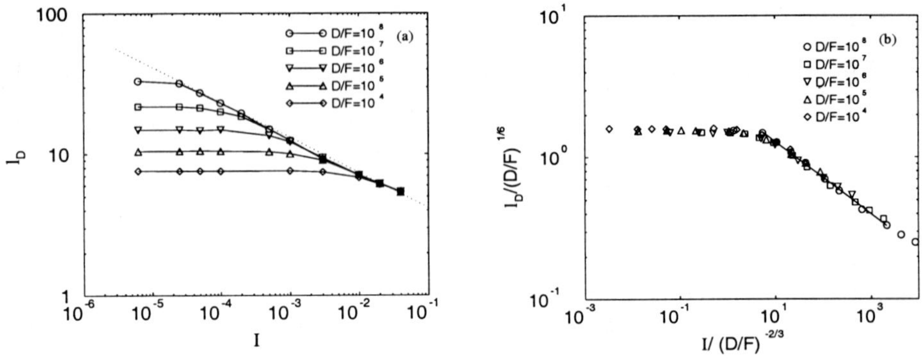

Figure 4. (a) Island distance as a function of the pulse intensity for different ratios of D/F. (b) Scaled data.

with

$$f(y)\begin{cases} = \text{const. for } y \ll 1 \text{ and} \\ \sim y^{-\nu} \quad \text{for } y \gg 1\,. \end{cases} \tag{12}$$

While the asymptotic power laws (9,10) were also obtained by Jensen and Niemeyer,[12] the scaling law (11) is new. In the previous work the pulses had a finite duration in contrast to the case presented here. This additional characteristic time spoiled the scaling.
Since for large I, the ω-dependence (via F (4)) has to vanish:

$$I_c^{\nu} \cdot \left(\frac{D}{F}\right)^{\gamma} = \text{const.} \tag{13}$$

and therefore:

$$I_c \propto \left(\frac{D}{F}\right)^{-\gamma/\nu} \tag{14}$$

For large pulse intensities we find in Fig. 4 (a) the slope $\nu = 0.248$, so that, together with $\gamma = 1/6$, we get:

$$\gamma/\nu \approx 2/3,$$

in accordance with the theoretical prediction (8).

Acknowledgements

We thank the Neumann Institute of Computing (NIC) in Jülich for providing computer time on the CRAY T3E for this project.

References

1. H. Jenniches, M. Klaua, H. Höchle, J. Kirschner, *Appl. Phys. Lett.* **69**, 3339 (1996).
2. H. Kallabis, L. Brendel, J. Krug, D. E. Wolf, *Int. J. Mod. Phys.* B **11**, 3621 (1997).
3. D. E. Wolf, *Adatom Diffusion and Epitaxial Growth*, in *Dynamics of Fluctuating Interfaces and Related Phenomena* (World Scientific, Singapore, 1997).
4. H. Kallabis, D. E. Wolf, *Phys. Rev. Lett.* **79**, 4854 (1997).
5. F. Westerhoff, *Musterpropagation im Lagenwachstum* (diploma thesis, Duisburg, 1998).
6. Harald Kallabis, *Theoretical Aspects of Crystal Growth* (Ph. D. thesis, Jülich, 1997).
7. F. Westerhoff, L. Brendel, S. B. Lee, D. E. Wolf, to be published.
8. J. Villain, A. Pimpinelli, L.-H. Tang, D. E. Wolf, *J. Phys. I France* **2**, 2107 (1992).
9. L.-H. Thang, *J. Phys. I France* **3**, 935 (1993).
10. S. Günther, E. Kopatzki, M. C. Bartelt, J. W. Evans, R. J. Behm, *Phys. Rev. Lett.* **73**, 553 (1994).
11. S. Stoyanov, D. Kashchiev, in *Current Topics in Material Science* **7**, 69, ed. E. Kaldis (North-Holland, Amsterdam, 1981).
12. P. Jensen, B. Niemeyer, *Surf. Sci.* **384**, L823 (1997).

PbS: FROM SOLIDS TO CLUSTERS

E. HOFFMANN AND P. ENTEL

Theoretische Tieftemperaturphysik Gerhard-Mercator-Universität Duisburg, Lotharstraße 1 47048 Duisburg, Germany

Semiconducting nanocrystallites like PbS exhibit electronic and optical properties greatly differing from those observed in the bulk material due to quantum size effects. By decreasing the diameter of PbS nanoparticles to about 1 nm the optical band gap increases significantly by a factor of 10 with respect to the bulk material. In this paper we investigate with *ab initio* methods the dependence of the optical band gap on the size of the system.

1 Introduction

The fundamental band gap of semiconductors like galena (PbS) is very sensitive against finite size effects. By varying the size and shape from bulk material to nanoparticles it is possible to change the optical band gap E_g from about 0.286 eV up to 2.8 or 5.2 eV.[1,2] Therefore, it is possible to build optical sensors with adjustable properties.

Galena is an ionic bounded system which crystallizes in the fcc based B1 structure. It is a narrow gap semiconductor with extrema of conduction and valence bands at the L point of the Brillouin zone. The fundamental band gap is a direct gap having at 4.2 K a width of 0.286 eV which increases up to 0.41 eV at room temperature.[3] If the size of PbS particles is reduced the absorption spectrum shows two effects. First the wedges of the absorption spectrum are smeared out and show long tails near the absorption edge[1,2,4-6] (see Fig. 1) which could be due to defect states, particle size distributions or indirect transitions. In order to determine the band gap energy, the transitions can be characterized by the Tauc[5] relation

$$\sigma \cdot E \; = \; A \cdot (E - E_g)^m,$$

where σ is the absorption coefficient and m=1/2, 2, 3/2 or 3 for allowed direct, allowed indirect, forbidden direct and forbidden indirect electronic transitions. Second the band gap shows a hyperbolic dependence on the particle size (see Fig. 2). Wang *et al.*[2] reported that E_g increases up to 2.32 eV for particles with a diameter of ~0.24 nm. Thielsch *et al.*[1] found a band gap maximum of 5.2 eV for a particle size of 1 nm. Additionally they found a photoluminescence transition with an energy of 2.85 eV for particles less than 2 nm in size. The smallest possible PbS particle is a dimer for which the lowest allowed dipole transition has the energy 2.337 eV (Table 1).

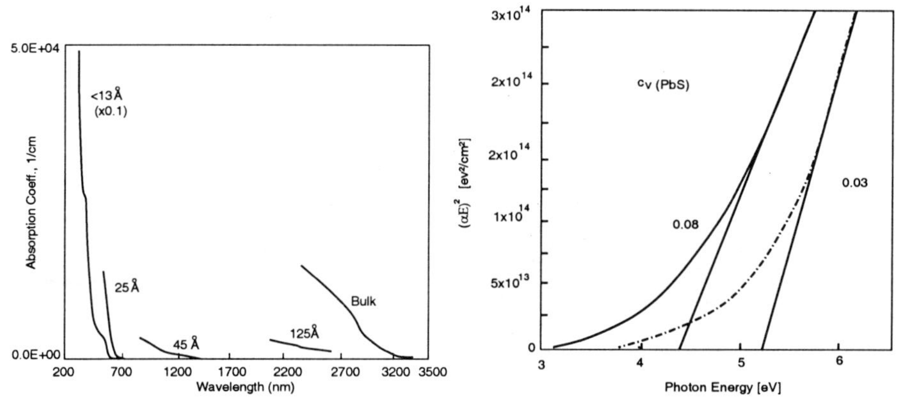

Figure 1. Left panel: The absorption spectra of PbS nanoparticles in a polymer film (see Wang *et al.*[2]) Right panel: Tauc's plots for the direct transition of PbS nanoparticles in a SiO_2 matrix, after Thielsch *et al.*[1]

With empirical models based on the exciton model, it is possible to describe the band gap behavior of nanoparticles. The models are based on the single-particle approximation and describe the electronic states in the nanoparticle as electrons and holes in a box of a given size (see Fig. 2). The only input required is then the effective mass of the valence electrons and holes. Wang[2] assumes an infinite potential and obtained a quantitatively good description for particles larger than 80 Å in diameter. Nosaka[9] found an improved agreement for particles down to 2 Å if the potential well of the exciton is reduced to 4.5 eV.

In this short communication we investigate in how for the band gap behavior of PbS nanoparticles can be described by *ab initio* methods. We have performed density functional calculations for the bulk system, a monolayer,

Transition	E (eV)
forbidden	1.846
$X \rightarrow A$	2.337
$X \rightarrow B$	2.709
$X \rightarrow C$	2.878
$X \rightarrow D$	3.163
$X \rightarrow E$	3.676
$X \rightarrow F$	5.923

Table 1. Electronic excitation spectrum of PbS dimers from the ground state after Joos and Saur.[7] The lowest transition is forbidden for dipole radiation, see Wang.[2]

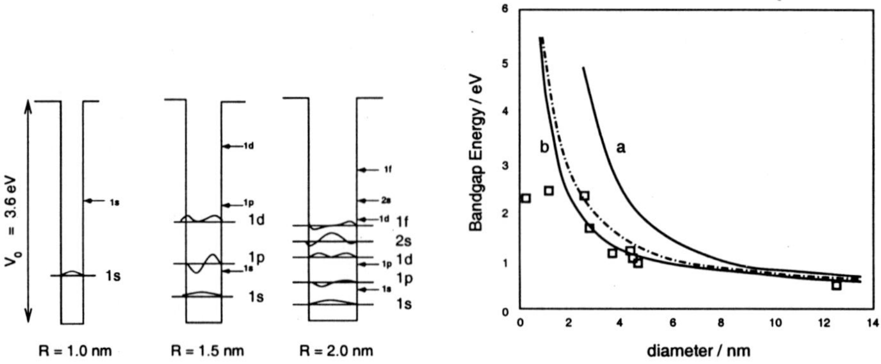

Figure 2. Left panel: Energy levels and wavefunctions of electrons having an effective mass of 0.19 m_e in a spherical well of finite depth. The arrows show the energy levels for $V_0 = \infty$, see Nosaka[9] for details. Right panel: Band gap energy of PbS nanoparticles calculated with the finite depth potential model,[9] a: $V_0 = \infty$, b: $V_0 = 4.5\,\text{eV}$. The dashed curve is calculated for the hyperbolic band model and the squares indicate the experimental results reported by Wang *et al.*[2]

and clusters with 8 and 18 atoms, and also for the dimer system. Nowadays density functional theory is a well established method which allows to calculate ground state properties of atoms, molecules, semiconductors and metals with high precision. For the investigation of band gaps the errors are much larger and usually of the order of 50% or more.[11] The reason for this large deviations is due to the fact that the Kohn-Sham eigenvalues are not the excitation energies of the electronic system,[12] nevertheless the agreement between the experimental and theoretical band structure inside the valence and conduction band is usually good. The description of band gaps can be improved by self interaction corrections (SIC)[13] or local mass approximations (LMA).[14]

2 Method of Calculation

All calculations presented here have been done in the framework of density functional theory employing the Vienna *ab initio* Simulation Package (VASP).[15,16] This method uses a plane wave basis set and ultrasoft Vanderbilt pseudopotentials,[17] allowing a smaller energy cutoff than in the case of normconserving pseudopotentials. Scalar relativistic effects are taken into account when generating the pseudopotentials. The exchange and correlation effects are treated in the local density approximation with the parameterization of Perdew and Zunger.[13] Because of the basis set the program uses periodic boundary conditions. For the plane wave expansion we use a cutoff energy of 198 eV. The Kohn-Sham Hamiltonian is solved via iterative matrix diagonalization schemes.[16] Low dimensional structures like monolayers or clusters are calculated by introducing vacuum layers in the unit cell. We usually used 10 Å vacuum, the use of 20 Å does not change the results. For the bulk and monolayer calculation we used the tetrahedron method with Blöchl correction,[18] while the dimer and cluster calculations are performed using the Γ point only. In this case we used a Gaussian smearing of the density of states with $\sigma = 0.1\,\text{eV}$. The eight atom cluster was initialized by using a simple cubic basis of the B1 structure: $(0,0,0)$, $(\frac{1}{2},\frac{1}{2},0)$, $(0,\frac{1}{2},\frac{1}{2})$, $(\frac{1}{2},0,\frac{1}{2})$; $(\frac{1}{2},0,0)$, $(\frac{1}{2},0,0)$, $(0,0,\frac{1}{2})$, $(\frac{1}{2},\frac{1}{2},\frac{1}{2})$. The initial eighteen atom cluster was constructed by doubling the lattice constant in each direction. All structures have been fully relaxed in the selfconsistent cycle leading to no significant deviations from the B1 positions. The partial density of states have been calculated by projections the wavefunctions onto a radial basis set up to $l_{\max} = 2$. As mentioned above the fundamental gap of PbS is related to a direct transition, therefore, we investigated only the density of states.

3 Results and Discussion

For solid galena in the B1 structure we found a lattice constant of 5.86 Å which is in good agreement with the experimental value of 5.94 Å at room temperature. For the band gap we obtained $E_g = 0.58\,\text{eV}$ which is by a factor of two larger than the experimental low temperature value of 0.286 eV. We would like to point out that in contrast to this the *ab initio* calculations of Wang *et al.*[2] yields metallic behavior. The reason is most probably not the failure of density functional theory, instead we believe that the main source of error is the use of only radial symmetric basis functions, because the packing

System	d (Å)	E_g (eV)	E_F (eV)
Bulk	3.0	0.58	
Monolayer	2.8	≈ 0	-3.3
Cluster (18)	2.7	1.44	-4.1
Cluster (8)	2.46	2.85	-4.8
Dimer	2.3	2.8,5.5	-5.8

Table 2. Ground state properties of Galena for different structures; d denotes the next neighbor distance, E_g the band gap derived from the density of states and E_F the Fermi energy with respect to the vacuum.

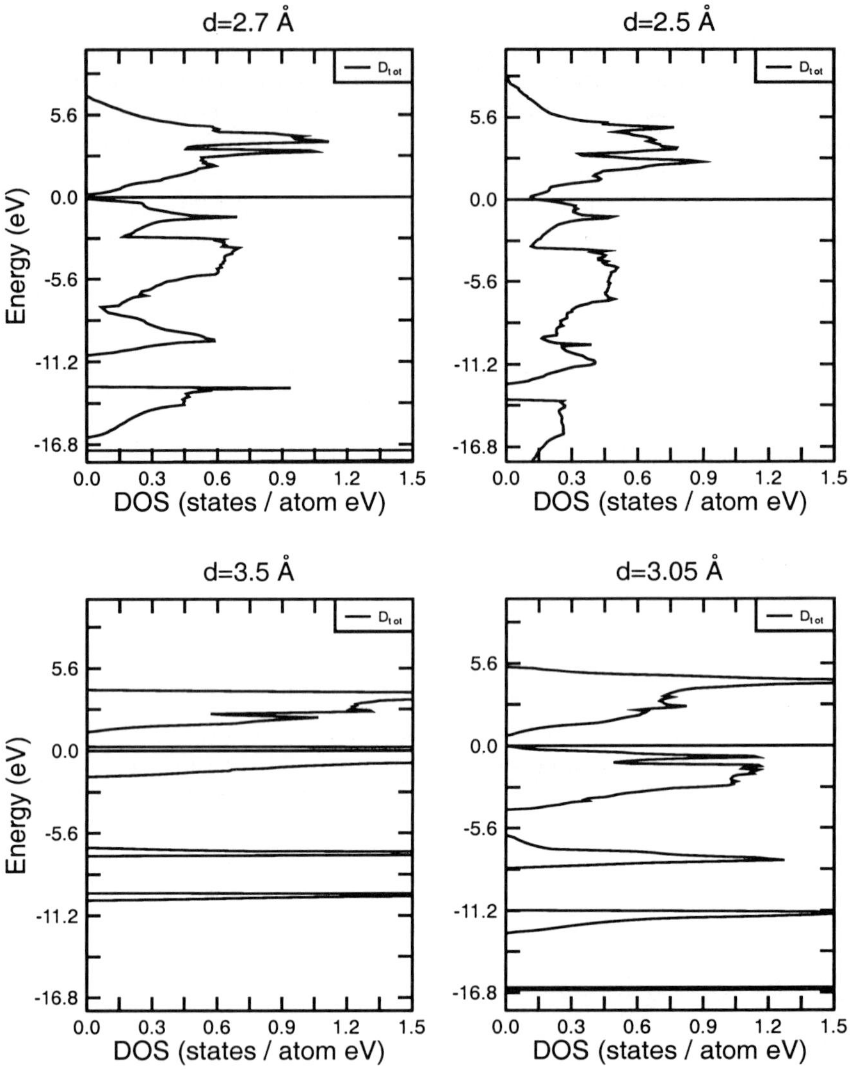

Figure 3. Density of states of solid PbS in the B1 structure. d indicates the nearest neighbor distance, for small d Galena becomes metallic. Zero energy denotes the Fermi level.

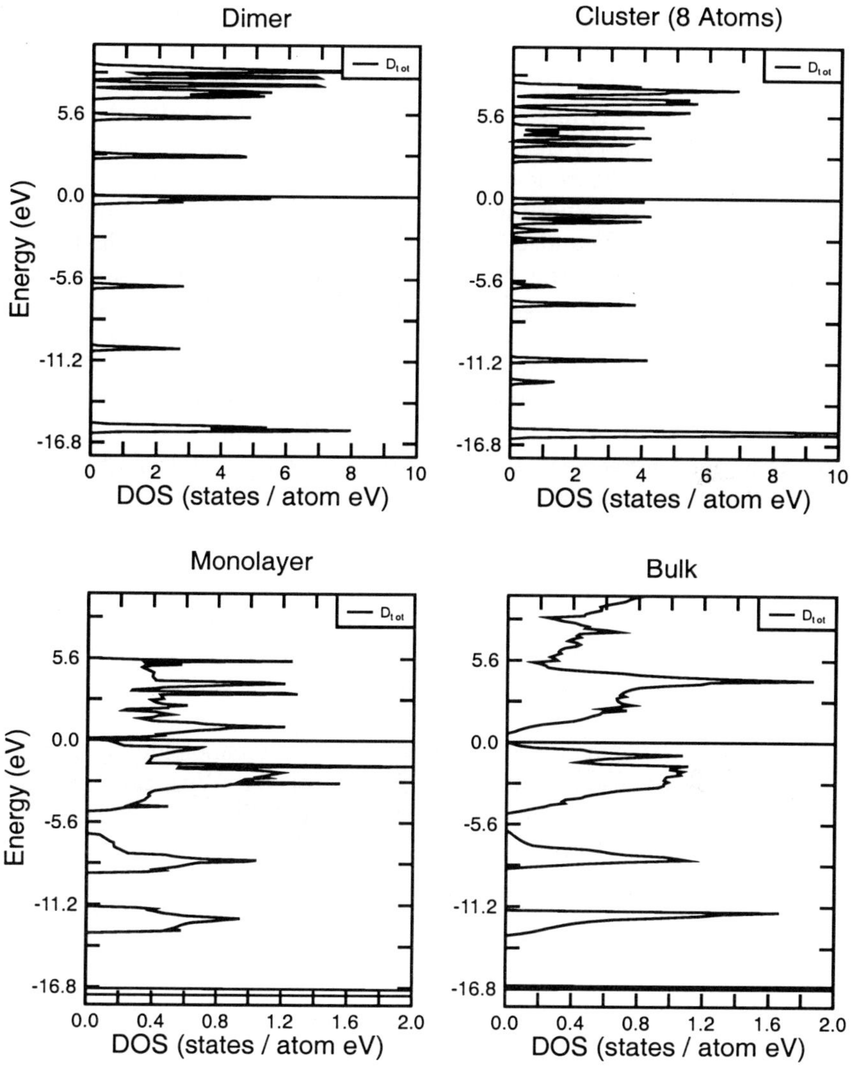

Figure 4. Density of states of PbS in reduced dimensions. All structures are fully relaxed, but there are no visible deviations from the B1 structure.

fraction of the B1 structure is only 0.52. Finally Mian *et al.*[6] reported an energy gap of 6.1 eV from Hartree-Fock calculations. The experimental pressure dependence of the band gap at room temperature is -9.1 μeV/bar. Assuming that the dependence is linear in pressure this, yields a critical pressure for the semiconductor-metal transition of 45.6 GPa. This is in remarkable good agreement with the theoretical value of 45 GPa.

In contrast to the experimental spectra of PbS (Table 1), where six transitions between 1.8 eV and 3.7 eV exist, we found only two eigenvalues (E = 2.8, 5.5 eV) between the Fermi and vacuum level. We believe that the main reason for this is the neglect of spin-orbit coupling. Because of the high atomic number of lead (82) relativistic effects should be noticeable. In our calculations (Fig. 4) the states at the Fermi level and the first excited states corresponds to threefold degenerate molecular orbitals. If we average the experimental values in order to cancel the splitting, we obtain $E_g = 2.66\,\mathrm{eV}$, which is in good agreement with the gap energy $E_g = 2.8\,\mathrm{eV}$ obtained from the nonrelativistic calculation.

If we increase the number of atoms to 8, which form the corner of a cube, the lattice constant is increased from 2.3 Å to 2.46 Å, but the value of the band gap does practically not change (see Table 2). In the case of the 18-atom cluster the band gap decreases to 1.5 eV. The decrease of the band gap with increasing particle size is stronger than observed in experiments (Fig. 2), where such a small band gap is expected for a particle diameter of about 40 Å.

Besides of nanoparticles thin films are also technologically important. Therefore, we have also investigated a free PbS layer. With the experimental bulk lattice constant the band gap amounts to 0.93 eV. If we relax the lattice constant to minimize the total energy, the lattice constant shrinks by about 7 % and the system becomes metallic. This agrees well with the pressure dependence of the band gap of the bulk material. From this calculation it seems to be possible to prepare metallic thin PbS films on a suitable substrate.

4 Conclusions

The electronic band gap of crystallized galena was calculated with a deviation of 0.3 eV with respect to the experimental value. The excitation spectra of PbS shows that relativistic corrections are not negligible. The result for the band gap in the atomic limit is in good agreement with the results of Wang *et al.*, but there is a discrepancy with the results of Thielsch *et al.* who found a band gap of 5.5 eV, which is in reasonable agreement with the semiempirical exciton model. In our calculation we found a second state of 5.5 eV above the Fermi level, which is in agreement with the result of Thielsch. In order to identify this energy difference with the band gap, the lower energy excitations must be forbidden. This could only be verified by analyzing the dipole radiation matrix elements. Another reason for the discrepancies between the two experiments may be the different methods of preparation the samples. In one experiment the nanoparticles are embedded in a polymer and in the other in SiO_2. Whether the different environment of the nanoparticle is responsible for the differences have to be investigated experimentally with accompanying theoretical calculations.

References

1. R. Thielsch, T. Böhme, R. Reiche , D. Schläfer, H.-D. Bauer, and H. Böttcher, *Nano Structured Materi.* **10**, 131 (1998).
2. Y. Wang, A. Suna, W. Mahler, and R. Kasowski, *J. Chem. Phys.* **87**, 7315 (1987).
3. Nimitz, *Landolt-Börnstein*, Vol III/17f (1983).
4. K. Kotani, *J. of Non-Crystalline Solids* **126**, 87 (1990).
5. J. Tauc and A. Menth, *J. of Non-Crystalline Solids* **8-11**, 569 (1972).
6. M. Mian, N.M. Harrison, V.R. Saunders, W.R. Flavell, *Chem. Phys. Lett.* **257**, 627 (1996).
7. Joos, A. Saur, *Landolt-Börnstein*, Vol I/3 (1951).
8. A.E. Sandström, *Handbuch der Physik* XXX, 78, (1951).
9. Y. Nosaka, *J. Phys. Chem.* **95**, 5054 (1991).
10. O.B. Maksimenko, A.S. Mishchenko, *Solid State Commun.* **92**, 797 (1994).
11. D. Vogel, P. Krüger, and J. Pollmann, *Phys. Rev.* B **54**, 5495 (1996).
12. L.J. Sham and A. Schüter, *Phys. Rev. Lett.* **51**, 1888 (1983).
13. J.P. Perdew and A. Zunger, *Phys. Rev.* B **23**, 5048 (1981).
14. G.E. Engel and W.E. Pickett, *Phys. Rev.* B **54**, 8420 (1996).
15. G. Kresse and J. Hafner, Phys. Rev. B **54**, 11169 (1994).
16. G. Kresse and J. Furtmüller, Phys. Rev. Lett. **45**, 566 (1996) and references therein.
17. D. Vanderbilt, *Phys. Rev.* B **41**, 7892 (1990).
18. P.E. Blöchl, O. Jepsen, and O.K. Anderson, *Phys. Rev.* B **49**, 16223 (1994).

MONTE CARLO STUDY OF THE MAGNETOELASTIC PROPERTIES OF FE-NI CLUSTERS

M. E. GRUNER, S. SIL,* AND P. ENTEL

Theoretische Tieftemperaturphysik, Gerhard-Mercator-Universität Duisburg, D-47048 Duisburg, Germany

1 Introduction

The search for new properties of materials composed of clusters (cluster-assembled materials) or of small nanoparticles has reached a state of technological importance. Moreover the evolution of the magnetic properties from the atom to the bulk via cluster growth is itself a fascinating and challenging problem. However, in spite of considerable effort spent on the investigation of the magnetic properties of metal clusters,[1-7] a coherent understanding of this subject has not yet been established. In this paper we study a feature of particular interest namely the magnetoelastic properties of small clusters composed of iron and nickel. It is known that alloys of these elements are characterized by outstanding anomalies regarding their structural and magnetoelastic properties on the iron-rich side of the phase diagram. Especially for compositions around 35 at.% Ni experiments reveal vanishing thermal expansion around room temperature, the so-called Invar effect, which has been the subject of numerous experimental and theoretical investigations since its discovery.[8-11] At slightly lower Ni concentrations, in addition to the magnetic anomalies, structural instabilities appear since the alloy undergoes a martensitic transformation from the fcc to the bcc phase.[12,13]

An early phenomenological explanation of the Invar effect was given by Weiss[14] who proposed that in fcc iron two different electronic states could coexist: One in which the electronic orbitals are filled without obeying Hund's rule characterized by the low magnetic moment (LM) and a smaller lattice parameter and the other in which Hund's rule is obeyed with high magnetic moment (HM) and a larger lattice parameter. Considering the HM state to be the fundamental state, one can explain the anomalous thermal expansion behavior by a gradual transition from the HM state to the LM state with increasing temperature. Thus the usual thermal expansion is compensated by reducing the volume as the fraction of iron atoms in the LM state

*Permanent address: *Department of Physics, Indian Institute of Science, Bangalore 560012, India*

increases. Considerations of this kind have also been supported by band structure calculations,[15-17] which prove the existence of two minima on the binding surface corresponding to LM and HM lattice constants. Furthermore the difference in energy between these minima is found to be well in the range of thermal fluctuations.

Assuming that the HM-LM aspect is the most important feature of Invar systems like Fe-Ni, we study the magnetoelastic properties of these alloys by using an Ising-like spin model with coupled magnetic and lattice degrees of freedom. For the bulk case this model has been investigated by Gruner et al.[18,19] with a constant pressure Monte Carlo method. In this work we will use the same finite-temperature Monte Carlo method in order to study the thermodynamic and magnetoelastic properties of small clusters of iron and nickel.

2 The model

In order to assign the LM and HM state to the Fe atom, we consider a modified spin-1 Ising model with $S_i = \pm 1$ for the HM state and $S_i = 0$ for the LM state. Furthermore we allow the statistical weight of the LM state to vary, which determines the fraction of LM spins in the high temperature limit. Since pure nickel does not exhibit the above mentioned anomaly in its thermal expansion curve, we can neglect the LM state for nickel atoms and represent the spin of nickel atoms by using Ising-spins with $S_i = \pm 1$. We take into account the magnetovolume effect through spin-dependent pair potentials of Lennard Jones type. For the sake of simplicity the pair potentials are assumed to be the same for iron and nickel. The total model Hamiltonian is then

$$H = -\sum_i D_i S_i^2 - \sum_{<ik>} J_{ik} S_i S_k + \sum_{<ik>} U(r_{ik}, S_i S_k), \tag{1}$$

where D_i are crystal field constants which separate the HM and LM states energetically. $J_{ij} = J_{\mathrm{Fe-Fe}}$, $J_{\mathrm{Fe-Ni}}$ and $J_{\mathrm{Ni-Ni}}$ are the exchange constants of the interaction between Fe-Fe, Fe-Ni and Ni-Ni atoms, respectively. The summation is performed over pairs of nearest neighbors. For the pair potentials we use

$$U(r_{ik}, S_i, S_k) = \begin{cases} 4\,\epsilon_{\mathrm{L}} \left(\left(\frac{d_{\mathrm{L}}}{r_{ik}}\right)^{12} - \left(\frac{d_{\mathrm{L}}}{r_{ik}}\right)^{6} \right), & S_i S_k = 0 \\ 4\,\epsilon_{\mathrm{H}} \left(\left(\frac{d_{\mathrm{H}}}{r_{ik}}\right)^{12} - \left(\frac{d_{\mathrm{H}}}{r_{ik}}\right)^{6} \right), & S_i S_k \neq 0 \end{cases}, \tag{2}$$

where $\epsilon_{L,H}$ denotes the lattice energy at the equilibrium distances $r_{ik} = 2^{1/6}d_{L,H}$ between two nearest neighbor atoms for LM and HM potentials, respectively.

3 Simulation method

Recent experimental investigations[6] indicate that fcc-Fe nanoparticles of size less than 100 nm do not undergo a martensitic transformation when the temperature is reduced to 4.2 K. This shows that at very low temperature the fcc lattice structure is stable for the case of small Fe-clusters. Therefore, it is reasonable to assume a fcc lattice structure for our systems (which is of the size 2...6 nm). In this work we performed a finite-temperature Monte Carlo simulation to study the Hamiltonian (1) using open boundary conditions. Our Monte Carlo routine consists of local update steps for spin and lattice degrees of freedom, respectively. First a new spin state is chosen using the usual Metropolis algorithm in order to simulate the spin configuration. Afterwards we include the vibrational degrees of freedom by moving the lattice site to a new trial position from a cubic region around the actual position. Again this new position is accepted or rejected according to the Metropolis criterion. In order to improve the convergence, the size of the cube is determined by the condition that about half of the proposed moves is accepted. For our calculation 160 000 to 320 000 lattice sweeps are performed for each temperature starting with the final configuration of the previous temperature. The first 40 000 steps are used to allow the system to reach equilibrium. Finally for each temperature we have averaged over 4 to 15 different random distributions with a fixed composition of iron and nickel atoms. Sample configurations of particles with different sizes are shown in Fig. 1.

In order to be able to compare the results with the simulations of bulk Fe-Ni alloys we used the same set of parameters when simulating nanoparticles. The magnetic parameters have been fixed to $J_{\mathrm{Fe-Fe}} = -0.72\,\mathrm{mRy}$, $J_{\mathrm{Fe-Ni}} = 0.96\,\mathrm{mRy}$ $J_{\mathrm{Ni-Ni}} = 0.40\,\mathrm{mRy}$ and $D_{\mathrm{Fe}} = -2.20\,\mathrm{mRy}$. We choose $D_{\mathrm{Ni}} \gg k_{\mathrm{B}}T$ in order to prevent nickel spins from switching to the LM state. Please note that the Fe-Fe exchange constant favors antiferromagnetic spin order while the others prefer parallel alignment of the spins. The elastic parameters are $d_H = 2.209$ Å, $d_L = 2.165$ Å and $\epsilon_H = \epsilon_L = 25.47\,\mathrm{mRy}$. Finally we set the statistical weight of the non-magnetic state to $g = 0.5$ in order to restrict the fraction of non-magnetic iron sites at infinite temperature to 20%. The choice of the parameters, which is based on results of band structure calculations,[17,20] has been explained in detail in Ref. 19.

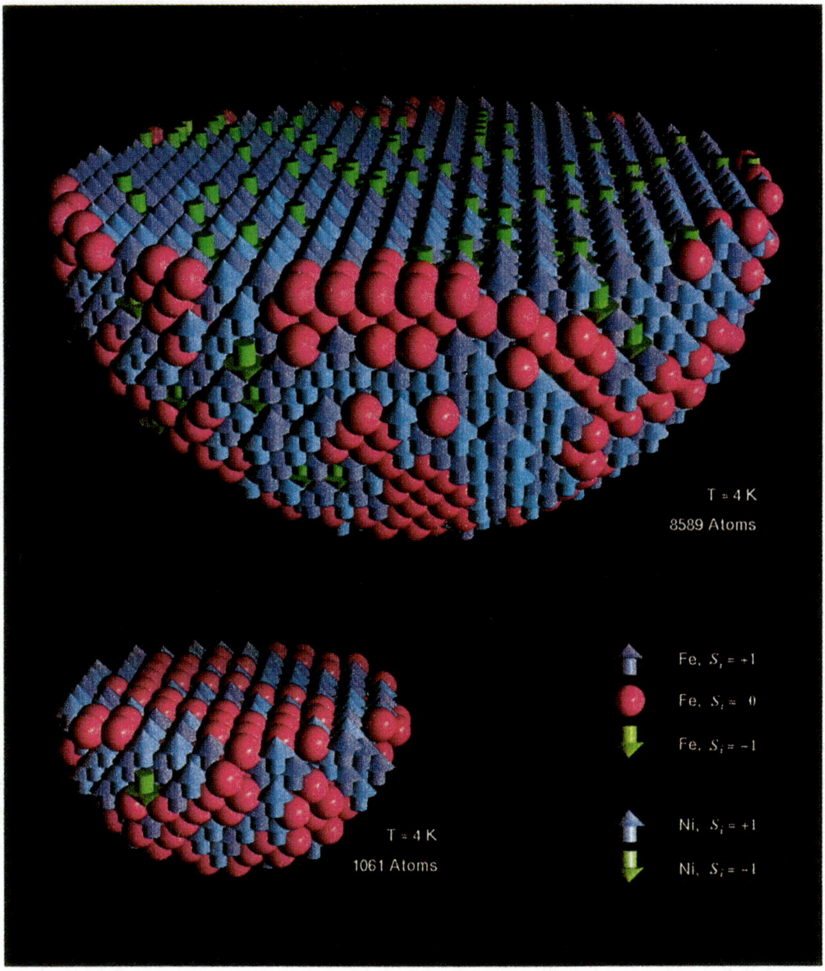

Figure 1. Sample configurations of $Fe_{65}Ni_{35}$ particles with 8589 atoms (upper panel) and 1061 atoms (lower panel) at $T = 4$ K. In order to illustrate the behavior of the inner part, the particle is cut through the center. The lighter arrows denote nickel atoms, the darker ones HM iron atoms with $S = \pm 1$. The spheres refer to LM iron.

4 Results

First we have investigated the change in the magnetoelastic behavior of particles at the classical Invar composition $Fe_{65}Ni_{35}$ when decreasing the particle

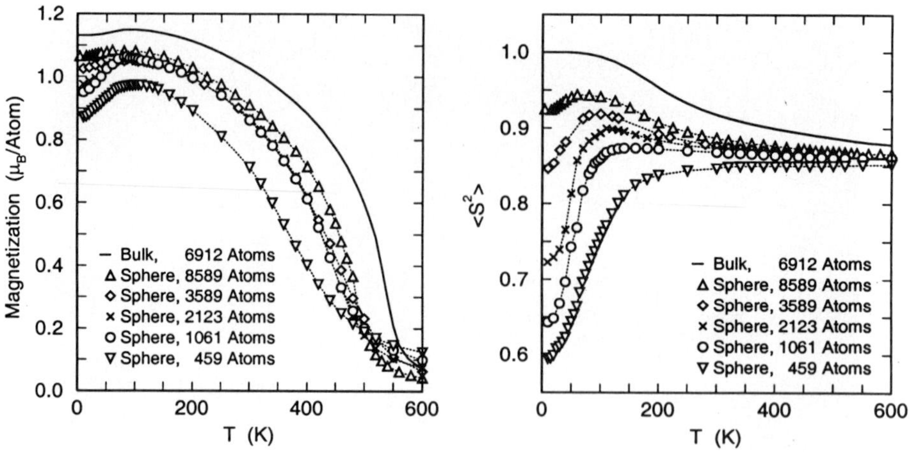

Figure 2. **Left:** Temperature dependence of the average magnetic moment for bulk and various particle sizes. **Right:** Temperature dependence of the averaged squared spin variable for different particle sizes.

size. The calculations were carried out for spherical particles ranging from 459 to 8589 atoms, which corresponds to diameters of 20 Å to 56 Å, respectively. In a second step we vary the concentration while keeping the particle size fixed to 1061 atoms ($\approx$ 28 Å in diameter).

4.1 Magnetoelastic behavior depending on the size of the particles

The left panel of Fig. 2 shows the dependence of the magnetization on the particle size. The magnetization is given by the thermal average of the sum over the individual spins, multiplied with a factor of 2.8 μ_B in case of an iron spin and 0.6 μ_B for a nickel spin, respectively. As expected we find a decrease of the onset of magnetic order for smaller particles. This effect is due to the increasing fraction of surface atoms which have a smaller coordination number than bulk atoms. At low temperatures we find a slight dip in the magnetization which becomes more eminent for smaller particles. In bulk and in large particles the large and negative Fe-Fe exchange interaction which is in competition with the ferromagnetic Fe-Ni and Ni-Ni interactions causes a fraction of iron atoms to prefer antiferromagnetical alignment to the main spin direction as can be seen in Fig. 1 (upper part), which leads to a reduction of the net magnetization. On the surface, however, another mechanism can be observed,

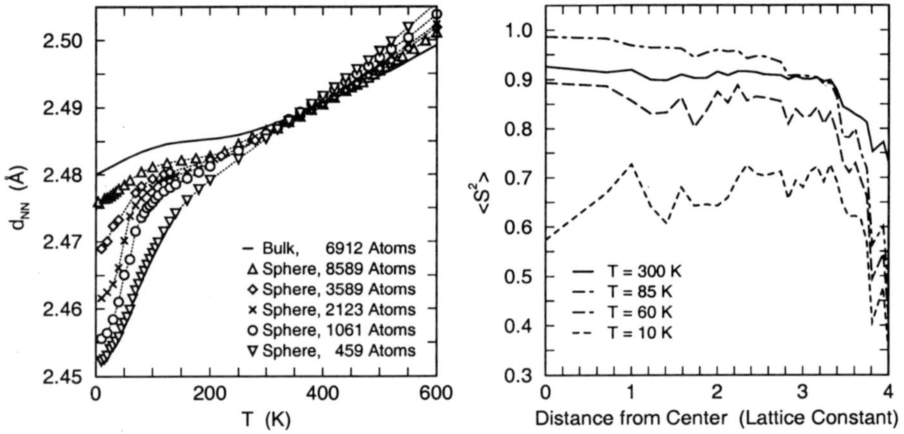

Figure 3. **Left:** Distance between nearest neighbors as a function of temperature. **Right:** Mean magnetic moment as a function of the distance to the center of the sphere at different temperatures. The values were averaged over particles with 1061 atoms but different distributions of the iron and nickel atoms.

which is dominating the behavior of small particles: In contrast to the bulk calculations, where periodic boundary conditions have been used, the reduced coordination on the surface allows for spatial relaxation of the atoms. Therefore, switching to the LM state, which comes along with potentials with a smaller equilibrium distance between the atoms causing energetically expensive lattice strains, becomes a feasible way for iron atoms on the surface to avoid magnetic frustrations induced by the competing exchange interactions.

This view is underlined by the right panel of Fig. 2 which shows the temperature dependence of the average of the squared spin variable giving the concentration of HM atoms in the system, which is also a measure for the mean magnetic moment of the atoms. In the bulk system we find that the fraction of LM atoms is vanishing at low temperatures, then gradually growing towards the high temperature limit of 13 %. Therefore, a growing number of elastic interactions appear with increasing temperature, which are described by the LM potentials characterized by a smaller lattice constant. This leads to a reduction of the mean neighbor distance compared to the pure HM case. For bulk $Fe_{65}Ni_{35}$ this reduction compensates the thermal expansion in a broad range of temperatures as it is observed in experiment (Fig. 3 left).

The Fe-Ni nanoparticles contain at low temperatures a considerable amount of nonmagnetic iron spins which is growing with decreasing particle size. With increasing temperature the mean magnetic moment is rising, eventually passing through a slight maximum before approaching the high-temperature limit as in the bulk case. The increase of the magnetic moment causes the Invar effect to vanish and – for small particles – accounts for a largely enhanced thermal expansion, i.e. a pronounced anti-Invar behavior.

In order to shed further light on the mechanisms causing the rapid change of the mean magnetic moment, a profile of the distribution of HM atoms with respect to the distance from the center for different temperatures is given in Fig. 3 (right). Here we have considered a cluster of size 1061 atoms. From the figure we observe that at very low temperature (around $T = 10$ K) a considerable amount of atoms on the surface and in the inner part of the cluster have switched to the LM state. As the temperature is increased to $T = 60$ K or $T = 85$ K the bulk atoms favor the HM state and the surface atoms remain in the LM state. At higher temperatures (around $T = 300$ K) due to thermal fluctuations the concentration of HM atoms decreases in the bulk, whereas it increases on the surface causing a reduction in the difference between surface and core moment. From the study of the variation of the profile of the distribution of HM atoms with the temperature it seems that for small clusters the lattice spacing and the spin states on the surface and in the bulk are correlated at very low temperature, where the correlation length is larger than the system size. When we increase the temperature the correlation length decreases and above some particular temperature it becomes smaller than the system size, which leads to an uncorrelated dynamics of spin and lattice degrees of freedom of the surface and the bulk. Thus for small clusters one can expect a sudden change in HM concentrations and lattice distortions in the low temperature region contrary to the bulk Invar alloys. A reasonable microscopic interpretation for this behavior is that for small particles the small distortions on the lattice spacing on the surface caused by iron atoms which have switched to the LM state penetrate the whole particle (Fig. 1 lower part), leading to a smaller energy due to a reduction of unsatisfied antiferromagnetic Fe-Fe couplings. In large particles and in bulk, however, there are fewer possibilities to achieve a distorted but energetically favorable configuration, so that the nickel atoms, which can not accustom to a reduced lattice constant by changing their moments, stabilize the HM state (Fig. 1 upper part). For increasing temperatures, thermal fluctuations and anharmonic lattice vibrations make the LM state less favorable due to its preferred smaller lattice spacing. This will happen first in the core part of the particle because of the higher coordination number of the atoms. On the surface, however, there are more possibilities for the Fe atoms to stay close enough to

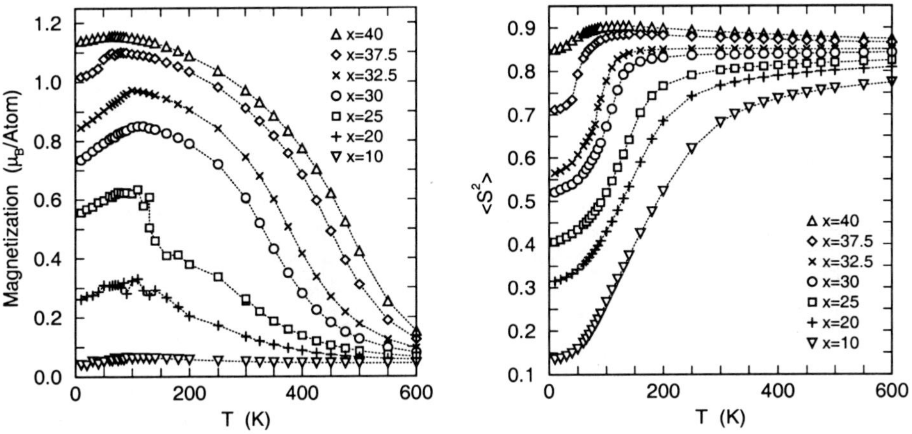

Figure 4. **Left:** Magnetization as a function of temperature for different compositions of $Fe_{1-x}Ni_x$ and fixed particle size of 1061 atoms. **Right:** Temperature dependence of the averaged squared spin variable for different nickel concentrations x.

their neighbors to avoid unsatisfied antiferromagnetic couplings by becoming nonmagnetic.

4.2 Composition-dependent properties of Fe-Ni particles

The magnetoelastic behavior of the Fe-Ni alloy shows a remarkable dependence on the Fe-Ni compositions (Fig. 4 left). For 40 at.% Ni Fig. 4 corresponds to the result expected for a pure ferromagnet and shows a dip in the magnetization at low temperature with the increase of the iron content because the anti-ferromagnetic Fe-Fe interaction destabilizes the ferromagnetic order. If the nickel content falls below 20 at.% ferromagnetism breaks down completely due to the almost complete loss of iron moment at low temperature (right panel of Fig. 4). At higher nickel concentration this loss of iron moment is less pronounced. Moreover from Fig. 4 we observe an increase of magnetic moment from a low value with the increasing temperature. The slope of this increase in moments seems to be maximum around the Invar concentration which is 35 at.%.

Again the behavior of the average moment is directly reflected in the change of the particle volume (Fig. 5 left) leading to a strong anti-Invar behavior over a wide range of temperatures, especially for low nickel concentrations. For higher nickel concentrations the specific heat (Fig. 5 right)

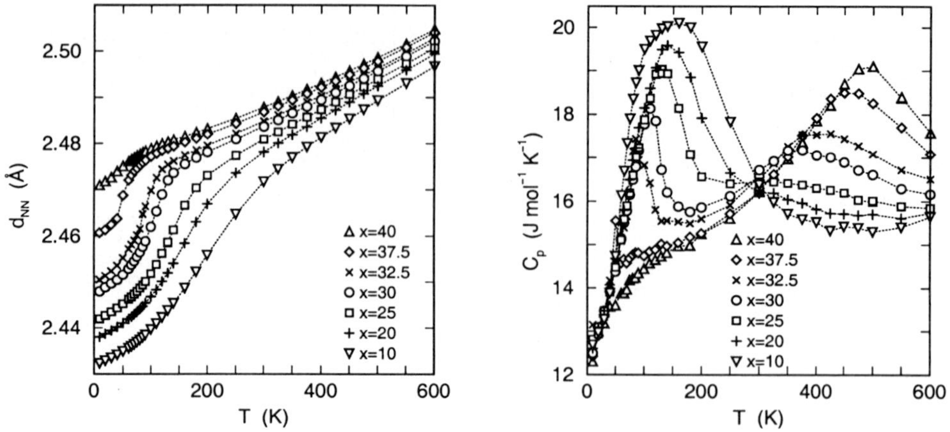

Figure 5. **Left:** Average nearest neighbor distance as a function of temperature for different compositions of $Fe_{1-x}Ni_x$. **Right:** Specific heat C_p as a function of temperature for different compositions of $Fe_{1-x}Ni_x$.

shows a pronounced maximum above $T = 400$ K, which corresponds to a breakdown of ferromagnetic order. With growing iron content this anomaly becomes increasingly washed out due to the weakening of ferromagnetism by the increasing number of antiferromagnetic Fe-Fe interactions in the particle and vanishes around 20 at.% Ni. At the same time a Schottky-type anomaly evolves below $T = 200$ K, which is located where the curve of the average particle moment and the average neighbor distance have their maximum slope. This maximum refers to the thermal excitation of HM atoms from the LM dominated ground state.

5 Conclusions

Our results underline that the transition from bulk to small particles or clusters account for a qualitative change of the properties of the material, which makes it extremely difficult to extrapolate the behavior of small particles from bulk properties. In the case of $Fe_{65}Ni_{35}$ the increasing surface to volume ratio causes the ground state of the particle to change so that the compensation of thermal expansion typical for Invar alloys turns into a largely enhanced volume expansion. However, our results imply that it is necessary to study particles of less than 10 nm in diameter in order to find a behavior that differs significantly from the bulk behavior. Unfortunately to the authors' knowledge, so far no experimental measurements of the magnetoelastic properties of Fe-Ni clusters of a few nanometers in diameter have been made, which would allow for a comparison with our theoretical predictions.

One crucial point of the calculations presented in this paper is, that a phenomenological model has been used to describe the anomalous behavior of Fe-Ni alloys. This approach presupposes the availability of a reasonable set of parameters. For bulk calculations these can be obtained from numerous experimental measurements and *ab initio* investigations. For particles and clusters, however, it can be expected that due to the reduced coordination of the surface atoms, the bulk parameters have to be modified in order to meet the situation on the surface correctly. *Ab initio* calculations of Fe-Ni surfaces can help to fix the surface parameters, but in order to account for magnetovolume effects, the possibility of a spatial relaxation of the surface atoms has to be included. This will make *ab initio* calculations computationally expensive, especially for disordered materials. Furthermore, regarding small clusters containing about 1000 atoms or less, our results let suspect that this relaxation takes place throughout the whole particle, which also makes it difficult to extrapolate from *ab initio* bulk and surface results the electronic properties of small clusters and particles.

Acknowledgements

This work has been supported by the DFG (Deutsche Forschungsgemeinschaft) through the SFB (Sonderforschungsbereich) 166, the SFB 445 and the Graduate College *Structure and Dynamics of Heterogeneous Systems*.

References

1. S. Kajiwara, S. Ohno, K. Honma, and M. Uda, *Phil. Mag. Lett.* **55**, 215 (1987).
2. Y. Zhou, M. Harmelin, and J. Bigot, *Mater. Sci. Engin. A* **124**, 241 (1992).
3. S. Kajiwara, S. Ohno, and K. Honma, *Phil. Mag. A* **63**, 625 (1987).
4. I.M.L. Billas, J.A. Becker, A. Châtelain, and W.A. de Heer, *Phys. Rev. Lett.* **71**, 4067 (1993).
5. I.M.L. Billas, A. Châtelain, and W.A. de Heer, *Science* **265**, 1682 (1994).
6. X. Zhao, Y. Liang and Z. Hu, *Nanostruc. Mater.* **7**, 313 (1996).
7. I.M.L. Billas, A. Châtelain, and W.A. de Heer, *J. Magn. Magn. Mater.* **168**, 64 (1997).
8. C.E. Guillaume, *C.R. Acad. Sci.* **125**, 235 (1897).

9. E.F. Wassermann, in *Ferromagnetic Materials*, Vol. 5, ed. K.H.J. Buschow and E.P. Wohlfarth (Elsevier, Amsterdam, 1990), p. 237.
10. M. Shiga, in *Materials Science and Technology*, Vol. 3B, ed. R.W. Cahn, P. Haasen, and E.J. Kramer (VCH, Weinheim, 1994), p. 159.
11. *The Iron-Nickel Alloys*, ed. G. Béranger, F. Duffaut, J. Morlet, J.-F. Tiers (Technique and Documentation, Paris, 1996).
12. L. Delaey, in *Materials Science and Technology*, Vol. 3B, ed. R.W. Cahn, P. Haasen, and E.J. Kramer (VCH, Weinheim, 1991), p. 339.
13. C.M. Wayman, H.K.D.H. Bhadeshia, in *Physical Metallurgy*, Vol. II, ed. R.W. Cahn and P. Haasen (North-Holland, Amsterdam, 1996), p. 1507.
14. R.J. Weiss, *Proc. Phys. Soc. London* **82**, 281 (1963).
15. P. Entel, E. Hoffmann, P. Mohn, K. Schwarz, and V.L. Moruzzi, *Phys. Rev. B* **47**, 8706 (1993).
16. E. Hoffmann, P. Entel, K. Schwarz, and P. Mohn, *J. Magn. Magn. Mater.* **140-144**, 237 (1995).
17. M. Schröter, H. Ebert, H. Akai, P. Entel, E. Hoffmann, and G.G. Reddy, *Phys. Rev. B* **52**, 188 (1995).
18. M.E. Gruner, P. Entel, *Comp. Mater. Sci.* **10**, 230 (1998).
19. M.E. Gruner, R. Meyer, and P. Entel, *Eur. Phys. J. B* **2**, 107 (1998).
20. R.F. Sabiryanov, S.K. Bose, and O.N. Mryasov, *Phys. Rev. B* **51**, 8958 (1995).

FREEZING AND MELTING OF CLUSTERS AND EXAMINATION OF THE POTENTIAL ENERGY SURFACE

K. L. NIERHOLZ, K.-P. NOS, R. MEYER AND P. ENTEL
Gerhard-Mercator-Universität – GH Duisburg,
Lotharstr. 1, 47057 Duisburg, Germany

This paper addresses the structural and thermodynamical properties of Ar_{55} clusters. A description of thermodynamical and structural properties is given. Caloric curves for the micro-canonical and the canonical ensemble have been obtained by molecular dynamics simulations and histogram methods. The structure of the clusters is investigated by defining basins of minima on the potential energy surface with a disconnectivity tree. Further, an approximation of the system's partition function is discussed.

1 Introduction

The technical development is reaching nanoscale regions where the material does not show conventional bulk behavior anymore. Hence, the investigation of systems containing $N = 10 \ldots 1000$ atoms or molecules where the assumption of the thermodynamical limes is no longer valid has become an important task. In this paper, we present some aspects of structure and thermodynamics of Ar_{55} clusters.

As clusters contain many surface atoms, many assumptions made in conventional thermodynamics are no longer valid. Clusters can be called solid-like at low temperatures and liquid-like at high temperatures, but in contrast to the bulk behavior an intermediate range of temperatures exists where clusters show both solid and liquid behavior.[1] As a consequence of this the heat capacity does not show a discontinuity but a smooth peak.[2] To explain and show this, we performed molecular dynamics simulations combined with histogram methods.

Moreover, we mapped the potential energy surface (PES) of clusters in order to characterize the preferred cluster structures and transitions between them. In order to do this, we calculated a selection of stationary points on the PES and connections between these. With the resulting data-base we discuss the connectivity of the icosahedral ground state and similar structures. Finally we discuss some aspects and problems of approximating the partition function with the set of minima. Another approach could be the investigation of collective motions of argon clusters.[3]

2 Methods

We simulated and computed 55-atom cluster making use of the pairwise-additive Lennard-Jones potential:

$$\Phi = \sum_{i=1}^{55} \sum_{j<i} 4\epsilon \left[\left(\frac{\sigma}{r_{ij}} \right)^{12} - \left(\frac{\sigma}{r_{ij}} \right)^{6} \right],$$

where $r_{ij} = |\mathbf{r}_i - \mathbf{r}_j|$ is the distance between the particles. We chose $\epsilon = 1.65324 \cdot 10^{-14}$ erg and $\sigma = 0.3405 \cdot 10^{-9}$ m in order to get reasonable results for argon.[4]

2.1 Thermodynamics of Ar_{55} by molecular dynamics simulations

Caloric curves for Ar_{55} have been calculated by molecular dynamics simulations employing the velocity form of the Verlet algorithm.[5] In order to be able to derive results valid for the canonical ensemble the Nosé-Hoover thermostat[6] has been implemented. A crucial point in the simulation of clusters is the definition of their volume V. Here we make use of the method of spherical approximation.[7] At each simulation step the greatest distance of an atom from the center of mass while still having at least 4 neighbors at a distance less than 1.5σ has been determined. The volume of the cluster has then been taken to be the volume of a sphere with a radius of this distance.

For the micro-canonical ensemble (constant energy), the mean kinetic energy for different total energies was determined. This was done by Gaussian fits to one-dimensional histograms over the kinetic energy. These histograms were calculated from a sampling over $3 \cdot 10^5$ time steps of 10^{-12} s after an equilibration run of 10^4 time steps. The temperature T is then defined by the equipartition theorem $T = 2\langle K \rangle / (k_B(3N-6))$ with the kinetic energy K and the number of atoms N, respectively.

From the simulations at constant temperature histograms of the (Φ, V) distribution of the system were derived. Every histogram was calculated during 500 simulation runs, each iterating over 10^4 (low temperatures) time steps of 10^{-12} s to 10^5 (high temperatures) time steps of 10^{-13} s. The reduction of the time step was necessary to prevent the cluster's evaporation. After norming, the histograms are proportional to a probability function $p(\Phi, V)$. E. g., the mean internal energy, the heat capacity and the mean volume can be calculated for the given temperatures.[7] Implementing the multiple histogram method,[8] the density of states $\Omega(\Phi, V)$ has been calculated. This is independent of T and the pressure p. Once $\Omega(\Phi, V)$ is obtained, thermodynamic quantities can be derived.

2.2 Mapping the Ar_{55} potential energy surface

An approach to investigate the $(3N-6)$-dimensional potential energy surface (PES) of the clusters is its description by some stationary points. A connected network of sequences (minima - transition states - minima) was constructed. At low temperatures only transition paths between different structures which correspond to the minimal energy pathes over first order saddle points should be possible. Because of that, saddle points were found with an eigenvector following method,[9] which started at low lying minima. Two minima connected unambiguously at $T = 0$ K have been determined by a steepest descent search ($\dot{\mathbf{r}} = -\nabla\Phi(\mathbf{r})$) starting at a previously found saddle point.

The computed data-set for Ar_{55} comprises 3037 minima and 5298 saddle points. We assume that at least the 20 lowest lying structures and the most important low energy transition states are found.

3 Results

3.1 Thermodynamical results

In Fig. 1(a), caloric curves for the micro-canonical and the canonical ensemble are shown. The micro-canonical internal energy exhibits an S-shaped curvature in the coexistence region where both, solid-like and liquid-like states can occur. This can be understood by considering the constant energy $U = \Phi + K$: The cluster switches between solid and liquid along the time scale. In solid-like states, the cluster's configuration has a lower configurational energy Φ. Switching to liquid-like states, it is increasing its configurational energy, hence the kinetic energy has to be reduced to keep the total energy constant. Because of this, the same mean kinetic energy can be obtained at different total energies.

The canonical caloric curve shows a smooth, monotonic change from solid to liquid. In Fig. 1(b), the corresponding heat capacity is presented, with a peak in the transition region. With increasing number of atoms N, this peak will become sharper and taller and finally, in the bulk limit, leading to a discontinuity for $c(T)$.

In Fig. 2 the (Φ, V) histogram for $T = 35.4$ K is shown. A double peak structure is observed giving evidence of the coexistence of the solid and the liquid phase. The tall peak at smaller volume and configurational energy represents the solid phase. Towards lower temperatures, this peak will dominate while the other one vanishes. On heating, the liquid peak at higher volume and configurational energy will enlarge. As this peak covers a larger area on the (Φ, V)-plane, it gets clear that the cluster's shape and structure fluctuates

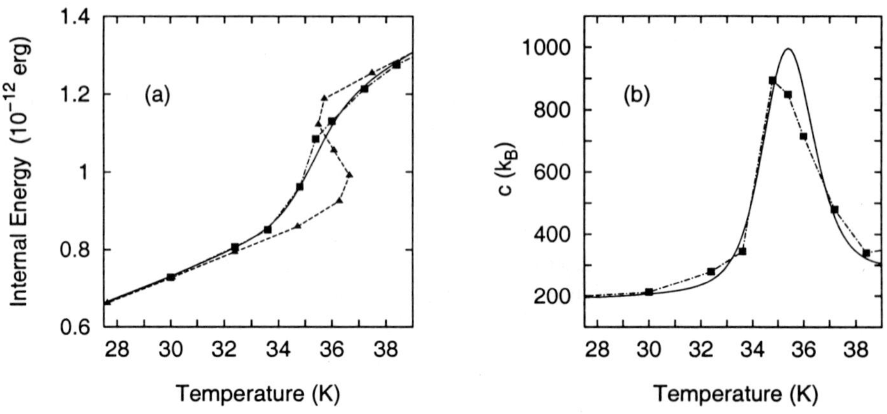

Figure 1. (a) Internal energy $U(T)$: Micro-canonical data (triangles/dashed line), canonical data obtained directly from the histograms (squares/dash-dotted line) and calculated from the density of states (solid line). (b) Heat capacity $c(T)$ for the canonical ensemble (see (a)).

strongly in the liquid phase. The small third peak at high volume but low energy represents solid, defective icosahedral structures.

With seven histograms at temperatures from 30 to 37.2 K in the phase change region, including that one in Fig. 2, the internal energy $U(T)$ and the heat capacity $c(T)$ represented by the solid lines in Fig. 1 (a) and (b), respectively, have been calculated. These give a smooth approximation to the data obtained by the histograms. An extension of this method in order to predict the system's behavior at nonzero pressures is planned. Calculations of caloric curves of systems with additional inert-gas atoms in order to simulate finite pressures are in progress.[10]

3.2 Structural Properties

One way to describe the shape and topography of the PES of clusters is a micro-canonical disconnectivity tree.[11] Every leaf of such a tree represents a local minimum, the relative height giving its potential energy. The nodes define basins, which consist of all connected minima and nodes, i.e. smaller basins with lower energy. These are placed on a ladder with discrete energy spacing ΔE. For each node at energy E_n saddle points below E_n exist. These transition states connect the lower minima or basins of the vertices which branch from the node. Each basin corresponds to a closed PES area which

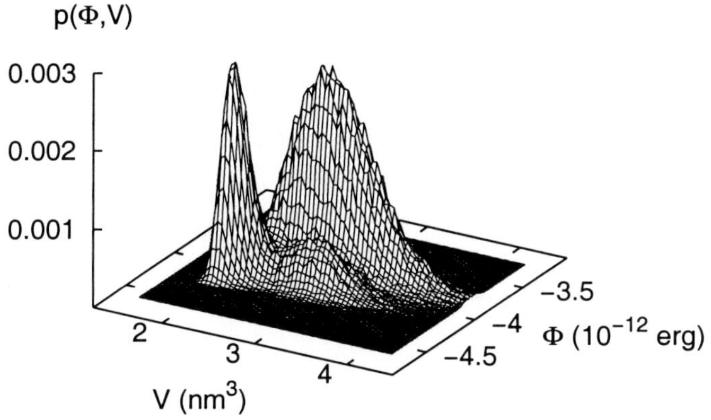

Figure 2. Normed histogram for $T = 35.4$ K over the volume V and the configurational energy Φ.

can be reached in a micro-canonical system. The given energy-step ΔE defines a kind of magnitude of the tree. Graphs with small ΔE emphasize the connectivity of local areas of the PES, greater energy-steps pronounce the general structure of the surfaces.

In Fig. 3 such a tree is shown for the 50 lowest found minima of Ar_{55}. An intermediate resolution was chosen ($\Delta E = 0.55\epsilon$). For symmetric configurations the class of their point group is given at the corresponding branch end.

The global minimum (icosahedron structure) is separated energetically from the next higher configurations, which are icosahedrons with one defect. Possibly it is these configurations which give rise to the small third peak in 2. Above them the enormous number of liquids like minima form a huge basin. The first of them, (the 12th-lowest structure) a twisted icosahedron with C_{5v} symmetry, is separated from the global minimum by a relatively high transition state. The central basin comprises at least 1306 minima before this one will become connected to it. At slightly higher energies some more of those degenerated icosahedron structures exist.

For kinetic and thermodynamic discussions it is necessary to know the

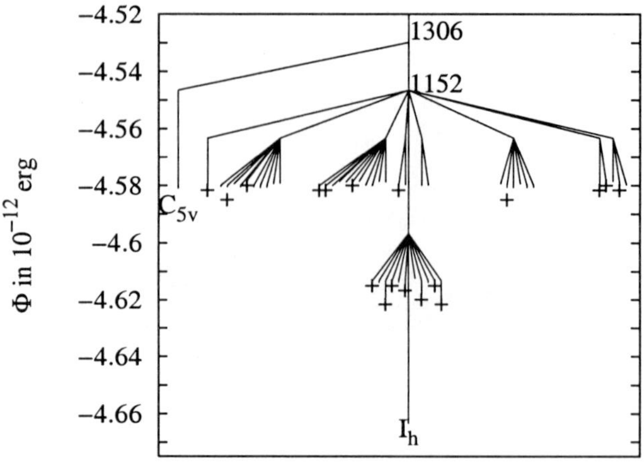

Figure 3. The micro-canonical disconnectivity graph for Ar_{55} with $\Delta E = 8.35 \cdot 10^{-15}$ erg. Only the 50 lowest minima are drawn. The real number of minima in the central basin is marked at the nodes. Configurations with C_s-symmetry are marked with a "+". All other of the 50 lowest minima have no symmetry properties, except the two with I_h and C_{5v}-symmetry.

partition functions for the stationary points. Its contribution around a stationary point $\mathbf{r}$ can be approximated within a harmonic approximation of the potential energy function $\Phi(\mathbf{r} + \delta\mathbf{r}) = \Phi(\mathbf{r}) + \frac{1}{2}\delta\mathbf{r}\ \Delta\Phi(\mathbf{r})\ \delta\mathbf{r}$ as:[12]

$$Z = \sum_{i}^{Minima} \frac{2N!}{h_s(i)} \prod_{j=1}^{3N-6} \frac{1}{\beta h \nu_j^i} \exp(-\beta\Phi(\mathbf{r_i})),$$

where ν_j^i are the vibrational frequencies, given by the positive eigenvalues of the Hessian matrix $\Delta V(\mathbf{r}_i)$. The multiplicity of the configuration i is included by $h_s(i)$, the order of the point group and N is the number of atoms.

With the knowledge of the potentials V, the point group orders and the mean vibrational frequencies $\bar{\nu}_i = \left(\prod_{j=1}^{(3N-6)} \nu_j^i\right)^{1/(3N-6)}$ of the minima, one can get a first impression of the kinetics of the PES. These mean vibrational frequencies are shown in Fig. 4 for Ar_{55}. Again the 12th lowest structure is remarkable (marked with an arrow), because its contribution to the partition function Z is much higher than that of the energetically neighboring minima, due to its low mean vibrational frequencies. However, at high temperatures the approximation of Z will become very bad, because only the few minima

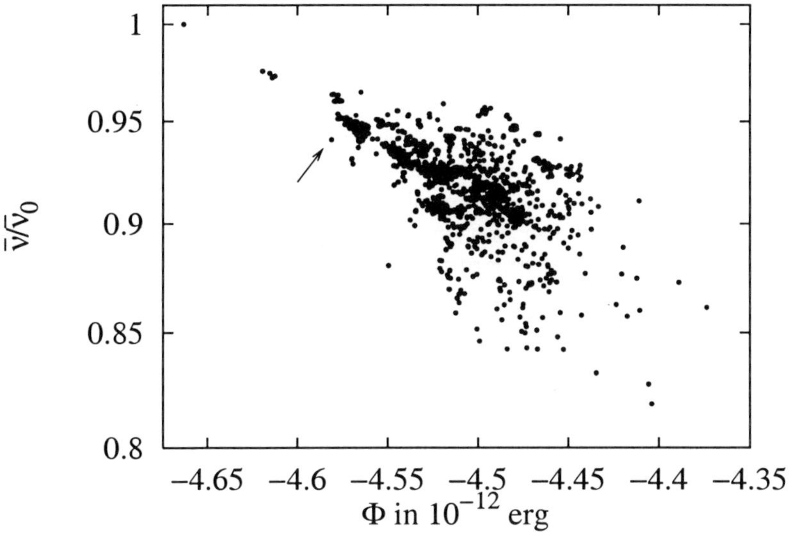

Figure 4. The mean geometrical vibrational frequencies of all accounted 3037 minima, $\bar{\nu}_0$ belongs to the icosahedron. The twisted icosahedron is marked with an arrow.

with very low vibrational frequencies dominate the partition functions behavior in this picture. It seems that the configurational space area occupied by these minima is largely overestimated in this simple theory.

In the next stage, rate constants between the minima will be computed by using, for example, the RRKM theory.[13] A master equation for the occupational probability of the structures can be solved to get an approximation of the structural dynamics of the system.[11,14,15]

4 Conclusions

The investigation of Ar_{55} clusters by molecular dynamics (MD) simulations has been performed successfully, partially by transferring methods originally developed for Monte Carlo simulations (MC)[2,7,8] to MD. With MD, a couple of problems, for example evaporation of cluster atoms, can be avoided or minimized. Caloric curves for the canonical ensemble have been obtained by calculating the density of states from (Φ, V) histograms. Simulations of nonzero pressure will be done soon Further it is planned to extend our calculations to simulations making use of embedded atom method potentials.

Classification of the PES of a system with separated regions is a powerful method to get an overall picture of cluster behavior: possible structures and their connectivity, kinetic behavior and thermodynamical properties can be explained. However, the quality of the data base is very sensitive to a correct selection of the essential minima and transition states. At temperatures of melting and fluid-like phases the simple reduction of the PES to regions around minima and transitions over first order saddle points becomes invalid. A more careful selection and weighting of the minima, combined with the inclusion of anharmonicities accounting to Z has to be used to model the correct melting behavior in this approach.[12,13,16] For low temperatures, both structural, and dynamic properties of clusters with different potentials or sizes can be easily compared and understood.

5 Acknowledgements

This work has been supported by the Sonderforschungsbereich 445 of the German Sience Council on *Nanoparticles from the vapour phase: Nucleation, structure and properties.*

References

1. J. Jellinek *et al.*, *J. Chem. Phys.* **84**, 2783 (1986).
2. P. Labastie and R. L. Whetten, *Phys. Rev. Lett.* **65**, 1567 (1990).
3. U. A. Salian, S. N. Behera and V. S. Ramamurthy, *J. Chem. Phys.* **105**, 3679 (1996).
4. M. F. Pas and B. J. Zwolinski, *Mol. Phys.* **73**, 471 (1991).
5. M. P. Allen and D. J. Tildesley in *Computer Simulations of Liquids* (Clarendon Press, Oxford, 1987).
6. S. Nosé, *Mol. Phys.* **52**, 255 (1984); W. G. Hoover, *Phys. Rev.* A **31**, 1695 (1985).
7. H.-P. Cheng *et al.*, *Phys. Rev.* A **46**, 791 (1992).
8. A. M. Ferrenberg and R. H. Swendsen, *Phys. Rev. Lett.* **61**, 2635 (1988); *Phys. Rev. Lett.* **63**, 1195 (1989).
9. D.J. Wales, *J. Chem. Phys.* **101**, 3750 (1994).
10. To be published.
11. O. M. Becker and M. Karplus, *J. Chem. Phys.* **106**, 1495 (1997).
12. K. D. Ball and R. S. Berry, *J. Chem. Phys.* **109**, 8541 (1998).
13. K. D. Ball and R. S. Berry, *J. Chem. Phys.* **109**, 8557 (1998).
14. R. E. Kunz and R. S. Berry, *J. Chem. Phys.* **103**, 1904 (1995).

15. R. E. Kunz, P. B. Blaudeck, K. H. Hoffmann and R. S. Berry, *J. Chem. Phys.* **108**, 2576 (1998).
16. J. P. K. Doye and D. J. Wales, *J. Chem. Phys.* **102**, 9659 (1995).

STRUCTURE AND DYNAMICS OF THIN FILMS ON SURFACES

INFLUENCE OF MAGNETISM FOR THE ALLOY FORMATION OF ULTRATHIN FILMS

G. BIHLMAYER, R. ABT AND S. BLÜGEL

Institut für Festkörperforschung, Forschungszentrum Jülich, D-52425 Jülich, Germany

E-mail: s.bluegel@fz-juelich.de

T. ASADA

Faculty of Engineering, Shizuoka University, Hamamatsu 432, Japan

In this article we review the present understanding of the interrelation between the energetic stability of surface alloys and magnetism, magnetic ordering, the choice of substrate and surface orientation. We discuss trends for 3*d*-alloys on a Cu(100) surface and specifically present *ab initio* results for the formation energy and the interdiffusion energy of Mn surface alloy films on the Cu(100), (110) and (111) surface and on the Ag(100) and Ni(100) surfaces. Results based on the density-functional theory elucidate the origin of the existence of thermodynamically stable surface alloys. Although in detail, the results depend on the selected substrate/alloy combinations, the trends presented here have a universal character. Recently, several of these surface alloys have been found experimentally. The large structural corrugation of the surface alloy atoms and their enhanced magnetic moment contribute to the stabilization of the alloys. This will be discussed as a result of a large magneto-volume effect. Total energy and force calculations are carried out using the full–potential linearized augmented plane–wave (FLAPW) method in film geometry.

1 Introduction

Surface alloying after deposition of one metal onto another has gained increasing interest in the last few years.[1] This is partly motivated by the desire to tailor the properties of metallic surfaces through intermixing in the near–surface region. Most of the studies so far have focused on the structure of surface alloys. Surprising results include the observation of surface alloying even for some metal–metal combinations which are practically immiscible in the bulk.

In a series of investigations Wuttig *et al.* have studied surface alloys formed after deposition of Mn on Cu(100),[2,3,4] Mn on Ni(100)[5] and Mn on Cu(110).[6] A particularly surprising result of these investigations is the observation of ordered Cu(100)c(2×2)Mn, Ni(100)c(2×2)Mn, and Cu(110)c(2×2)Mn surface alloys, stable over a large temperature range and characterized by a considerable atomic corrugation in the ordered surface alloy layer. Interesting enough, no ordered bulk alloy exists for the Cu-Mn system. Early total energy calculations have shown that the corrugation[7,8] of the Cu(100)c(2×2)Mn surface alloy layer is due to the formation of a large magnetic moment of the Mn atoms. Several recent experiments confirmed the formation of these ordered surface alloys[9,10] as well as the formation of these large magnetic moments.[11,12,13,14,15,16,17] From this we can conclude that we are dealing here with ordered two-dimensional magnets, which are structurally well-defined and stable over a large temperature range. In search of an understanding of the correlation between volume, magnetism and surface alloy formation, Mn deposited on Pd(100)[18] and Ag(100)[19,20] has been investigated experimentally as well as theoretically.[21]

In this paper we shortly review the present understanding on the interrelation of the formation and stability of the Mn surface alloy with the magnetic properties. We first report on a set of systematic first-principles electronic-structure investigations of the alloy formation and stability of $3d$ transition-metal surface alloys on Cu(100). From these systematic calculations we are able to unravel the underlying chemical and magnetic trends leading to the surface alloy formation. The results show a clear trend: i) Magnetism acts against interdiffusion. At low coverage, where the magnetism of Cr and Mn is very large, interdiffusion is suppressed. ii) Magnetism acts unfavorably for clustering and promotes alloy formation. For Mn we find a strong tendency for alloy formation. From i) + ii) we expect for the case of the Cu substrate a thermodynamically stable, two dimensional surface alloy: Cu(100)c(2×2)Mn.

For the rest of this paper we concentrate exclusively on the investigation of Mn surface alloys. To elucidate the dependence of the magnetic properties on the coordination number and the occupied volume, we carry out case studies such as Cu(100)c(2×2)Mn, Cu(110)c(2×2)Mn, and Cu(111)c(2×2)Mn to explore the role of the surface orientation, Ag(100)c(2×2)Mn to investigate effect of the size of the in–plane lattice constant, and Ni(100)c(2×2)Mn in order to see changes due to a magnetic substrate.

This work is based on total energy differences of different structural and magnetic configurations. By comparing the total energies of different configurations we extract the total energy differences which are interpreted as an indication of the stability of the surface alloy against clustering or interdiffusion into the bulk. Although what occurs in real growth would be strongly governed by the energy barriers of each process, we believe that the examination of such total energy aspects will give a certain indication of how systems evolve.

This article is organized as follows: after a short discussion of the computational details, we first investigate the possibilities for alloy formation of $3d$

transition–metals (TM) on the Cu(100) surface. This gives a nice overview of how different magnetic properties influence the stability of surface alloys. Then we vary the substrate orientation for the Mn/Cu system, to see how coordination and relaxation effects may alter these observations. In a final section, we will investigate the influence of the substrate on the alloy–formation and we choose the Mn/Ni(100) and Mn/Ag(100) as examples for magnetic and nonmagnetic substrates.

2 Computational details

The results for the $3d$ surface alloys were obtained with the full–potential linearized augmented plane–wave method (FLAPW) in film geometry,[22,23] as implemented in the computer code FLEUR, based on density functional theory in the local spin–density approximation (LSDA) or the generalized gradient approximation (GGA). For LSDA we used the exchange–correlation potential of von Barth and Hedin,[24] but with the parameters as chosen by Moruzzi, Janak, and Williams.[25] Recently, we have extended the calculations including the GGA[26] with the version of Perdew and Wang.[27,28] All calculations use consistently the theoretical lattice constant of Cu, $a_0 = 6.65$ a.u. for LSDA[29] or $a_0 = 6.82$ a.u. for GGA,[26] respectively, determined by means of a total energy minimization with a bulk FLAPW code.

The ultrathin films were modeled by seven- to nine-layer films consisting of five or seven layers of substrate and one monolayer film on each side of the film. Throughout this paper all total energy results presented are calculated using 70-80 basis-functions per atom. The force calculations appear to be a bit more critical with respect to the cut-off parameters and we used 110 basis-functions. The forces were minimized down to a maximum force of 3 meV/a.u. per atom. For some of the films, the ground state atom positions were determined by total energy minimization.[7]

3 3d - c(2×2) surface alloys on Cu(100)

The $c(2 \times 2)$ substitutional surface alloys can be thought of as a checkerboard type monolayer film consisting of $3d$ metal atoms, where every second transition–metal atom is substituted by a substrate atom, Cu in this particular example (cf. Fig. 1). The surface unit cell is classified by the notation $c(2 \times 2)$ and we call these surface alloys Cu(100)c(2×2)TM. In the following, we will not consider any relaxational effects – these will be discussed in the next section for the Mn/Cu surface alloys. All results were obtained in the GGA, for a comparison with LDA results see Asada and Blügel.[26]

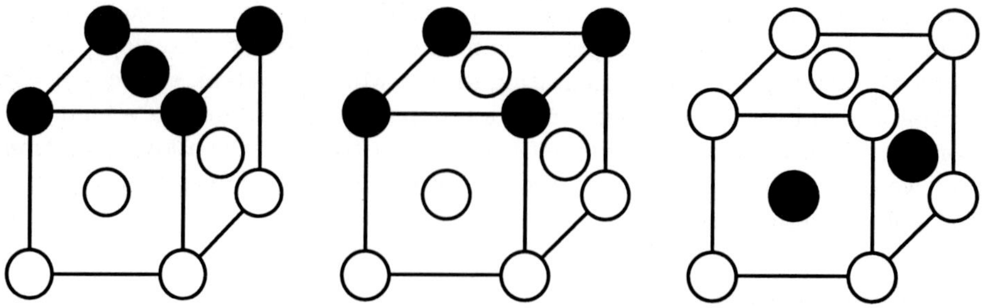

Figure 1. Left: schematic representation of a $3d$ monolayer (filled circles) on a Cu(100) substrate (open circles). Middle: a substitutional surface alloy in a c(2×2) arrangement on the same substrate. Right: an interlayer of $3d$ transition-metal atoms in Cu(100).

Within the surface alloy layer the $3d$ transition-metal atoms are not nearest but next nearest neighbors to each other. Thus the overlap between the $3d$ wave-functions is substantially reduced, effectively being the one of reduced coordination number. Therefore each transition metal atoms can be considered, crudely in the first approximation, as surface isolated $3d$ impurities on Cu(100) substrate. Thus the magnetic moments of the $3d$ transition-metal atoms should be enhanced considerably. This is particularly true for the transition–metal atoms around the center of the $3d$ series: V, Cr, Mn. For V mono-layers on Cu(100) no magnetism was found, while for Cu(100)c(2×2)V the V moment is close to $2\mu_B$. For Cr and Mn the magnetic moments increase to $3.5\mu_B$ and $3.8\mu_B$, respectively. In comparison to the moments of the c(2×2) anti-ferromagnetic phase of the full coverage mono-layers on Cu(100) this amounts to an increase of $0.5\mu_B$ for Cr and $0.4\mu_B$ for Mn.

3.1 Stability against clustering

To assess the existence of the c(2×2) surface alloy we first investigate the stability of the surface alloy against the phase separation. Phase separation of a surface alloy is synonymous with atom rearrangement such as to obtain the clean Cu(100) substrate surface and the clustering of the former $3d$ transition–metal constituent surface alloy atoms. The clustering is modeled by the formation of monolayer films. The stability of the surface alloy against the phase separation or the clustering is investigated in terms of the formation energy

$$\Delta E_F = E_{\mathrm{Cu(100)c(2\times2)TM}} - \frac{1}{2}\Big(E_{\mathrm{TM/Cu(100)}} + E_{\mathrm{Cu(100)}} \Big)\,, \qquad (1)$$

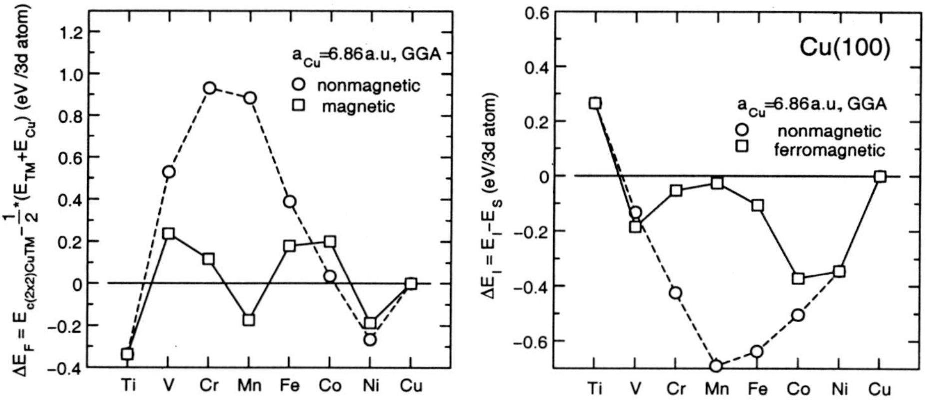

Figure 2. Left: The formation energy per $3d$ atom for the Cu(100)c(2 × 2)TM surface alloys. Right: The interdiffusion energy per $3d$ atom for the Cu(100)c(2 × 2)TM surface alloys. Dashed lines connect the nonmagnetic solutions, full lines connect the magnetic ones.

which is determined by the total energy difference between the systems containing the surface alloy ($E_{\mathrm{Cu(100)c(2\times2)TM}}$) and the average of the total energies determined by the $3d$ mono-layers on Cu(100) ($E_{\mathrm{TM/Cu(100)}}$), and the clean Cu(100) surface ($E_{\mathrm{Cu(100)}}$). Negative ΔE_F means that the surface alloy is energetically favored against the phase separation. The ground state magnetic configurations for each $3d$ monolayer system were determined as either ferromagnetic or anti-ferromagnetic (or nonmagnetic). But ferromagnetism is assumed for the magnetic configuration of the c(2×2) alloy geometry.

On the left side of Fig. 2 we show for all $3d$ transition–metal elements the formation energy ΔE_F derived either from the ground state magnetic or from the nonmagnetic calculations. For the latter we rename the formation energy by ΔE_F^P. We recognize the parabolic behavior of ΔE_F^P originates from the same behavior of the cohesive energy of the nonmagnetic $3d$ metals, since the energy is considered mostly from the disappearance of in-plane nearest neighbor $3d$ bondings. The quantity, ΔE_F^P, is distinctly positive from V to Fe, indicating that the phase separation or clustering is expected for these four systems if we suppress the formation of local magnetic moments. The phase separation energy ($-\Delta E_F$) is largest for Cr and Mn with about

0.9 eV/TM atom. This means, hypothetical nonmagnetic Cu(100)c(2 × 2)Cr and Cu(100)c(2 × 2)Mn surface alloy films may never be thermodynamically stable and surface alloys may only be expected for Ti and Ni systems. For Ti, since all Ti systems remain nonmagnetic, the formation energy ΔE_F^P is equal to the ground state formation energy ΔE_F.

We next observe that magnetism brings about a remarkable change in the behavior of ΔE_F. Moving from V to Mn, ΔE_F decreases monotonically to a minimum at Mn with its formation energy of about -160 meV/atom. From Mn to Fe and Co the formation energy again increases and changes its sign. For Ni, the effect of magnetism is tiny. Owing to the substantial magnetic effect the formation energy from V to Fe drops significantly and there is caused a pronounced dip in the formation energy with the minimum at Mn. However, the formation energy for V, Cr, Fe and Co is still positive, showing no possibility of surface alloy formation for these elements. Only for Mn the strong inclination of phase separation at non-magnetism has been altered to a distinct tendency towards surface alloy formation. Thus, as far as can be seen from the formation energy, we can expect the existence of the surface alloy for Ti, Mn, and Ni.

3.2 Stability against wetting by the substrate

We discuss in this subsection the stability of the 3*d* surface alloy against interdiffusion into the substrate by comparing the total energy of the surface alloy, E_S, with the one, E_I, of the system with an interlayer c(2×2) alloy layer buried as a subsurface layer in Cu(100). We define the difference $\Delta E_I = E_I - E_S$ as the interdiffusion energy, where negative ΔE_I indicates that the system is unstable against the interdiffusion. The impact of magnetism for the stability of the surface alloy against interdiffusion is analyzed by performing both the spin-polarized and the nonmagnetic calculation for the two structural configurations. The magnetic configurations are assumed to be ferromagnetic as done for the c(2×2) structure in the calculation of formation energy of the previous subsection.

On the right side of Fig. 2 we show the interdiffusion energy ΔE_I deduced from both nonmagnetic and magnetic calculations. We first discuss this quantity for the nonmagnetic case. Moving from Ti to Cu ΔE_I shows clearly a parabolic behavior with the minimum at Mn. We consider that this shape reflects the surface energy dependence of the nonmagnetic 3*d* metals and that each value may be regarded roughly as the energy of a 3*d* atom-Cu bond plus a Cu dangling bond subtracted by the one of the Cu-Cu bonds plus a 3*d* atom dangling bond. For Ti, ΔE_I is positive, which may show that the too delocalized Ti wave-function causes an energy increase if it exists as a bulk impurity in Cu.

When we lift the constraint of non-magnetism by carrying out spin-polarized calculations, all systems except Ti become magnetic. We observe that, with the exception of V, a remarkable increase of ΔE_I that acts against wetting is caused for Cr, Mn and Fe. For Ni the magnetic contribution is small with the value of ΔE_I being largely unfavorable for the stability of the surface alloy. The significant changes by magnetism should be considered as coming not only from the magnetic energy enhancement but also from the elastic energy contribution caused by the associated volume expansion. However, we recognize that the magnetic effect is still not strong enough to bring about a positive value for Mn. ΔE_I for Mn is -40 meV/atom, showing that Mn surface alloy is marginally unstable against interdiffusion. But we regard this as being able to be reconciled. We think over the fact that, with the present structure model of the buried c(2×2) alloy layer for 3*d* bulk impurities, interactions remain between the second and further nearest neighbor Mn atoms. We have to properly subtract those further interactions to obtain the quantity more reliable. If we do this by use of Hoshino *et al.*'s results,[30] we obtain a value of $+60$ meV/atom showing a distinct stability of the surface alloy against the interdiffusion. The inclusion of the relaxation effect is expected to make an interdiffusion energy more positive giving a further intensified tendency toward surface layer formation for Mn. It is because the elastic energy could be released more efficiently for the surface layer than for the buried layer. In the next section it is shown that the relaxation energy gain for the Mn c(2×2) surface alloy amounts up to some 90 meV/atom. It is expected that this large relaxation energy on the surface alloy layer contributes much toward the stability of the surface alloy against both the phase separation and the interdiffusion.

In the previous subsection, arguing about the formation energy, we identified three candidates, Ti, Mn, and Ni for the formation of a c(2×2) surface alloy. Now, from the interdiffusion energy point of view, it has been found that Ni can not remain a candidate any more. Thus, to summarize at this point, Mn is an exclusive candidate for the magnetically stabilized surface alloy, whereas Ti may be the one for a nonmagnetic surface alloy.

4 Mn/Cu surface alloys: influence of the surface orientation

Up to this point, we investigated magnetically stabilized surface alloys only at the Cu(100) surface. But magnetism is a quantity, which depends strongly on coordination number, symmetry and on the surface orientation. Thus the existence, the formation and the properties of surface alloys at other surfaces is a open question. To reveal the systematics that govern surface alloying for Mn on Cu single crystal surfaces we study now, after discussing relaxational effects on the Cu(100) surface, also Mn on Cu(110) and Cu(111).

4.1 The (100) surface

All results given above have been obtained neglecting any interlayer relaxations. Since the Cu(100)c(2 × 2)Mn system is just at the edge of being a magnetically stabilized surface alloy, we have improved the investigations by including the interlayer relaxation for all 3 subsystems: ferromagnetic Cu(100)c(2 × 2)Mn, Cu(100), and the anti-ferromagnetic Mn monolayer on this substrate. We performed within LDA a structural optimization by total energy minimization[7] for ferromagnetic (F) and of (hypothetical) nonmagnetic (P) Cu(100)c(2 × 2)Mn surface alloys (S).

The structural optimization for the Cu(100)c(2 × 2)Mn surface alloy is shown in Fig. 3, which is based on minimizing the total energy $E_S\{P/F; \Delta z_{\mathrm{Mn},F/P}; \Delta z_{\mathrm{Cu},F/P}\}$. We included for the nonmagnetic as well as the ferromagnetic case two degrees of freedom. These were the relaxations Δz of Mn and the top Cu atom along the surface normal. This is sufficient as it is shown by a LEED[7,3] analysis. Two important points are to be seen upon comparing the results of the nonmagnetic with the ferromagnetic surface alloy: i) the ferromagnetic solution is lower than the paramagnetic one by about 1.4 eV/Mn atom; and ii) in the nonmagnetic case, the position of the Mn and Cu atoms remains nearly at the ideally terminated Cu site and no corrugation of the surface plane is found, ($\Delta z_{\mathrm{Mn},P} = \Delta z_{\mathrm{Cu}} = -0.5\%\, d_{\mathrm{Cu}}$), whereas the ground state solution was found for magnetic Mn showing a large outwards buckling of $\Delta z_{\mathrm{Mn},F} = +11.5\%\, d_{\mathrm{Cu}}$ and Cu showing a small inwards buckling of $\Delta z_{\mathrm{Cu},F} = -2.5\%\, d_{\mathrm{Cu}}$, all relaxations are given in relative units ($\Delta d/d_{\mathrm{Cu}}$ in %) with respect to the interlayer distance of the substrate (d_{Cu}). This large buckling is in agreement with the experiment,[3] and we can therefore conclude at this point that i) the CuMn surface alloy is magnetic and ii) the buckling motion of Mn is caused by its magnetism. On the other hand, the small inwards relaxation and the absence of the corrugation found for the nonmagnetic solution is in line with the classical arguments of atomic radii as conveyed after the investigation of the CuAu and CuPd surface alloy.[31,32] The inwards relaxation is smaller than the value for the clean Cu(100) surface which we determined the same way to $\Delta z_{\mathrm{Cu}(100)} = -1.5\%$. Thus nonmagnetic Mn pushes the entire CuMn plane up by avoiding any corrugation.

The buckling of the CuMn layer is a nice example of a giant magneto-

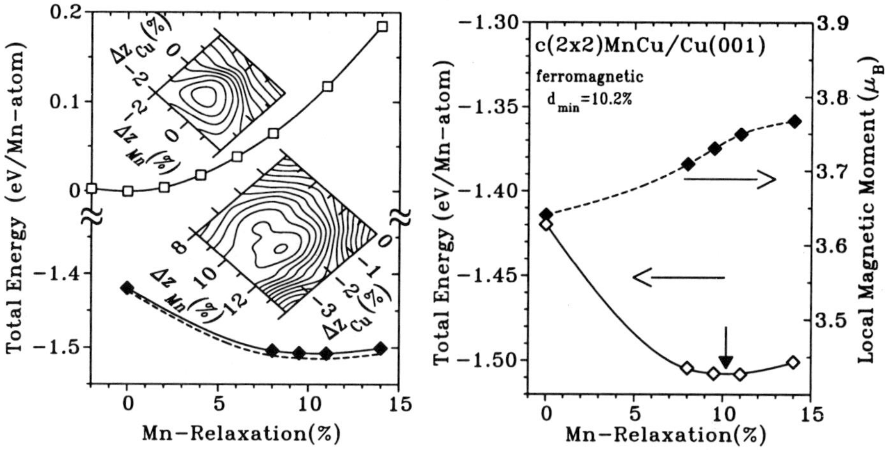

Figure 3. Left: Total energy per Mn atom of Cu(100)c(2×2)Mn vs. the buckling relaxation $\Delta z_{\rm Mn}$ of Mn in relative units with respect to the theoretical interlayer spacing of Cu, $d_{\rm Cu}$ = 1.76 Å. The open squares represent the nonmagnetic and the solid diamonds the ferromagnetic results. The solid lines (for Cu atoms fixed at the ideally terminated positions $\Delta z_{\rm Cu} = 0$) and dashed line (the top Cu atom is always at its optimally relaxed position) are the fitting polynomials. The upper (lower) inset shows the contour plot of the nonmagnetic (ferromagnetic) total energy with respect to the buckling of Mn and Cu. The minimum, which determines the optimal structure is found in the inner circle. The contour interval is 1 meV. The energy of the nonmagnetic solution at 0% relaxation was chosen as the origin of the total energy scale. Right: Total energy (open diamonds) per Mn atom of Cu(100)c(2×2)Mn measured at the left vertical scale, and the local magnetic moment (solid diamonds connected by dashed line) measured at the right vertical scale as function of the buckling relaxation $\Delta z_{\rm Mn}$ of Mn in relative units with respect to the theoretical interlayer spacing of Cu, $d_{\rm Cu}$ = 1.76 Å. Cu atoms are fixed at the ideally terminated position $\Delta z_{\rm Cu} = 0$. The solid line describing the energy is due to a least square fit to a quadratic polynomial. The vertical arrow indicates the minimum energy interlayer distance d_{min} as determined from a least square fit to a quadratic polynomial. The horizontal arrows indicate the proper vertical scale.

volume effect, where the volume of the Mn atom increases with the magnetism and exerts even a pressure on the Cu surface atom, which is why $|\Delta z_{{\rm Cu},F}| > |\Delta z_{{\rm Cu},P}|$. From a microscopic point of view the double occupancy of the Mn bonding states is lifted due to magnetism and half of the Mn bonding states are promoted to anti-bonding states, which make the Mn atom effectively larger and cause the large magneto-buckling effect. In the end, this volume dependence of the Mn atom on the magnetism is a complicated many-atom interaction and therefore this buckling is very difficult to describe or predict

Table 1. Comparison of the experimentally (LEED) and theoretically determined atom relaxations in relative units (%) of the bulk Cu(110) interlayer distance. (P) and (F) indicate nonmagnetic and ferromagnetic calculations, respectively, and Energy and Force indicate whether the relaxations are determined by static total energy minimization or the use of the force method. d_{ij} is defined as the distance from the lower atom in layer i to the highest atom in layer j. ΔZ_{kl} is the height of atom l in layer k. A positive value for the buckling means that the atom is displaced outwards.

Technique	ΔZ_{11}	ΔZ_{31}	Δd_{12}	Δd_{23}	Δd_{34}
LEED	17.2	0.8	−5.7	0.9	−1.0
Energy (P)	−0.6		−5.8		
Energy (F)	12.7		−5.7		
Force (F)	16.3	3.6	−11.3	1.9	−2.2

by hard sphere type of models with element specific atomic radii.

4.2 The (110) surface

On this open surface, our LDA *ab initio* calculations predict an even larger magnetic moment for Mn. For the unrelaxed surface alloy a magnetic moment of $M(\Delta z = 0) = 3.77\mu_B$ was found. The buckling increases the magnetic moment to $M(\Delta z_{\mathrm{Mn}} = 7\%, \Delta z_{\mathrm{Cu}} = -5.7\%) = 3.82\mu_B$. Including multilayer relaxation we find no significant change of these results. This moment is larger than for the Cu(100)c(2×2)Mn surface alloy, for which we found within LDA moments of $M(\Delta z = 0) = 3.64\mu_B$ and $M(\Delta z_{\mathrm{Mn}} = 11.5\%, \Delta z_{\mathrm{Cu}} = -2.5.\%) = 3.75\mu_B$[7]. We attribute this increase in the local moment of the Cu(110) based surface alloy to the more open structure of the fcc(110) surface. At the Cu(100) surface each surface atom has 8 nearest neighbor atoms, at the (110) there are only 7 neighbors. This reduces the hybridization and the local magnetic moment increases (towards the atomic limit, which would be $5\mu_B$). Since Mn is already in the limit of being a strong ferromagnet[16] and the moment is already in the saturated limit, the increase is rather small. Thus we conclude that the magnetism of Mn at the Cu(110) surface is larger than on the Cu(100) surface, which is consistent with the larger spin-polarization energy E_{sp} calculated and the larger buckling (in relative units) of the surface alloy atoms found.

The results of the structural optimization for the Cu(110)c(2×2)Mn surface alloy are collected in Tab.1 together with the experimental LEED data from the previous section. The results of the static total energy minimization are shown in Fig.4(a) and (b). We first discuss the nonmagnetic calculations

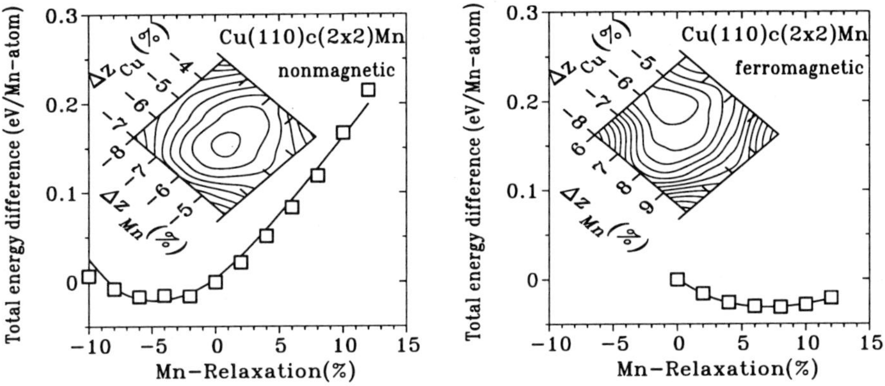

Figure 4. Theoretical total energy difference per Mn atom of Cu(110)c(2 × 2)Mn vs. the buckling relaxation Δz_{Mn} of Mn in relative units with respect to the theoretical interlayer spacing of Cu $d_{\mathrm{Cu}(110)} = 1.244$ Å. The left and right figures show the nonmagnetic and the ferromagnetic results, respectively. The origin of the energy scale of the right side is -1.55eV per Mn atom lower than in the left drawing. The solid lines (surface Cu atoms fixed at the ideally terminated position $\Delta z_{\mathrm{Cu}} = 0$) are the fitting polynomials. The insets show the contour plots of the total energy difference with respect to the buckling of Mn and Cu. The minimum, which determines the optimal structure is found in the inner circle. The contour interval is 1 meV. All relaxations Δz are measured from the ideal bulk terminated surface coordinates. Cu atoms, with the exception of the surface atom, are fixed at the ideal bulk positions.

summarized in Fig.4(a). If we allow for relaxations of the Mn and Cu surface alloy atoms we lower the total energy by 29meV/Mn atom and we find that their equilibrium positions, $\Delta z_{\mathrm{Mn},P} = -5.8\%$ and $\Delta z_{\mathrm{Cu},P} = -5.2\%$, are nearly identical to the relaxed atom positions of the plain Cu(110) surface, which was -5.3% taking only the Cu surface atom into account. Thus, we end up with a small buckling of about $\Delta z_{11,P} = \Delta z_{\mathrm{Mn},P} - \Delta z_{\mathrm{Cu},P} = -0.6\%$ (all relaxation given in relative units ($\Delta z / d_{\mathrm{Cu}(110)}$ in %) where $d_{\mathrm{Cu}(110)}$ is the ideal Cu(110) interlayer spacing). The Mn relaxation is nearly unaffected by the relaxation of the surface Cu atom. Keeping the Cu atom at the bulk terminated position, the Mn atom relaxes by $\Delta z_{\mathrm{Mn},P} = -5.2\%$.

When we lift the constraint of non-magnetism and switch from a local-density calculation to a local-spin-density calculation, the system becomes magnetic and the total energy is lowered by the magnetic energy or the spin-polarization energy, $\Delta E_{sp} = E_S(F) - E_S(P)$, respectively. The energy lowering is large and amounts to $\Delta E_{sp}(\Delta z = 0) = 1.6$ eV/Mn atom for Cu and Mn atoms at bulk terminated atom positions. Relaxing the atomic positions of the surface Cu and Mn atoms (see Fig.4(b)) reduces the total energy further by 51meV. Again Cu relaxes inwards by $\Delta z_{\mathrm{Cu},F} = -5.6\%$ but Mn relaxes outwards by $\Delta z_{\mathrm{Mn},F} = 7.1\%$. Thus we find a large buckling in the surface alloy layer of $\Delta z_1 = \Delta z_{\mathrm{Mn},F} - \Delta z_{\mathrm{Cu},F} = 12.7\%$. This in line with the large experimental buckling of 17.2% discussed above.

In Tab 1 we also show the results of a calculation where all layers are allowed to relax. The equilibrium geometry was determined by minimization of the forces on the atoms. The theoretically determined buckling of 16.2% reproduces the experimental situation. The calculations predict a large magnetic moment for Mn of $M = 3.82\mu_B$. A hypothetical nonmagnetic Cu(110)c(2 × 2)Mn surface alloy shows no buckling (< 1%), proving that the buckling is due to a large magneto-volume effect of Mn.

4.3 The (111) surface

The stability of the (100) and (110) surface alloys is attributed to a magnetic stabilization that was enhanced in these alloys: the lower coordination and the missing nearest-neighbor magnetic atoms lead to an enhancement of the magnetic moment as well as to an outward relaxation of these atoms that reduces the coordination even more. No straightforward extension of the insights gained from these surfaces to the (111) surfaces of fcc crystals is possible: the close packing of the (111) layers will reduce the substantial relaxations observed in the more open surfaces and hexagonal symmetry of the surface causes new possibilities of magnetic ordering of the atoms. The preferred magnetic ordering of Mn on an Ag (100) surface in an alloy as well as in a monolayer was found to be anti-ferromagnetic.[36,37] Therefore, we have to include the possibility of the formation on non-collinear structures on the hexagonal (111) surfaces. Alloys of high Mn content could form interesting combinations of magnetic and structural ordering.

In the case of the (100) and the (110) Cu surfaces, the formation of a surface alloy was observed experimentally and theoretically. To investigate the possibilities of surface-alloy formation on the (111) surface, we calculate[33] within the GGA the ordered Mn/Cu surface alloys depicted in Fig. 5 as nine layer films containing seven layers of Cu(111) substrate. Note, that our total energy calculations neglect the kinetic effects that may stabilize or prevent the formation of the alloy.

We define the formation energy of an alloy with $x\%$ Mn as $\Delta E_F = E^{\mathrm{alloy}} - [xE^{\mathrm{Mn/Cu(111)}} + (1-x)E^{\mathrm{Cu(111)}}]$, where all energies have to be taken

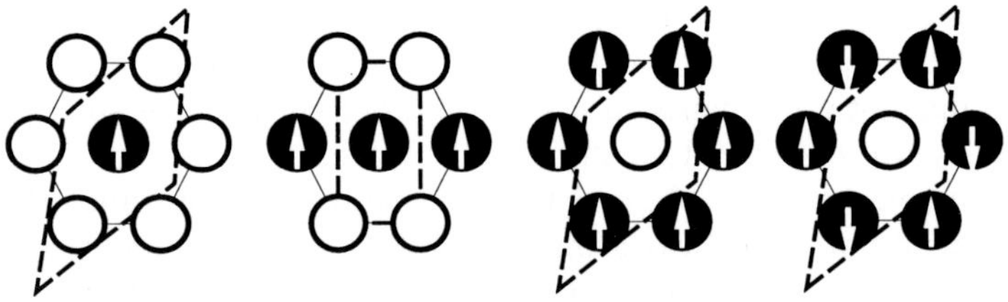

Figure 5. Surface units cells of the calculated Mn/Cu surface alloys on Cu(111): open circles represent Cu atoms, black circles the Mn atoms. The magnetic configuration of the surface alloys with 33, 50 and 66 % Mn is indicated by white arrows. The actual units cells are outlined by thick dashed lines.

Table 2. Magnetic moment μ, relaxation of the Mn atoms ΔZ_{Mn}, relaxation of the first two layers Δd_{12} and energy of formation ΔE_F for ferro (F) and anti-ferromagnetic (AF) Mn/Cu alloys.

		μ [μ_{B}]	$\Delta Z_{\mathrm{Mn}}[a.u.]$	Δd_{12} [%]	ΔE_F [meV]
33 %	F	3.68	0.39	−0.4	−38.0
50 %	F	3.53	0.29	−0.2	3.8
66 %	F	3.55	0.25	0.5	112.6
	AF	3.41	0.28	−1.2	0.1

per surface atom. For $E^{\mathrm{Mn/Cu(111)}}$ we have to take the magnetic ground state configuration, i.e. (within the collinear subspace) the anti-ferromagnetic overlayer.[34] To eliminate the effects from differently sized unit cells, we calculated $E^{\mathrm{Cu(111)}}$ and ferromagnetic $E^{\mathrm{Mn/Cu(111)}}$ for a (2×1) and $(\sqrt{3}\times\sqrt{3})$R30° cell and compared only energies of unit cells of the same size. The alloy atoms are allowed to relax perpendicular to the surface but again no in-plane relaxation was taken into account.

The magnetic moments, relaxations and the formation energies of the configurations are listed in Table 2. The 33% Mn alloy can be compared to the Mn/Cu(100) and (110) surface alloys, since there are no nearest-neighbor Mn-Mn interactions. Indeed, the outward relaxation of the Mn atom, ΔZ_{Mn}, of 0.39 a.u. is the same as for the Mn in the (100) oriented 50% surface alloy and almost identical to the (110) oriented one (0.38 a.u.).

One has to be careful when comparing these results, since the latter two were obtained in the local density approximation (LDA). This may underestimate the relaxations, as it was shown to yield smaller magnetic moments.[26] In the GGA, the magnetic moment of the (100) surface alloy is 3.84 μ_{B}, while in LDA it is only 3.64. The LDA value for the (110) alloy is 3.82 μ_B and our LDA calculation yields 3.51 μ_B for the 33 % (111) alloy so that the general rule, that magnetic moments in a close packed surface are smaller than on open ones, is still conserved on the (111) surface.

The moments of the 50% and 66% alloys are smaller, since the number of nearest-neighbor (NN) Mn atoms are 2 and 3, respectively, and the Mn overlayer (6 NN Mn atoms) has an even smaller moment of 3.15 μ_B. Also the outward relaxation of the Mn atoms gets smaller which fits into the picture of the magneto–volume effect. As the Mn atoms move outwards in the 33% and 50% alloy, the first two Cu layers resume almost the same distance as in the clean Cu(111) surface.

From the formation energies of Table 2 we see, that the 33 % alloy is stable against clustering and the value $E_F = -38$ meV indicates stability also at room temperature. The 50 % (2 × 1) arrangement is on the border of stability while the ferromagnetic 60% $(\sqrt{3} \times \sqrt{3})$R30° structure is highly unstable. But when we switch from this structure to the anti-ferromagnetic configuration we gain 112 meV and almost get a stable surface alloy. This large stabilization is not unexpected, for the energy difference between the ferro- and anti-ferromagnetic Mn overlayers was already 215 meV per Mn atom. There, four ferromagnetic NN Mn-Mn couplings were changed to anti-ferromagnetic coupling, while in the case of the 66 % surface alloy only three bonds were changed.

This suggests, that the 50 % (2 × 1) alloy could gain up to 70 meV when the Mn chains couple anti-ferromagnetic and this would be the most stable surface alloy. Indeed, we calculated this AF 50 % (2 × 1) alloy and found a stabilization energy (as compared to the ferromagnetic alloy) of −71 meV. Thus, the formation of this anti-ferromagnetic chains of Mn on a Cu (111) surface is - from a thermodynamic point of view - highly favorable. Experimentally, only evidence for the 33 % (111) alloy has been found,[35] and the STM pictures indicate that the Mn atoms indeed form a $(\sqrt{3} \times \sqrt{3})$R30° structure.

5 Mn (100) surface alloys: influence of the substrate

In order to shed some light onto the volume dependence on the surface alloy formation we also performed *ab initio* total energy calculations for the Mn/Ag system.[37] Finally, we present results for the Ni(100)c(2 × 2)Mn surface alloy to extend out studies also to magnetic substrates.

5.1 Mn/Ag (100) surface alloys

We calculated the total energies and the magnetic moments for 0.5 ML and 1 ML Mn on Ag(100). The configurations include Mn monolayers in the p(1×1) ferromagnetic and the c(2×2) anti-ferromagnetic configuration at different layers from the surface, and 0.5 ML Mn arranged as c(2×2)-surface alloy with ferromagnetic and p(2×2) anti-ferromagnetic configurations at different layers from the surface. (The Ag(001)p(2×2)Mn surface alloy is chemically equivalent to Ag(001)c(2×2)Mn, but with an anti-ferromagnetic arrangement of the Mn atoms.)

On the Ag substrate we find large magnetic moments for Mn of about 4 μ_B, irrespective of the structural configuration. A closer look reveals that the moments for Mn monolayers are smaller (3.93 and 3.95 μ_B for the ferromagnetic and anti-ferromagnetic ordered monolayers respectively) than for the Ag(001)c(2×2)-configurations (4.15 and 4.24 μ_B, again for ferromagnetic and antiferromagnetic order) and smaller for monolayers located in inner layers ($\approx$ 3.7 μ_B).

The anti-ferromagnetic Ag(001)p(2×2)Mn is 37 meV lower in energy than the ferromagnetic Ag(001)c(2×2)Mn surface alloy. The anti-ferromagnetic energy gain here is smaller than for the monolayers due to the neglect of direct in–plane *3d*-*3d* hybridization or nearest neighbor Mn atoms, respectively. For the c(2×2) surface alloy the formation energy is only 28 meV, indicating that this alloy is on the verge of stability. This value has to be compared to the formation energy of the unrelaxed Cu(100)c(2×2)Mn surface alloy that is almost 80 meV higher. Thus the larger magnetic moments, caused by the larger lattice constant of silver, lead to better stabilization of the Ag(001)c(2×2)Mn surface alloy. The p(2×2)-anti-ferromagnetic state has a formation energy of -19 meV and is, therefore, stable even within LDA.

Looking at the layer dependence of the formation energy we find a large energy gain by wetting of the c(2×2)-alloy by Ag atoms. Among the c(2×2)-alloys the subsurface configuration is the most stable one. Comparing the c(2×2)-alloys with monolayers, both for the subsurface configuration, we find the monolayer layer is the most stable configuration. Assuming that the energy difference between the p(1×1)-ferromagnetic and the c(2×2)-anti-ferromagnetic configuration for the monolayer and the c(2×2)-ferromagnetic and p(2×2)-anti-ferromagnetic configuration for the c(2×2)-alloy film both calculated for the layers on Ag(001) hold also for the subsurface layer, the c(2×2)-anti-ferromagnetic monolayer in the subsurface calculation is the most stable configuration and 61 meV lower in energy than the p(2×2)-anti-ferromagnetic alloy film in the subsurface layer.

5.2 Mn/Ni (100) surface alloys

A theoretical structure optimization has also been performed for the ferromagnetic Ni(001)c(2 × 2)Mn surface compound. The structural model and the computational parameters are those stated in the subsection 4.1. Again two degrees of freedom, the relaxations Δz_{Mn} of Mn and Δz_{Ni} of Ni along the surface normal have been taken into account. The bulk Ni lattice constant ($a_{\mathrm{Ni}} = 3.43$ Å) or the interlayer spacing ($d_{\mathrm{Ni}} = a_{\mathrm{Ni}}/2 = 1.715$ Å), respectively, has been determined first by a FLAPW method.[38] The results are shown in Fig. 6. We find a large outwards relaxation of Mn of $\Delta z_{\mathrm{Mn},F} = +9.0\%$, and an inwards relaxation of Ni of $\Delta z_{\mathrm{Ni},F} = -2.0\%$ adding up to a total corrugation of $\Delta z_1 = 11\%$. Omitting the relaxation of the Ni constituent atom ($\Delta z_{\mathrm{Ni},F} = 0$) of the NiMn surface compound, a Mn relaxation of only $\Delta z_{\mathrm{Mn}} = 6.4\%$ was found (cf. Fig. 6). We recall a similar behavior for the Cu based surface compound. Neglecting the Cu relaxation ($\Delta z_{\mathrm{Cu},F} = 0$) the Mn relaxation changed from $\Delta z_{\mathrm{Mn},F} = 11.5\%$ to $\Delta z_{\mathrm{Mn},F} = 10.2\%$. This is a typical example of many–atom interactions not contained in the classical model of atomic radii.

Two different spin configurations have been taken into account. One of which is the ferromagnetic (F) coupling of Mn with Ni and the other is the anti-ferromagnetic (AF) one. Both magnetic configurations have been found as stable magnetic solutions. At $\Delta z_{\mathrm{Mn}} = \Delta z_{\mathrm{Ni}} = 0$ the ferromagnetic coupling was found to be 120 meV per Mn atom lower in energy than the anti-ferromagnetic state. Thus the coupling is the same as for the full–coverage Mn monolayer on Ni(001). A comparison shows that the energy difference between the two magnetic states in the surface alloy is three times larger than in the monolayer. On the other hand the relaxation energy is about 35 meV per Mn atom, practically the same as for the complete Mn monolayer. Therefore, from the difference in the energy scale between the relaxation energy and the magnetic coupling energy we may expect for the surface alloy no low-spin to high–spin state transition. In fact the magnetic moment in a Mn atom of the NiMn surface compound, $M_{S,F}(\Delta z_{\mathrm{Mn}} = \Delta z_{\mathrm{Ni}} = 0) = 3.41\mu_B$, $M_{S,AF}(\Delta z_{\mathrm{Mn}} = \Delta z_{\mathrm{Ni}} = 0) = 3.36\mu_B$, are large, nearly as large as for the Mn atom in the CuMn surface alloy, and the local moment of Mn in the NiMn surface alloy is already in the high–spin state. This result is a late proof that the low–spin to high–spin transition is due to the direct $d-d$ hybridization of Mn in presence of the ferromagnetic Ni substrate. With increasing corrugation the Mn magnetic moment increases. A closer look reveals that the increase is twice as large as for the Mn moment in the Cu surface alloy. At the same time the Mn moment in the MnNi alloy is smaller than the MnCu alloy to

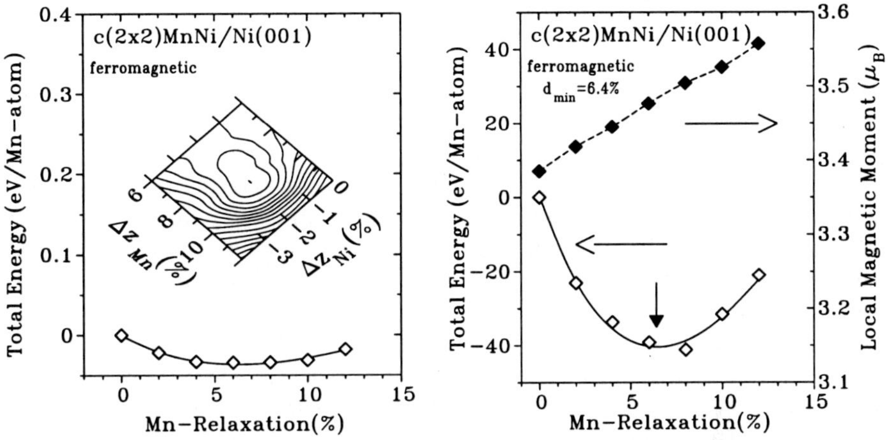

Figure 6. Left: Total energy per Mn atom of Ni(100)c(2 × 2)Mn vs. the buckling relaxation Δz_{Mn} of Mn in relative units with respect to the theoretical interlayer spacing of Ni, $d_{\mathrm{Ni}} = 1.72$ Å. The open diamonds represent results for the ferromagnetic coupling of Mn to Ni. The solid line describes the total energy, for Ni atoms fixed at the ideally terminated positions $\Delta z_{\mathrm{Ni}} = 0$, in terms of a quadratic fitting polynomial. The inset shows the contour plot of the total energy with respect to the buckling of Mn and Ni. The minimum, which determines the optimal structure is found as a point in the inner contour. The contour interval is 1 meV. The origin of the energy scale is the total energy at $\Delta z_{\mathrm{Mn}} = \Delta z_{\mathrm{Ni}} = 0\%$ relaxation. Right: Total energy (open diamonds) per Mn atom of Ni(100)c(2 × 2)Mn measured at the left vertical scale, and the local magnetic moment (solid diamonds connected by dashed line) measured at the right vertical scale as function of the buckling relaxation Δz_{Mn} of Mn in relative units with respect to the theoretical interlayer spacing of Ni, $d_{\mathrm{Ni}} = 1.72$ Å. Ni atoms are fixed at the ideally terminated position $\Delta z_{\mathrm{Ni}} = 0$. The solid line describing the energy is due to a least square fit to a quadratic polynomial. The origin of the energy scale is the total energy at $\Delta z_{\mathrm{Mn}} = \Delta z_{\mathrm{Ni}} = 0\%$ relaxation. The vertical arrow indicates the minimum energy interlayer distance d_{min} as determined from a least square fit to a quadratic polynomial. The horizontal arrows indicate the proper vertical scale.

begin with. Obviously, the d bands of Mn and Ni overlap. This leads to a $d - d$ hybridization which is comparatively larger than in Cu. This reduces the Mn moments in the MnNi alloy slightly and increases the cohesion, makes the corrugation smaller although the lattice constant of Ni is smaller as for Cu. The Mn magnetic moment increases by outwards relaxation due to the *spd* dehybridization and in difference to the Cu based surface alloy also due to a reduction of the $d - d$ hybridization with Ni.

6 Summary and Outlook

Based on total-energy arguments we investigated the possibilities of surface alloys. Exemplary for $3d$ transition-metal monolayers on noble metal substrates, detailed calculations for $3d$ monolayers on Cu(100) have been performed, showing clear trends: i) Magnetism acts against interdiffusion. At low coverage, where the magnetism of Cr and Mn is very large, interdiffusion is suppressed. ii) Magnetism reduces clustering and promotes alloy formation. For Mn we find a strong tendency for alloy formation. From i) and ii) we expect for the case of the Cu substrate a thermodynamically stable, two dimensional surface alloy: Cu(100)c(2×2)Mn. Although in detail, the results depend on the Cu(001) substrate, the trends presented above have an universal character. We investigated in detail the energetics, magnetic and structural properties of Cu(100)c(2×2)Mn, Cu(110)c(2×2)Mn, Cu(111)c(2×2)Mn, and Ni(100)c(2×2)Mn, and Ag(100)c(2×2)Mn surface alloys. We found in all cases large magnetic moments for Mn accompanied with a large outwards relaxation of the Mn atoms being a result of a large magneto-volume effect. For Ag(100)c(2×2)Mn the alloy formation might be kinetically possible but energetically not favored due to the large lattice constant of Ag. In addition we found a nonmagnetic surface alloy Ti/Cu(001), which is not yet confirmed experimentally.

While in this article we have shown the possibility of magnetically stabilized surface alloys, many questions remain unanswered. For example, do additional magnetic surface alloys exist? What is their magnetic order, how large is their critical temperature T_c for the magnetic transition, which surface alloys at one monolayer range will form multilayer alloy films? Do additional structural phases as a function of the coverage exist? Is there a structural order–disorder transition? What is the termination of the surface alloys at step edges? Which microscopic growth mechanisms are involved in the kinetic processes responsible for the formation of these surface alloys? More questions can easily be asked. We hope we will answer some of these questions in future.

Acknowledgements

Work was supported by the DFG under Grants BL444/1-1, the Program 'Verbundforschung Synchrotronstrahlung' of the BMBF, and the TMR Networks, Contract Nos: EMRX-CT96-0089 and FMRX-CT98-0178.

References

1. For a collection of presently known surface alloys see reference: U. Bardi, *Rep. Prog. Phys.* **57**, 939 (1994).
2. T. Flores, M. Hansen, and M. Wuttig, *Surf. Sci.* **279**, 251 (1992).
3. M. Wuttig, C.C. Knight, T. Flores, and Y. Gauthier, *Surf. Sci.* **292**, 189 (1993).
4. M. Wuttig, B. Feldmann and T. Flores, *Surf. Sci.* **331-333**, 659 (1995).
5. M. Wuttig, T. Flores, and C.C. Knight, *Phys. Rev.* B **48**, 12082 (1993).
6. C. Ross, M. Wuttig, Y. Gauthier, G. Bihlmayer, and S. Blügel, *Phys. Rev.* B **57**, 2607 (1998).
7. M. Wuttig, Y. Gauthier, and S. Blügel, *Phys. Rev. Lett.* **70**, 3619 (1993).
8. M. Wuttig, S. Junghans, T. Flores, and S. Blügel, *Phys. Rev.* B **53**, 7551 (1996).
9. H.P. Noh, T. Hashizume, D. Jeon, Y. Kuk, H.W. Pickering, and T. Sakurai, *Phys. Rev.* B **50**, 2735 (1994).
10. R.G.P. van der Kraan, H. van Kempen, *Surf. Sci.* **338**, 19 (1995).
11. W.L. O'Brien, J. Zhang, and B.P. Tonner, *J. Phys.: Condens. Matter* **5**, L515 (1993).
12. W.L. O'Brien and B.P. Tonner, *J. Appl. Phys.* **76**, 6468 (1994).
13. W.L. O'Brien and B.P. Tonner, *Phys. Rev.* B **51**, 617 (1995).
14. A.B. Hayden, P. Pervan, and D.P. Woodruff, *J. Phys.: Condens. Matter* **7**, 1139 (1995).
15. D. Schmitz, O. Rader, C. Carbone, W. Eberhardt, *Phys. Rev.* B **54**, 15352 (1996).
16. O. Rader, W. Gudat, C. Carbone, E. Vescovo, S. Blügel, R. Kläsges, W. Eberhardt, M. Wuttig, J. Redinger, and F.J. Himpsel, *Phys. Rev.* B **55**, 5404 (1997).
17. O. Rader, E. Vescovo, M. Wuttig, D.D. Sarma, S. Blügel, F.J. Himpsel, A. Kimura, K.S. An, T. Mizokawa, A. Fujimori, and C. Carbone, *Europhys. Lett.* **39**, 429 (1997).
18. D. Tian, R.F. Lin, and F. Jona, *Solid State Commun.* **74**, 1017 (1990).
19. P. Schieffer, C. Krembel, M.C. Hanf, and G. Gewinner, *Phys. Rev.* B **55**, 13884 (1997).
20. P. Schieffer, C. Krembel, M.C. Hanf, and G. Gewinner, *Surf. Sci.* **398**, 332.
21. W. Kim, W. Kim, S.-J. Oh, J. Seo, J.-S. Kim, H.-G. Min, and S.-C. Hong, *Phys. Rev.* B **57**, 8823 (1998).
22. E. Wimmer, H. Krakauer, M. Weinert, and A.J. Freeman, *Phys. Rev.* B **24**, 864 (1981).

23. M. Weinert, E. Wimmer, and A. Freeman, *Phys. Rev.* B **26**, 4571 (1982).
24. U. von Barth and L. Hedin, *J. Phys. C.* **5**, 1629 (1972).
25. V.L. Moruzzi, J.F. Janak, and A.R. Williams, *Calculated Electronic Properties of Metals* (Pergamon, New York, 1978).
26. T. Asada and S. Blügel, *Physica* B, **237-238**, 359 (1997).
27. P. Perdew and Y. Wang, *Phys. Rev.* B **33**, 8800 (1986).
28. P. Perdew and Y. Wang in *Electronic Structure of Solids* (Akademie Verlag, Berlin, 1991), Vol. 11.
29. S. Takizawa (PhD-Thesis, University of Tokyo, Japan, 1990).
30. T. Hoshino, W. Schweika, R. Zeller, and P.H. Dederichs, *Phys. Rev.* B **47**, 5106 (1993).
31. Z.Q. Wang, Y.S. Li, C.K.C. Lok, J. Quinn, and F. Jona, *Solid State Commun.* **62**, 181 (1987).
32. S.C. Wu, S.H. Lu, Z.Q. Wang, C.K.C. Lok, J. Quinn, Y.S. Li, D. Tian, and F. Jona, *Phys. Rev.* B **38**, 5363 (1988).
33. G. Bihlmayer, P. Kurz, and S. Blügel, in preparation.
34. P. Kurz, G. Bihlmayer, and S. Blügel, submitted to *J. Appl. Phys.* .
35. J. Schneider, A. Rosenhahn, and K. Wandelt, *Appl. Surf. Sci.* **142**, 68 (1999).
36. S.C. Hong, M. Kim, and A.J. Freeman, *J. Appl. Phys.* **83**, 7016 (1998).
37. R. Abt, and S. Blügel, *Phil. Mag.* B **78**, 659 (1998).
38. C. Blaas, private communication.

TRANSPORT AND MAGNETIC PROPERTIES OF DISCONTINUOUS $CO_{80}FE_{20}/AL_2O_3$ MULTILAYERS

J.B. SOUSA[A], G.N. KAKAZEI[A], YU.G. POGORELOV[A], S. CARDOSO[B], P.P. FREITAS[B], A.M.L. LOPES[A], M.M. PEREIRA DE AZEVEDO[C,A]

[a] *IFIMUP and Departamento de Física, Universidade do Porto, 4150 Porto, Portugal;*

[b] *INESC, Rua Alves Redol, 9-1, 1000 Lisbon, Portugal;*

[c] *Departamento de Fsica, Universidade de Aveiro, 3810 Aveiro, Portugal*

We report on the preparation of multilayered $Co_{80}Fe_{20}(t)/Al_2O_3(30$ Å) films by ion-beam sputtering, on their static magnetization, and on tunnel magnetoresistance (TMR) in the current-in-plane (CIP) and current-perpendicular-to-plane (CPP) geometries. It is found that $Co_{80}Fe_{20}$ layers are discontinuous in Al_2O_3 matrix at thicknesses $t \leq 18$ Å. In the CIP case, the thickness t can be optimized to give maximum TMR (~6.5 % for $t = 10$ Å) or maximum initial slope of TMR with field (~24 %/kOe for $t = 13$ Å) at room temperature. Magnetization data show that ferromagnetism onsets at $t \geq 13$ Å, indicating that magnetic percolation precedes electrical percolation in this system. The temperature dependence of TMR was found quite different for the two geometries: fairly strong in the CIP case and weak in the CPP case. To explain these features we propose a model taking into account the significant differences specifics of short-range magnetic correlations within and across the layers.

1 Introduction

Discontinuous metal-insulator (DMI) multilayers are a new class of granular magnetoresistive (MR) films which combine the known merits of the cermet films[1,2] (easy preparation, high chemical/electrical stability, insensitivity to pinholes) with higher MR sensitivity to low and moderate magnetic fields. Research on DMI multilayers started with $(CoFe/HfO_2)$n and $(Co/SiO_2)_n$,[3,4] leading to room temperature MR values (both in CPP and CIP geometries) up to 4 %, after optimized annealing. Further optimization of layer thicknesses and suitable choice of metal/insulator materials could make these systems a useful supplement (at not too low applied fields) to the high sensitive TMR tunnel junctions.[5,6] In this work we investigate the transport and magnetic properties of as-deposited $Co_{80}Fe_{20}(t)/Al_2O_3(30$ Å) multilayers with different $Co_{80}Fe_{20}$ layer thickness (10 Å $\leq t \leq$ 25 Å) over a wide temperature (10 - 300 K) and magnetic field (0 - 10 kOe) range. A significant result of this study is the observation of a crossover from granular-like to continuous-like behavior, which takes place at *different* values of magnetic layer thickness (t) for electric and magnetic properties. This can indicate that magnetic percolation has a different mechanism than common structural percolation. Also, comparison between CIP and CPP MR behavior reveals a significant difference in their temperature dependence, which we attribute to different magnetic correlations along the tunneling paths followed by electrons in each case.

2 Preparation and structural characterization

The preparation routine for our DMI samples was chosen by the following reasons. $Co_{80}Fe_{20}$ is a soft magnetic alloy with high spin polarization.[7] The choice of Al_2O_3 was motivated by a recent report on sputtered Co-Al-O cermet films with enhanced MR (˜10% at room temperature[2]). Finally, the ion beam sputtering technique was chosen as a more flexible method for preparation of nanostructures than common magnetron sputtering, due to the independent control of beam parameters (ion density and kinetic energy) and deposition pressure.[8] All the films were deposited on glass substrates using Xe-ion beam sputtering acting alternatively on two separate targets, and having a rotating substrate table. Alumina was sputtered from an Al_2O_3 target and $Co_{80}Fe_{20}$ from a mosaic target (pieces of Fe on a Co plate). The base pressure was 8×10^{-8} Torr. All the films contained 10 bilayers. The nominal thickness of the $Co_{80}Fe_{20}$ layers was varied from 10 to 25 Å and that of the Al_2O_3 layers was fixed at 30 Å. The thickness values were determined from known ion source calibrations. The sample for CPP measurements (t = 14 Å) was deposited between two crossed Ta(70 Å)/Cu(100 Å)/ Ta(70 Å) electrodes with a contact area of 0.5 mm2, using a succession of appropriate shadow masks.

Characterization of these multilayers by low angle X-ray diffraction revealed great improvement of interface quality for $Co_{80}Fe_{20}$ layer thickness above 18 Å, with appearance of extra Bragg peaks and pronounced Keissing fringes. This is consistent with our further conclusion that thin magnetic layers are discontinuous (isolated granules), while for $t \geq 18$ Å we have continuous ferromagnetic layers. However, even for 12 Å thick $Co_{80}Fe_{20}$ layers we still observe three low angle diffraction peaks, indicating the periodicity of the multilayer structure.

3 CIP geometry

The room temperature electrical resistance decreases sharply with the increase of $Co_{80}Fe_{20}$ layer thickness, from 16 MΩ for t = 10 Å to 22 Ω for t = 25 Å (see Fig. 1). The lower value is consistent with the estimated resistance for a stack of ten $Co_{80}Fe_{20}$ continuous layers, each 25 Å thick. Such behavior suggests a

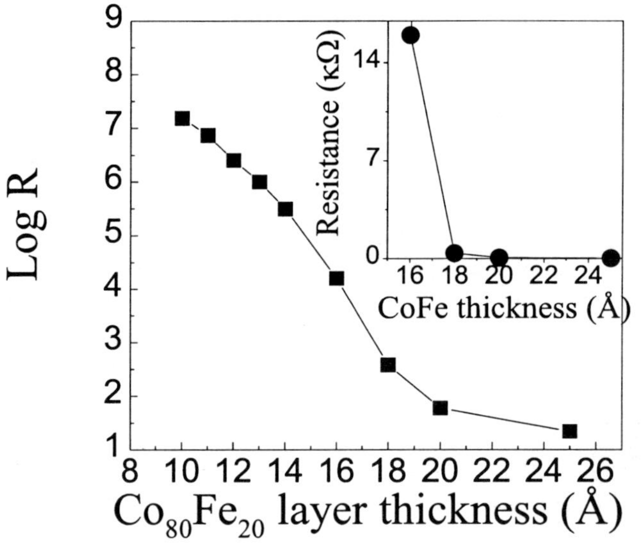

Figure 1. Resistance vs metal layer thickness.

change from dielectric to metallic regime. This is directly confirmed by the temperature dependence of the resistance, changing from $dR/dT < 0$ at $t \leq$ 18 Å (Fig. 2a) to the metallic behavior ($dR/dT > 0$) above 18 Å of $Co_{80}Fe_{20}$ thickness (Fig. 2b).

However, a slight upturn is still observed at the lowest temperatures (< 50 K, Fig. 2b), indicating the presence of a remnant tunnel contribution, likely associated with residual bridges in a non-ideal metallic network within the $Co_{80}Fe_{20}$ layers. The magnetoresistance for all the samples, measured at room temperature and in magnetic fields up to 10 kOe is displayed in Fig. 3.

The onset of MR response correlates well with the transition from metallic to dielectric (tunneling) conductance at $t \simeq 18$ Å. Then MR significantly increases with decreasing thickness t, reaching a maximum saturation value

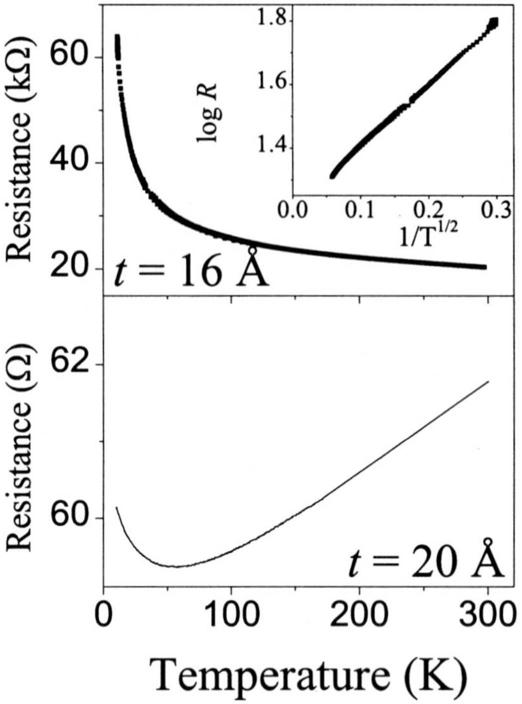

Figure 2. Temperature dependence of electrical resistance at different CoFe layer thicknesses. a) At $t < 18$ Å activated behavior is extended over whole temperature range. Inset: a linear $\ln R$ vs $T^{-1/2}$ plot permits to determine the activation energy $C \simeq 0.5$ meV. b) At $t > 18$ Å activated behavior only holds at lowest temperatures, changing for metallic behavior at higher temperatures.

(at 10 kOe), MR_{max} of 6.5% at $t = 10$ Å. To our knowledge, this is the highest value observed in DMI multilayers at room temperature. The low field behavior of the magnetoresistance, characterized by the sensitivity coefficient $S = dMR/dH$ is shown in Fig. 4.

This sensitivity depends strongly on the $Co_{80}Fe_{20}$ layer thickness, and displays a fairly sharp maximum for an optimum thickness $t* \simeq 13$ Å, reaching MR ~1% in 40 Oe. Notice that $t*$ does not correspond to maximum MR

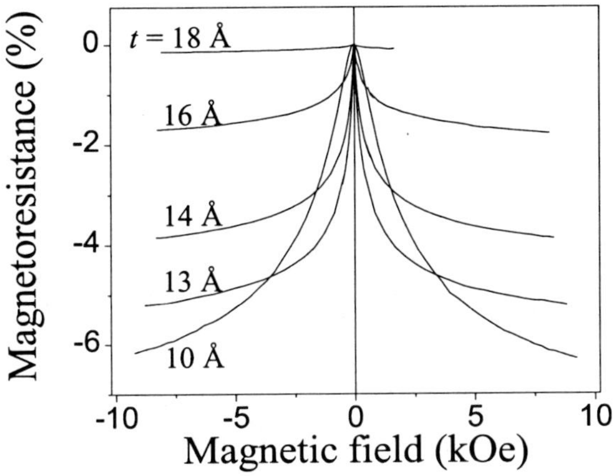

Figure 3. Field dependences of room temperature MR at different thicknesses of ferromagnetic layer.

itself. In order to clarify the physical mechanisms underlying the peculiar thickness dependence of the MR sensitivity, we performed room temperature VSM magnetization measurements. With growing thickness, the magnetization curves $M(H)$ display a crossover from superparamagnetic to ferro-like behavior just at $t_m \simeq t*$, where the remanence first appears, as shown in Fig. 4. We remark that this occurs when the $Co_{80}Fe_{20}$ layer is still discontinuous, as inferred, for instance, from the tunnel behavior of $R(T)$ data.

The $MR(H)$ curves change only weakly with temperature down to about 100 K. Below this temperature (which can be associated with the blocking temperature T_b, as confirmed by SQUID measurements) we observe the onset of hysteresis and coercivity for $MR(H)$ in the 16, 14 and 13 Å samples, which increase as temperature decreases. As illustrated in Fig. 5 for the t = 14 Å sample at T = 15 K, the hysteresis extends up to 2 kOe and the coercive field H_c reaches 250 Oe. This hysteretic regime is accompanied by a strong enhancement of the saturation MR value, also seen in Fig. 5.

Within our accessible experimental range, the highest saturation MR value of ~12% was observed for the t = 13 Å sample at 15 K (huge resistance values precluded low temperature MR measurements for t < 13 Å).

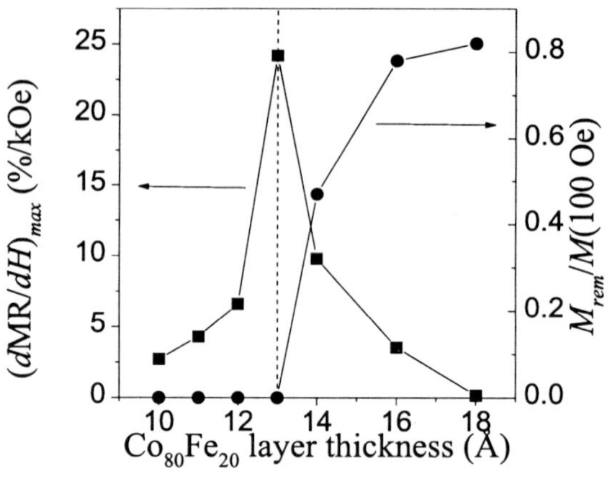

Figure 4. Maximum slope of MR in field and normalized magnetization remanence vs layer thickness.

4 CPP geometry

We prepared one sample for CPP measurements with $t = 14$ Å which demonstrates, alike the CIP case, the temperature dependence with $dR/dT < 0$, consistent with thermally assisted tunneling. However the $\log R\tilde{}T^{-1/2}$ dependence is not observed in CPP geometry. This difference from CIP is due to the large voltage existing between consecutive layers in CPP geometry, leading to the so-called high electrical field regime.[9] Correspondingly, the I-V characteristics are nonlinear, indicating a non-ohmic transport mechanism. For similar experimental conditions, Coulomb blockade was observed in CoFe(15 Å)/HfO_2(40 Å) and Co(15 Å)/SiO_2(50 Å) DMI multilayers at temperatures below 40-50 K.[3,10] However, in our CoFe(14 Å)/Al_2O_3(30 Å) sample we did not find clear evidence of such phenomenon even at $T = 10$ K (as seen from persisting finite slope[3] of dI/dV vs V, inset to Fig. 6). We estimated the granule charging energy $E_c\tilde{}0.1$ meV, from the fitted value of activation energy $C\tilde{}3$ meV (see section V) for CIP transport in 14 Å CoFe layers within the standard theory.[11] Thus, the expected Coulomb blockade effects in our system should be restricted to very low temperatures ˜1 K.

The maximum room temperature MR value found in CPP geometry was about

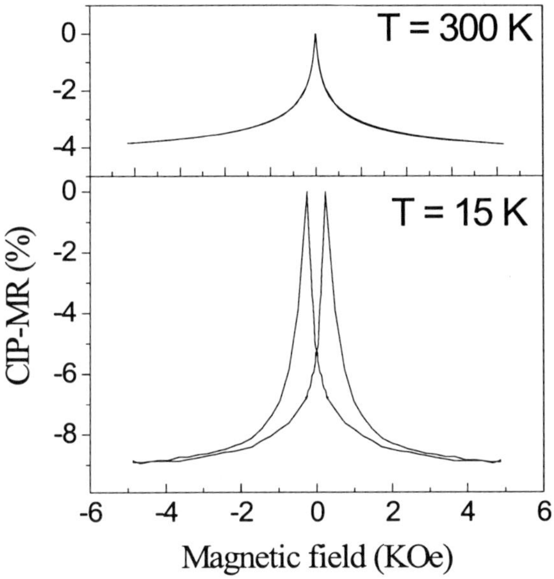

Figure 5. Field dependencies for CIP-MR in $Co_{80}Fe_{20}$(14 Å)/Al_2O_3(30 Å) sample at room temperature (upper panel) and at low temperature (lower panel).

2.5% in the sample with t = 14 Å. We notice that MR considerably decreases with increasing bias voltage. A similar tendency has been observed in spin junctions and Co/SiO_2 discontinuous multilayers.[6,10] The MR ratio at fixed current changes quite weakly with temperature in the whole range down to 10 K. Also MR reveals hysteresis at low temperatures like the CIP case, within almost the same range and with similar coercive fields.

5 Discussion

A quantitative analysis has been done on the thermally activated $R(T)$ dependence. As shown in the inset of Fig. 2a for the t = 16 Å sample in the CIP geometry, the data agrees well with the standard dependence $\log R = 2(C/k_B T)^{1/2}$ characteristic of the low electric field tunneling regime,[11] where C is the activation energy. This is expected to occur in a granular layer ($CoFe/Al_2O_3$), since the potential drop at each tunnel bridge in a series of

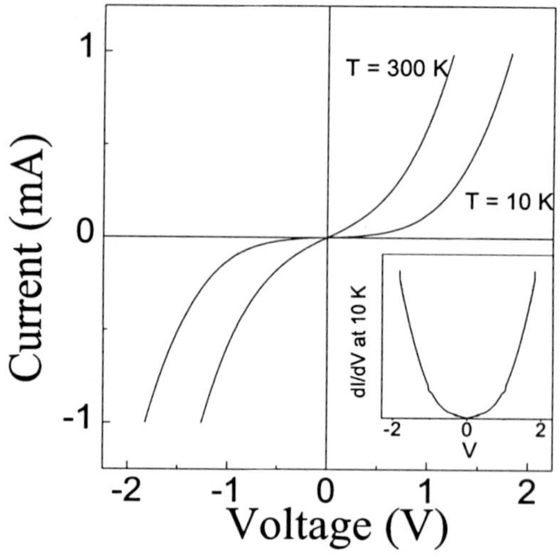

Figure 6. I-V characteristics of $t = 14$ Å sample in CPP geometry at different temperatures. Inset: the finite slope of differential resistance vs voltage corresponds to absence of Coulomb blockade at $T = 10$ K.

~10^6 bridges per cm is extremely small (see below). Correspondingly, the I-V characteristics for CIP should be strictly linear for all the samples in the whole temperature range, as indeed observed. Also, electrical breakdown is virtually excluded in our films, unlike tunnel junctions where the external voltage is applied to a single bridge.

For our samples with $t < 18$ Å, we obtain C ranging from 0.5 meV at $t =$ 16 Å (inset of Fig. 3a) to ~8 meV at $t = 12$ Å. The intergrain distance-to-size ratio s/d was estimated from this data, using the Sheng-Abeles model.[9] Unexpectedly small values s/d~0.01 were always obtained (for the low t samples, d can be roughly compared with a typical value of ~40 Å reported for Co clusters in Al_2O_3;[3] while s should not be less than one atomic spacing). This pronounced difference from typical values of s/d~0.1 in cermet systems[9] can be attributed to the planar structure of the discontinuous layers.

For t > 13 Å, as is directly seen from the magnetization data, ferromagnetic behavior starts to occur within the discontinuous magnetic layers. This is also associated with an anomalous break in the MR vs t curve (not shown) and with a pronounced peak in the low-field MR sensitivity (Fig. 4). We interpret this phenomenon as due to the establishment of long-range correlations between sufficiently close but still disconnected magnetic granules coupled by dipolar interactions and conventionally we call it *magnetic percolation.*[12] Further analysis is needed to conclude whether it corresponds to a specific phase transition.

For $t \geq 18$ Å, electrical percolation occurs among the $Co_{80}Fe_{20}$ grains, leading to the onset of metallic conduction. This is indicated by the ascending $R(T)$ curve (Fig. 1) and by the null TMR effect (Fig. 3). Further support is given by the qualitative change of the X-ray diffraction pattern, with the appearance of extra Bragg peaks for $t \geq 18$ Å.

An interesting result is the much stronger temperature dependence of MR for CIP than for CPP geometry. To analyze it, we consider the spin-dependent tunnel conductance[13] $G\tilde{\ }1 + P^2 \cos\theta$, where P is the spin polarization of the ferromagnetic material and θ is the angle between magnetic moments of neighbor granules. Assuming that tunnel transitions in DMI are mostly in series for CIP and mostly in parallel for CPP, we obtain the respective (saturation) MR ratios:

$$MR_{\mathrm{CIP}} = P^2 \left(1 - \langle \cos\theta \rangle_{\perp}\right) / \left(1 - P^2 \langle \cos\theta \rangle_{\perp}\right) \tag{1}$$

$$MR_{\mathrm{CPP}} = P^2 \left(1 - \langle \cos\theta \rangle_{z}\right) / \left(1 + P^2\right) \tag{2}$$

Here the averages $\langle\rangle_{\perp,z}$ are respectively for neighbor granules in the same layer and across layers, taken at $H = 0$ (or at $H = H_c$), and the terms $\tilde{\ }P^4$ and higher are dropped in Eq. (1). Of course, MR can increase with the polarization P at lower temperatures, but this effect should affect both geometries and it is expected to be small. However an important temperature dependence results from the magnetic correlation factor cos. At room temperature, the in-layer value $\langle \cos\theta \rangle_{\perp}$ is rather high, due to the strong in-plane dipolar coupling. But it should decrease considerably below the blocking temperature T_b (when a strong short-range disorder is produced at magnetization reversal near the coercive field, e.g. due to the increasing anisotropy fields), leading to the increase of MR_{CIP} up to P^2. From the measured value ~9% (Fig. 5), we obtain $P \simeq 0.3$, a reasonable figure. The absence of such increase for MR_{CPP} is likely due to the fact that correlation $\langle \cos\theta \rangle_{z}$ between granule moments in different (almost uncoupled) layers is random and hence less sensitive to the change of ordering in each layer with temperature.

6 Conclusions

We found that TMR in discontinuous $Co_{80}Fe_{20}(t)/Al_2O_3$(30 Å) multilayers depends strongly on the magnetic layer thickness t. The maximum negative MR value at room temperature ~6.5% (at 10 kOe) was observed for $t = 10$ Å, whereas the maximum initial dMR/dH value (~24 %/kOe) was observed for $t = 13$ Å. The crossover from high MR (discontinuous layers) to almost vanishing MR (continuous layers) was found at t~18 Å, while the transition from paramagnetic to ferromagnetic behavior was found to occur around t ~13 Å. This suggests that the processes of magnetic and electrical percolation are essentially different, the former one occurring at lower t. Another difference is observed in the temperature dependence of TMR for different geometries. CIP-MR increases up to 3 times with cooling below the blocking temperature T_b~100 K, whereas CPP-MR remains almost temperature insensitive. This dissimilarity is explained by the difference of short-range magnetic correlations between neighbor granules in the same layer and in adjacent layers.

This work was supported by the program PRAXIS XXI through the projects 3/3.1/MMA/1787/95, P/CTM/11242/98 and the grants BPD/3866/96 (G.N.K.), BPD/14226/97 (Yu.G.P.) and BD/11533/97 (S.C.)

References

1. J. I. Gittleman, Y. Goldstein, S. Bozowski, *Phys. Rev.* B **9**, 3891 (1972).
2. S. Mitani, K. Takanashi, K. Yakushiji, and H. Fujimori, *J. Appl. Phys.* **83**, 6524 (1998).
3. S. Sankar, B. Dieny, A. E. Berkowitz, *J. Appl. Phys.* **81**, 5512 (1997).
4. S. Sankar, A. E. Berkowitz, D. J. Smith, *Appl. Phys. Lett.* **73**, 535 (1998).
5. T. Miyazaki, N. Tezuka, *J. Magn. Magn. Mater.* **139**, L231 (1995).
6. J. S. Moodera, L. R. Kinder, T. M. Wong, R. Meservey, *Phys. Rev. Lett.* **74**, 3273 (1995).
7. D. I. Bardos, *J. Appl. Phys.* **40**, 1371 (1969).
8. H. R. Kaufman, R. S. Robinson, W. E. Hughes, in *Characteristics, Capabilities and Applications of Broad-Beam Sources* (Commonwealth Scientific Corporation, 1987).
9. B. Abeles, P. Sheng, M. D. Coutts, and Y. Arie, *Adv. Phys.* **40**, 407 (1975).

10. B. Dieny, S. Sankar, M. R. McCartney, D. J. Smith, P. Bayle-Guillemaud, A. E. Berkowitz, *J. Magn. Magn. Mat.* **185**, 283 (1998).
11. P. Sheng, B. Abeles, Y. Arie, *Phys. Rev. Lett.* **31**, 44 (1973).
12. G. N. Kakazei, A. F. Kravets, N. A. Lesnik, M. M. Pereira de Azevedo, Yu. G. Pogorelov, J. B. Sousa, *J. Appl. Phys.* **85**, 5654 (1999).
13. J. Inoue, S. Maekawa, *Phys. Rev.* B **53**, 11927 (1996).

A MOLECULAR-DYNAMICS STUDY: NUCLEATION PROCESSES AND STRUCTURAL TRANSFORMATION OF THIN IRON FILMS ON CU(001)

K. KADAU, R. MEYER, AND P. ENTEL

Theoretische Tieftemperaturphysik, Gerhard-Mercator-Universität Duisburg, 47048 Duisburg, Germany

E-mail: kai@thp.uni-duisburg.de

We have studied the structure of monolayers of iron on a Cu(001) substrate. By means of molecular-dynamics simulations in combination with embedded-atom potentials for the description of the interactions between the atoms, we have investigated the growth of bcc domains in the iron film with decreasing temperature. It turns out that the bcc domains become larger with decreasing temperature, but even at lowest temperatures there remains a significant amount of fcc iron. In contrast to this, cylindrical iron islands transform nearly completely to the bcc-structure at low temperatures. This different behavior is discussed in the frame of experimental deposition techniques which strongly influences the final structure of the grown films.

1 Introduction

Many experiments have been made on ultra-thin iron films on copper substrates, because it is possible to grow iron in the fcc structure at room temperature on Cu(001) substrates. These experiments show that with increasing thickness the iron film begins to transform to the bcc structure which is the bulk equilibrium-structure of iron at room temperature and below.[1,2] Experiments also reveal that the decrease of temperature can lead to structural changes from fcc to bcc in the iron film.[3] Recent experiments of Kirschner *et al.*[4] show that there is a significant difference between the structure and magnetic properties of iron films grown by means of thermal deposition (TD) and pulsed-laser deposition (PLD). They found that films grown by PLD are more stable in the fcc structure than films grown by TD. Whereas PLD grown films show layer-by-layer growth, TD grown films show rougher surfaces with islands of 6 nm to 40 nm in diameter.

We present here results of molecular-dynamics simulations of a perfect 15 monolayer (ML) iron film on a Cu(001) substrate. We cooled down the film from room temperature to 1 K. Furthermore we show results of a 15 ML high cylindrical iron island on a Cu(001) substrate.

2 Method

2.1 Model

The method of molecular-dynamics (MD) simulations which we used in this work, is based on the numerical solution of Newtonian equations of motion for a coupled system of atoms. In a way this method is like a microscope with atomic resolution with which one can follow the atomic movements on the fs to ps scale. The potential which is used to describe the interatomic forces, plays an important role for the reliability of such simulations. In this paper we used the embedded-atom method (EAM) proposed by Daw and Baskes in 1984.[5,6] In the EAM the atoms are supposed to be embedded in the electron density of the surrounding atoms, on which the so-called embedding energy depends. To the embedding energy an ordinary two-particle term, in this case a screened Coulomb-potential, is added so that the total energy of the system can be written as:

$$E = \sum_{i} F_{\alpha_i}(\rho_i) + \sum_{i} \sum_{j \neq i} \frac{Z_{\alpha_i}(r_{ij}) Z_{\alpha_j}(r_{ij})}{r_{ij}}, \tag{1}$$

where the indices i and j denote individual atoms; $\alpha_i, \alpha_j = \mathrm{Fe}, \mathrm{Cu}$ refer to the species of the corresponding atoms, and r_{ij} is the distance between atoms i and j. The function F_{α_i} is the embedding energy of atom i depending on the electron density of the surrounding atoms,

$$\rho_i = \sum_{j \neq i} \rho^{at}_{\alpha_j}(r_{ij}), \tag{2}$$

which is the sum of the individual contributions $\rho^{\mathrm{at}}_{\alpha_j}$ of all atoms j except i. The second term corresponds to a screened Coulomb potential with effective charges Z_{α_i}.

In our simulations the functions F_{α_i} and Z_{α_i} are represented by cubic spline functions. These spline functions have been fitted to the experimental data of lattice constant, cohesive energy, vacancy formation energy, elastic constants and some phonon frequencies of the elements iron and copper. Details of these calculations and the parameters of the resulting functions for iron can be found elsewhere.[7] It should be noted that in the case of bulk iron our EAM potential is able to describe the structural transition from the bcc structure to the fcc structure with increasing temperature very accurately.[7]

2.2 Simulation procedure

We apply periodic boundary conditions to our simulation cell except for the z-direction. This allows to simulate a film-like system with two free surfaces, whereby the copper surface at the bottom of our system is not of interest in this work. Furthermore we used the Nosé-Hoover thermostat in order to control the temperature and the Parrinello-Rahman scheme to allow the simulation cell to change its size and shape.[9,10] We integrated Newtonian equations of motion numerically by using the velocity Verlet algorithm[8] with a time step of 1.5 fs. Results are based on the mean atomic configurations. In order to distinguish between fcc and bcc structure in the simulations, we made use of the different coordination number in the fcc and the bcc lattice.[11] Simulations have been made for two types of cells described below (see also Fig.1).

Complete iron film
In this case our simulation cell is cubic and contains approximately 40000 atoms initially arranged on a fcc lattice, whereas the uppermost 15 ML are Fe atoms and the other atoms are Cu. In order to take into account the fact that in a real growth process the occurrence of defects is inevitable, we allow for 2 percent vacancies in the whole system. Starting the simulations at room temperature, we cool down the system to 1 K in steps of 5 K making runs with a length of 0.375 ps for each temperature.

Iron island
Here we chose a simulation cell which contains approximately 60000 atoms with fcc structure arranged in such away as to have a 15 ML thick cylinder of iron atoms with a diameter of 7.5 nm on the copper substrate. In this case the system is simulated at a temperature of 50 K for 20 runs each over a time interval of 0.45 ps.

3 Results

When cooling the complete iron film from 300 K to 1 K, the amount of bcc structured atoms in the iron film increases. Figure 2 shows the structure of the iron film slice-wise at 150 K and 1 K, where the blue color marks the regions with bcc structure, whereas orange is used for the fcc structure. It

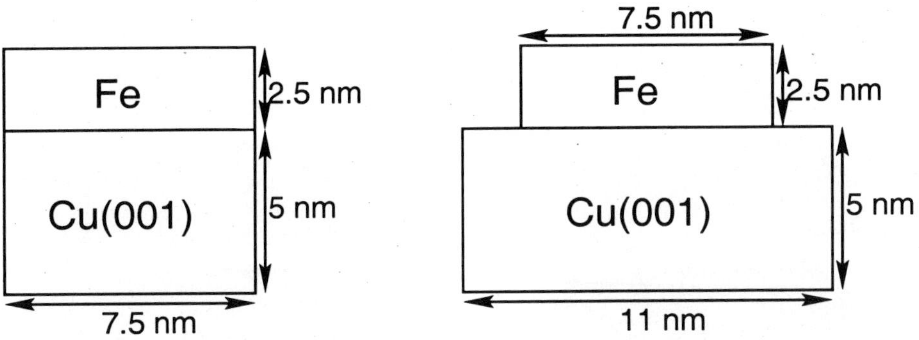

Figure 1. Sketch of the dimensions of the simulation cells: Complete iron film on Cu(001) (left panel) and iron island on Cu(001) (right panel). The sketch is a 2-dimensional cut through the simulation cells.

should be mentioned that the surface atoms in this case are always regarded as bcc atoms because of their low coordination number. At higher temperatures fluctuations of bcc nuclei can be observed as well as the formation of stable bcc nuclei which are pinned at the initial defects. At higher temperatures there is no special preference for the nuclei being near the surface (left upper slice in Fig. 2) or near the copper substrate (left lower slice in Fig. 2). At lower temperatures there is a growth of some preferred nuclei and the amount of bcc structured atoms decreases coming closer to the copper substrate which is shown in the right panel of Fig. 2. The transformation from fcc to bcc in the iron film is not complete even at lowest temperatures due to lateral stresses in the iron film. This leads to a very rough bcc-fcc interface which might have fractal properties.

In Figure 3 the structure of the cylindrical iron island is shown slice-wise for a temperature of 50 K. In contrast to the previous case now practically the whole iron island transforms to the bcc structure which is again marked by blue color. In this case only the layers close to the copper substrate retain their fcc structure due to lattice adaption in the first layers close to the substrate. Here the lack of lateral stresses, as compared to the previous situation, leads to the more pronounced fcc to bcc transformation with a more or less plane fcc-bcc interface. Just as for the case of the iron film[12] we observed twinned bcc-Fe in the Pitsch orientation to the copper substrate.

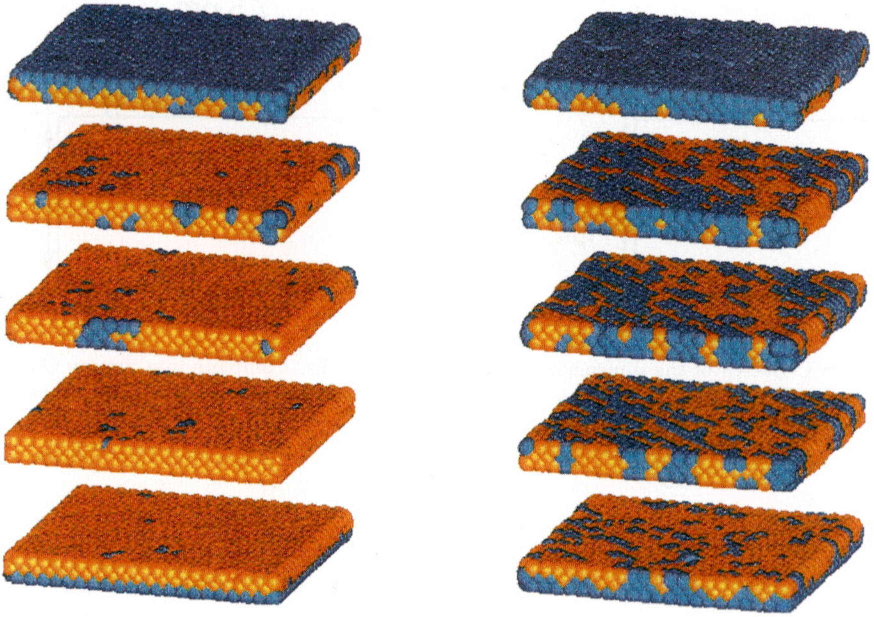

Figure 2. Structure of the iron film at 150 K (left panel) and 1 K (right panel). Nucleation of bcc-domains (blue) in the fcc environment can be seen. In order to show nuclei and the domain structure in more detail the film is decomposed into slices. The Cu substrate is not shown in this figure.

4 Summary and Discussion

The investigation of a 15 ML iron film on Cu(001) by means of molecular-dynamics simulations has shown, in agreement with experiments, that the amount of bcc-Fe increases with decreasing temperature. We observed fluctuations of bcc-nuclei as well as pinned bcc-nuclei which grow and form domains when decreasing the temperature.

For the simulated iron island of cylindrical form we found that, in contrast to the iron film, nearly the whole island transforms to the bcc structure at low temperatures. The question is what provokes the difference in the simulations between the film and the island and how does this behavior compare with experimental findings? Kirschner *et al.*[4] observed that iron films grown by PLD show a layer by layer growth which leads to a more or less flat surface, whereas iron films grown by TD methods show a more structured 3D-island like surface. These experiments reveal that the flat iron film is more stable in

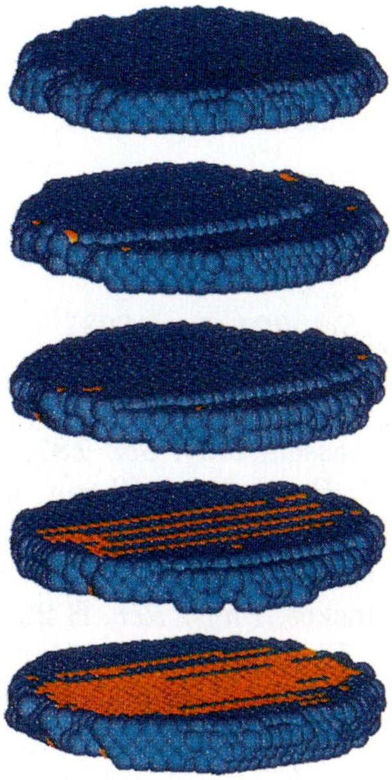

Figure 3. The structure of an iron island shown slice-wise: Nearly the whole island has transformed into the bcc structure (blue). The bottom layers close to the copper substrate are still in the fcc structure (orange). The Cu substrate is not shown in this figure.

the fcc structure than the structured iron film. We initiatively conclude that in the experiments the lack of lateral stresses in the 3D-island growth regime leads to a more pronounced fcc to bcc transformation.

In both types of simulation - for the iron film and the iron island - we found a significant amount of fcc-structured atoms close to the copper substrate. Keune *et al.* investigated the structure of the 8 layers close the copper substrate in a 42 ML iron film at room-temperature using Mössbauer spectroscopy.[13] They observed that about 25 percent of these layers retain their fcc structure due to the stabilizing feature of the underlying substrate. Our simulations essentially confirm this picture and show two types of possible distributions of the fcc-structured atoms: The first one is a rough bcc-fcc interface as observed for simulations of the iron film, the second one is a more flat bcc-fcc interface as observed for simulations of cylindrical iron islands.

5 Acknowledgements

This work has been supported by the German Science Council (Deutsche Forschungsgemeinschaft) through the *Graduate College Structure and Dynamics of Heterogeneous Systems.*

References

1. M. Wuttig *et al.*, *Surf. Sci.* **291**, 14 (1993).
2. F. Scheurer, R. Allenspach, P. Xhonneux, E. Courtens, *Phys. Rev.* B **48**, 9890 (1993).
3. M. Wuttig and J. Thomassen, *Surf. Sci.* **282**, 237 (1993).
4. P. Ohresser, J. Shen, J. Barthel, M. Zheng, Ch. V. Mohan, M. Klaua, and J. Kirschner, *Phys. Rev.* B **59**, 3696 (1999).
5. M. S. Daw and M. I. Baskes, *Phys. Rev. Lett.* **50**, 1285 (1983).
6. M. S. Daw and M. I. Baskes, *Phys. Rev.* B **29**, 6443 (1984).
7. R. Meyer and P. Entel, *Phys. Rev.* B **57**, 5140 (1998).
8. M. P. Allen and D. J. Tildesley, *Computer Simulation of Liquids* (Clarendon Press, Oxford, 1987).
9. W. G. Hoover, *Phys. Rev.* A **31**, 1695 (1985).
10. M. Parrinello and A. Rahman, *Phys. Rev. Lett.* **45**, 1196 (1980).
11. K. Kadau, R. Meyer, and P. Entel, *Surf. Rev. Lett.* **6**, 35 (1999).
12. K. Kadau and P. Entel, *J. Magn. Magn. Mater.* **198-199**, 531 (1999).
13. W. Keune, A. Schatz, R. D. Ellerbrock, A. Fuest, K. Wilmers, and R. A. Brand, *J. Appl. Phys.* **79**, 4265 (1996).

ELECTRONIC STRUCTURE OF SI(111): SN-($\sqrt{3} \times \sqrt{3}$) R 30°: AN AB INITIO STUDY

R. SPETTMANN AND P. ENTEL
Theoretische Physik, Gerhard-Mercator-Universität, 47048 Duisburg, Germany

K. SCHROEDER AND S. BLÜGEL
Institut für Festkörperforschung, Forschungszentrum Jülich, 52425 Jülich, Germany

Different coverages of Sn on Si(111) have been investigated by means of *ab initio* molecular dynamics simulations. The resulting structure, density of states and Schottky barrier have been determined for the relaxed configurations and are compared with experimental results.

1 Introduction

1.1 Motivation

Ultra-thin metal films on semiconductor surfaces are nowadays a matter of growing interest because of their potential technological application. A detailed theoretical comprehension of the physical properties of the interface for different substrate covering is therefore very important. As model system we have investigated Sn coverages on Si, such as possibly obtained experimentally by a molecular beam epitaxy growth process. Experimental information of local lattice properties of α-Sn or β-Sn films is available by ^{119}Sn-Conversion-Mössbauer spectroscopy (see B. Roldán Cuenya in this volume). Here the isomeric shift gives information about the chemical binding properties at the interface. Important aim of the experimentalists is to epitaxially stabilize α-Sn in the diamond structure on Si(111). The present paper contains first accompanying *ab initio* molecular-dynamics simulations of this growth process.

1.2 Method

We used a program for *ab initio* calculations, which was developed by S. Blügel, P. Richard[1,2] and B. Engels[3,4] at the Research Center of Jülich. The program is based on the Car-Parinello-method and on the density-functional theory in the local density approximation with norm-conserving pseudopotentials[5] of the Kleinman-Bylander[6] form and a plane wave basis set. The Kohn-Sham Hamiltonian is solved by an iterative diagonalization scheme for the eigenvalue problem.[7,8,9]

1.3 Components

The semiconductor Si in the diamond structure is used as substrate with Sn as the adsorbate. There are two different modifications of Sn:

- α-Sn, a semiconductor with a diamond structure and
- β-Sn, a metal with a tetragonaly body centered structure.

Table 1 lists some bulk properties while Fig. 1 illustrates lattice structures of the components.

Table 1. Some bulk properties of Si, α-Sn and β-Sn

	nn distance	lattice constant	bandgap (T=0 K)
Si	$r_{nn} = 2.34$ Å	$a = 5.43$ Å	$E_g = 1.17$ eV
α-Sn	$r_{nn} = 2.81$ Å	$a = 6.49$ Å	$E_g = 0.10$ eV
β-Sn	$r_{nn} = 3.02$ Å	$a = 5.83$ Å $c = 3.18$ Å $c/a = 0.55$	–

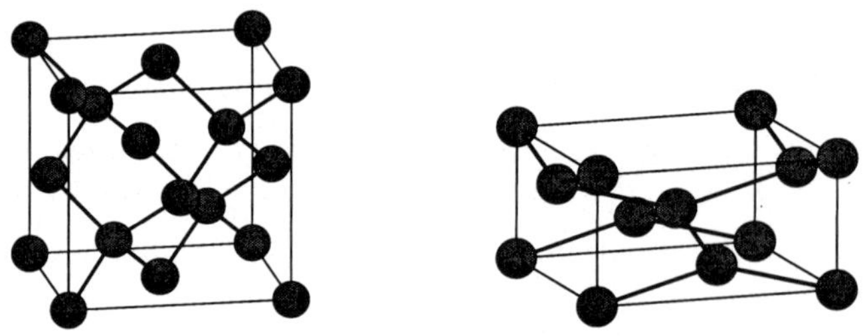

Figure 1. Diamond (left) and β-Sn structure (right).

1.4 The unit cell

The side-view of the unit cell in the unrelaxed state in the left panel of Fig. 2 shows a double layer structure of diamond crystal cut in the (111) direction. Double layer means that in the direction orthogonal to the surface, the distance between two layers alternates by the ratio of 6 to 2. This leads to a much stronger bonding of the two layers forming the double layer. The unit cell is built up in such a way that inversion symmetry exists. The inner layers are fixed, all other layers can be freely moved. All systems have been relaxed before analyzing the structure and Schottky barrier.

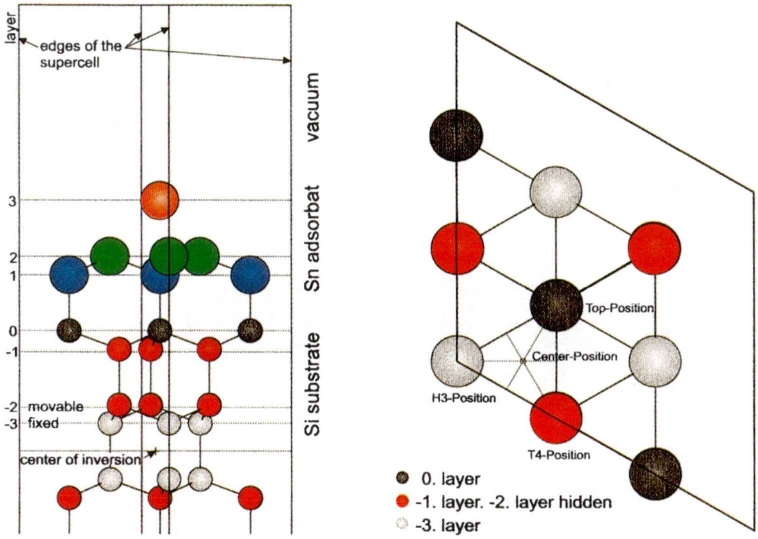

Figure 2. The unrelaxed unit cell in side- and top-view.

The top-view (Fig. 2 right) shows the symmetry of the chosen reconstruction. The $(\sqrt{3} \times \sqrt{3})$ R 30° reconstruction was built up from the 1×1 structure by increasing the angles between the basis vectors by 30° and stretching them by $\sqrt{3}$. This leads to three atoms per layer compared to one atom per layer for the 1×1 cell. This reconstruction was suggested by recent experimental investigations.[10]

2 Structure

Many different systems have been analyzed. In this article we concentrate on those systems with full monolayers of Sn.

System, each on 2 DL Si	N_{Si}	N_{Sn}
2 ML(Sn)	24	12
4 ML(Sn)	24	24
1 ML(Si) + 1 ML(Sn)	36	0

A complete (full) monolayer consists of three atoms per unit cell. Side-views of the analyzed full monolayer systems are shown in the in the following figures. All figures show the atoms which belong to the cell at the start of the relaxation process. Atomic positions in bulk Si are plotted in thin lines.
The aim of the present investigations was to find out whether we can observe growth processes of the diamond structure of the α state of Sn induced by the substrate. The one-monolayer system was built by filling up a half double layer of Si atoms with a monolayer of Sn, i.e. both layers form a mixed Si/Sn double layer. The alternative is to build up a one-monolayer system by placing the Sn atoms on a full double layer. This leads, however, to an extremely instable system.

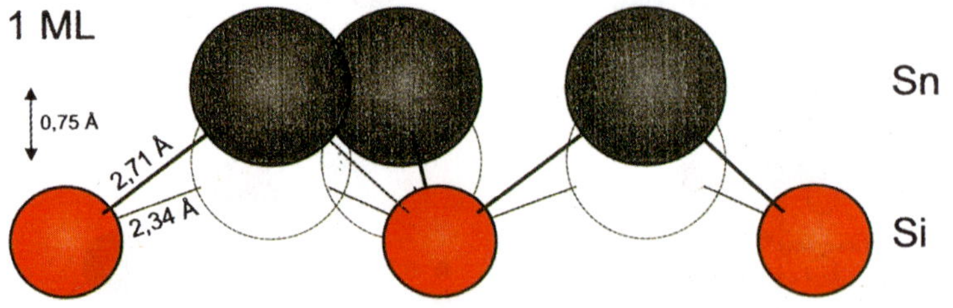

Figure 3. One monolayer Sn on a half double layer Si.

The relaxed one-monolayer system shown in Fig. 3 is very symmetrical. The adsorbate atoms just relax in z direction by enlarging the distance from the substrate compared to the Si bulk system.

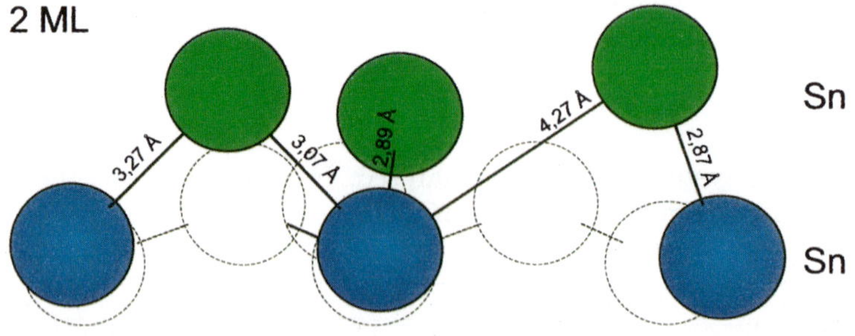

Figure 4. Two monolayers of Sn on a full double layer of Si

The two-monolayer system in Fig. 4 shows a much larger distortion than the one-monolayer system. The atoms relax in the z direction as well as in the $x-y$ plane in order to compensate the ca. 20% larger lattice constant of Sn (α-state) compared to the lattice constant of Si. The nn distances of the adsorbate atoms are larger than in bulk α-Sn. As largest system we analyzed a four-monolayer system.

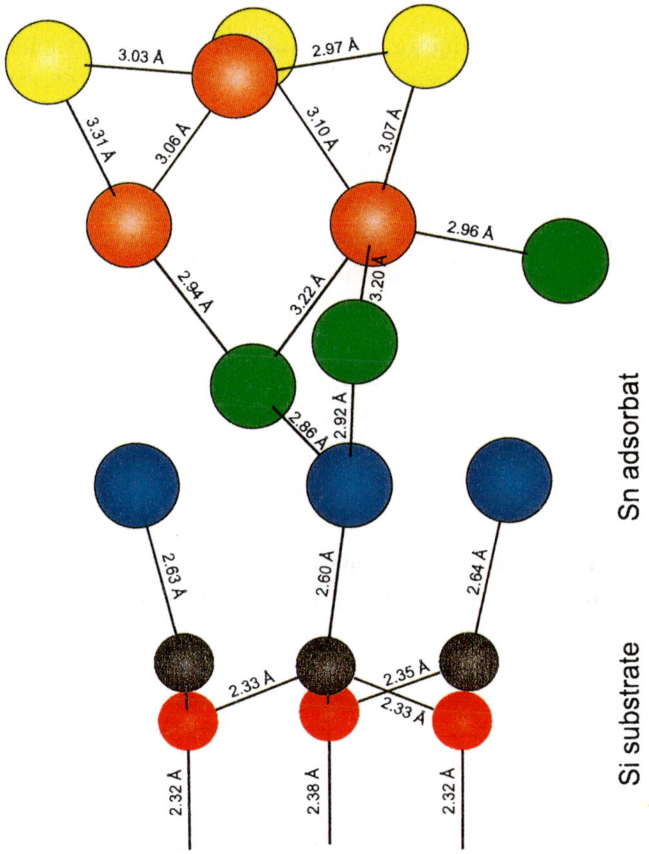

Figure 5. Four monolayers of Sn on a full double layer of Si.

The first Sn layer of the four-monolayer system (Fig. 5) is grown symmetrically in the diamond structure, but the other layers show strong distortions. The nn distances of the outer layers are similar to β-Sn.

Our calculations show that growing Sn in the diamond structure is energetically not very favorable and can, therefore, hardly be observed in epitaxial growth processes.

3 The silicon/tin Schottky contact

3.1 Schottky contacts

If a metal and a semiconductor are contacted their Fermi energies are assimilated, leading to band bending near the interface on the semiconductor side.[11,12,13] The n-type Schottky barrier Φ_{Bn} is given by the difference between the lower edge of the conduction band, E_C of the bulk semiconductor and the Fermi energy of the connected system, E_F. The p-type Schottky barrier Φ_{Bp} is given by the difference between the upper edge of the valence band E_V of the unconnected semiconductor and the Fermi energy of the connected system. The sum of the n- and p-type Schottky barrier gives the bandgap of the semiconductor, E_g.

$$\begin{aligned} \Phi_{Bn} &= E_C - E_F, \\ \Phi_{Bp} &= E_F - E_V, \\ E_g &= \Phi_{Bp} + \Phi_{Bn}. \end{aligned} \tag{1}$$

3.2 Density of states

In order to determinate the Schottky barriers we had to check whether the adsorbate of the analyzed systems are really metallic and whether the metal/semiconductor interface is also the Si/Sn interface. In order to achieve this we have calculated the density of states (DOS) for each layer. For this purpose the local DOS of each atom is determined and the DOS of all atoms of a specific layer are then combined. Fig. 6 shows the DOS of the uppermost Si layer (layer 0 in Fig. 2) in the upper diagram and the lowest Sn layer (layer 1) in the lower one for the two-monolayer system. We found a clear bandgap around the Fermi energy on the Si side of the interface and no bandgap on the Sn side, which indicates that our systems contains metal/semiconductor interfaces.

3.3 The Si/Sn Schottky contact

We calculated Φ_{Bn} for all systems after complete relaxation. The resulting barrier height as a function of the layer thickness is plotted in Fig. 7.

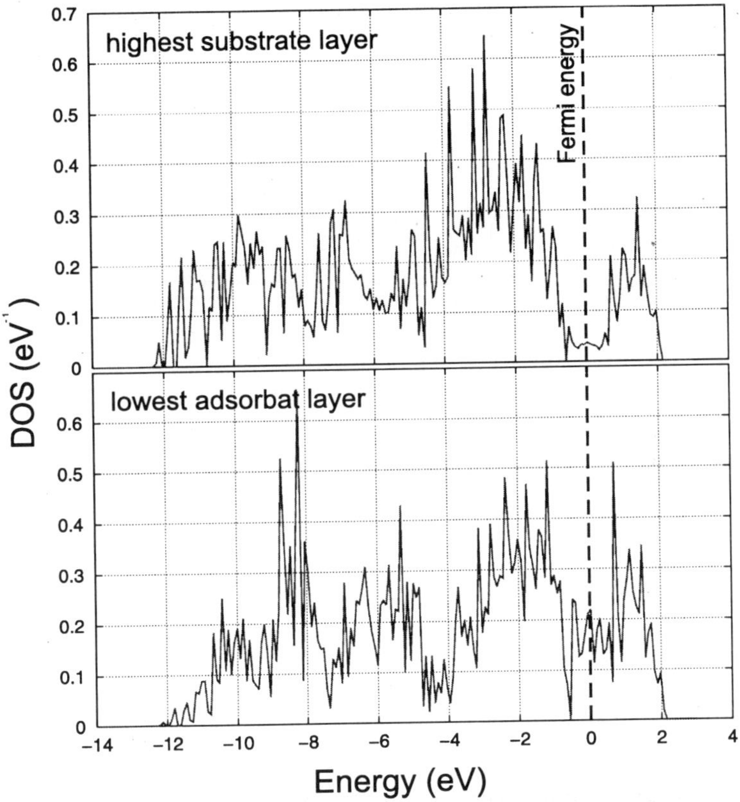

Figure 6. Local DOS for the interface layers of the 2 ML system.

For comparison with experiment we use values measured by Schmitsdorf and Mönch.[14] They found for the p-type Schottky barrier of a many-monolayer system:

$$\Phi_{\mathsf{Bn}} = 0.679 \text{ eV}. \tag{2}$$

With Eq. (1) and the bandgap of Si of $E_g = 1.17$ eV at $T = 0K$ we obtain

$$\Phi_{\mathsf{Bp}} = \mathsf{E_g} - \Phi_{\mathsf{Bn}} = 0.491 \text{ eV}. \tag{3}$$

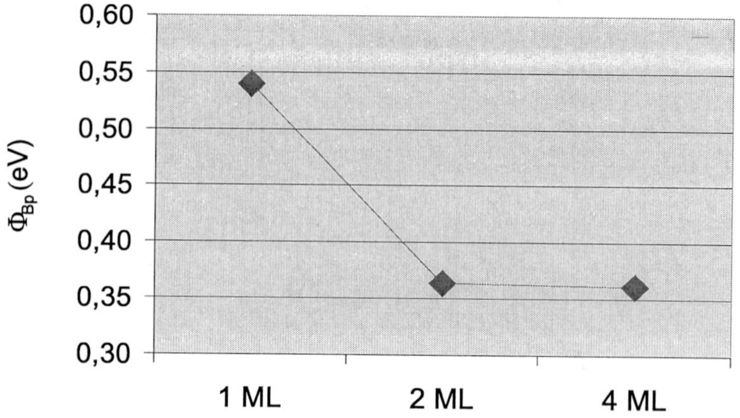

Figure 7. The n-type Schottky barrier Φ_{Bn}. for one, two and four monolayers of tin

For comparison we used the value $\Phi_{Bp} = 0.361$ eV from the four-monolayer system, showing that the calculated value differs by 26% from the experimental one.

4 Summary

We found that growing of Sn on a silicon substrate leads to increasing distortions with increasing coverage. Already a few monolayers of Sn show this effect. Of interest is, however, that Sn as an adsorbate behaves metallically. The Schottky barrier of the Si/Sn system was determined from our calculations. A comparison with experimental values results in a deviation of 26%.

5 Acknowledgements

This work has been supported by the German Science Council (Deutsche Forschungsgemeinschaft) through the *Graduate College Structure and Dynamics of Heterogeneous Systems.*

References

1. P. Richard, *Iterative Lösung der Schrödinger-Gleichung für große Systeme* (Master Thesis, RWTH Aachen, 1993).

2. Peter Richard, *Banddiskontinuitäten an III–V Halbleitergrenzflächen: Verspannungsabhängigkeit und Einfluß von Zwischenschichten* (PhD Thesis, RWTH Aachen, 1996).
3. B. Engels, *Konstruktion von glatten normerhaltenden Pseudopotentialen* (Master Thesis, RWTH Aachen, 1992).
4. B. Engels, *Ab initio Berechnung der (110) Oberfläche von III–V Halbleitern: Simulation von Rastertunnelmikroskopie–Aufnahmen* (PhD Thesis, RWTH Aachen, 1995).
5. D. Vanderbilt, *Optimally smooth norm–conserving pseudopotentials*, *Phys. Rev.* B **32**, 8412 (1985).
6. L. Kleinman and D.M. Bylander, *Efficacious forms for model pseudopotentials*, *Phys. Rev. Lett.* **48**, 1425 (1982).
7. W.E. Davidson, *The iterative calculation of a few of the lowest eigenvalues and corresponding eigenvectors of large real–symmetric matrices*, *J. Comput. Phys.* **17**, 87 (1975).
8. B. Liu, in *Report on Workshop: Numerical Algorithms in Chemistry: Algebraic Methods* (Lawrence Berkley Lab., Univ. of California, 1978).
9. N. Kosugi, *Modification of the Liu–Davidson Method for Obtaining One or Simultaneously Several Eigensolutions of a Large Real–Symmetric Matrix*, *J. Comput. Phys.* **55**, 426 (1984).
10. D.T. Wang, N. Esser, M. Cardona, and J. Zegenhagen, *Epitaxy of* Sn *on* Si(111), *Surf. Sci.* **343**, 31 (1995).
11. W. Mönch, *On the present understanding of Schottky contacts* in *Festkörperprobleme* Vol. **26** (Vieweg & Sohn, Braunschweig, 1986), p. 67.
12. *Electronic structure of metal–semiconductor contacts*, in *Perspectives in condensed matter physics*, Vol. **1**, ed. W. Mönch (Kluwer, Dordrecht, 1990).
13. W. Mönch, *Semiconductor Surfaces and Interfaces* (Springer, Berlin, 1993).
14. R.F. Schmitsdorf, T.U. Kampen, and W. Mönch, *Explanation of the linear correlation between barrier heights and ideality factors of real metal-semiconductor contacts by laterally nonuniform schottky barriers*, *J. Vac. Sci. Technol.* B **15**, 1221 (1997).

IN-SITU RHEED AND STM INVESTIGATIONS ON THE EPITAXIAL GROWTH OF FE/PD MULTILAYERS ON PD(001)-BUFFERED SAPPHIRE

T. STEFFL, H. MÜHLBAUER, B. RELLINGHAUS AND G. DUMPICH
Experimentelle Tieftemperaturphysik, Gerhard-Mercator-Universität Duisburg, Lotharstr. 1, Duisburg, D-47048, Germany
E-mail: dumpich@ttphysik.uni-duisburg.de

We present the preparation of (001) oriented (Fe/Pd) multilayer films on sapphire substrates. The samples are grown on specifically prepared Pd buffer layers, which are pre-seeded with a thin Fe layer. Structural and topographical investigations utilizing in-situ RHEED and STM and ex-situ XRD clearly evidence epitaxial growth along the [001] direction throughout more than 30 Fe/Pd bilayers.

1 Introduction

Epitaxial growth of $(Fe/Pd)_n$ multilayers (ML) has attracted considerable attention due to the possibility to control the crystal structure of the Fe layers via the growth process. This is of particular interest, since the magnetic properties of Fe are expected to strongly depend on its structure.[1] Most important for the epitaxial growth of ML are the structure and topography of the underlying substrate and usually, buffer layers are used to induce epitaxial growth in a particular orientation. In the case of $(Fe/Pd)_n$ ML it has been shown that for (111) oriented Pd layers, Fe grows in the bcc-phase in (110) orientation.[2] Investigating the growth of Pd/Fe/Pd trilayers we have found that on (111) oriented Pd buffer layers (deposited at elevated temperatures onto sapphire), bcc-Fe grows according to the Kurdjumov-Sachs relationship.[3] Within this growth mode, the formation of domains with different in-plane orientations is unavoidable, thereby preventing subsequent epitaxial growth of $(Fe/Pd)_n$ ML. On the other hand, epitaxial growth has been observed for $(Fe/Pd)_n$ ML prepared on Pd(001)-buffered MgO. In this case, Fe grows in the bcc-phase obeying the epitaxial relationship Fe(001)[110] || Pd(001)[100].[4,5]

In the present paper, we focus on in-situ RHEED and STM investigations of epitaxially grown $(Fe/Pd)_n$ ML on Pd(001)-buffered sapphire. We have evidence for epitaxial growth along the [001] direction throughout more than 30 Fe/Pd bilayers with the Fe layers growing in a tetragonally distorted phase.

2 Experimental

Fe/Pd multilayers are deposited by alternately evaporating Fe and Pd from independently controlled electron beam guns onto sapphire substrates. We use 99.99% Fe and 99.95% Pd ingots (Goodfellow) as primary materials. The base pressure of the UHV preparation chamber is 2×10^{-8} Pa, and during deposition, it rises to 2×10^{-6} Pa. The typical deposition rate is 0.04 nm/s. Deposition rate and layer thickness are monitored by quartz crystal oscillators, which are calibrated utilizing an optical interference method. Epi-ready polished $\alpha - Al_2O_3$ (sapphire) single crystals, which are cut parallel to the c-axis [a-plane, $(11\bar{2}0)$], are used as substrates. Before mounting the substrates to the sample holder they are cleaned in acetone and destilled water.

The $(Fe/Pd)_n$ ML stacks are grown onto Pd buffer layers, which are pre-seeded with Fe. The thickness and growth temperature of both, the Fe seed layer and the Pd buffer layer have been varied in a wide range to find an optimized set of parameters for ML growth along the [001] direction. The resulting procedure is as follows. A thin Fe layer with a thickness $t_{Fe,seed} \simeq 0.5$nm is deposited onto the sapphire substrate at temperatures $300K \leq T_{seed} \leq 700K$. A Pd film with a thickness $t_{Pd} = 30$nm is then evaporated at temperatures $T_{buf} \simeq 700K$. To avoid interdiffusion between adjacent layers, the alternating Fe and Pd layers are deposited at room temperature (RT). The ML stack is always terminated by a Pd layer to prevent the film from oxidation when being exposed to ambient conditions e.g. for ex-situ investigations.

The growth of the films is monitored by means of in-situ RHEED measurements (Reflective High Energy Electron Diffraction). Furthermore, the sample can be transfered to a second UHV chamber equipped with a beetle type scanning tunneling microscope (STM) for topographical investigations without breaking the UHV. Both, RHEED and STM measurements are carried out at various stages of the ML growth. Ex-situ characterization is done by means of X-ray diffraction (XRD) experiments utilizing Co-k_α radiation ($\lambda = 0.17902$nm).

3 Results and discussion

Quality and texture of the Pd buffer layer are crucial factors, since they control the orientation of subsequently grown layers. Thus, before presenting the results of our investigations on the growth of the $(Fe/Pd)_n$ ML, we firstly report on the structural properties of single Pd films.

In Fig. 1 we present the RHEED patterns obtained from the surface of a Pd film with thickness $t_{Pd} = 30$nm. The film is grown at $T_{Pd} = 670$K on

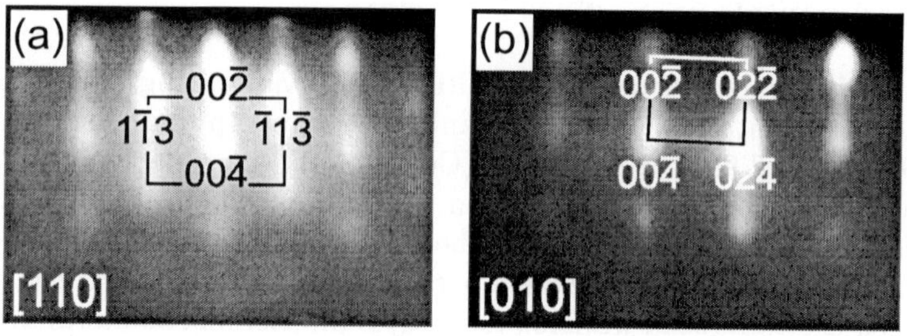

Figure 1. RHEED patterns of a single Pd film with a thickness $t_{Pd} = 30nm$ on a sapphire substrate pre-seeded with Fe. The pattern in fig. (b) is obtained after rotating the sample azimuthally by 45° with respect to the orientation in fig. (a).

an α-$Al_2O_3(11\bar{2}0)$ substrate pre-seeded with Fe ($t_{Fe,seed} = 0.5nm$) at $T_{seed} = 300K$. The occurrence of diffraction spots instead of streaks indicate volume diffraction and thus a certain roughness of the sensed surface. From the symmetry of the pattern in Fig. 1(a) and by measuring the distances and angles between the spots the orientation of the film is determined to be [110]. Azimuthal rotation of the sample by 45° results in the observation of the pattern shown in fig. 1(b). It represents diffraction along the [010]-direction of an fcc crystal. Additional RHEED investigations on the relative orientation of the Pd film with respect to the substrate reveal that Pd grows with its [110] direction parallel to the [0001]-direction of the $\alpha - Al_2O_3$ (not shown here, epitaxial relationship: Pd(001)[110]$\|$ α-$Al_2O_3(11\bar{2}0)$[0001]).

To confirm our findings from RHEED we present in fig. 2 the result of an ex-situ XRD measurement of the Pd film. Intensity is plotted on a logarithmic scale vs. the double diffraction angle 2Θ. Besides some narrow Bragg peaks originating from the sapphire substrate (labeled $\bullet$), we observe strong (002) and (004) reflections of the fcc-Pd. There are also the (111) and (222) Bragg peaks present in fig. 2, but their intensities are two orders of magnitude smaller than those of the (002) and (004) peaks, respectively. As a consequence and in very good agreement with the results from RHEED, the Pd film is found to be strongly textured with the (001) plane laying parallel to the substrate surface.

In the following, we exemplarily present topographical and structural results obtained from a $(Fe_{2.5nm}Pd_{4nm})_{32}$ ML grown on a $Pd_{30nm}/Fe_{0.5nm}$ buffered $\alpha - Al_2O_3(11\bar{2}0)$ substrate. Deposition temperatures of the seed and buffer layers are $T_{seed} \simeq 600K$ and $T_{buf} = 670K$, respectively. In Fig. 3 we present a set of STM images taken at three different stages during the growth

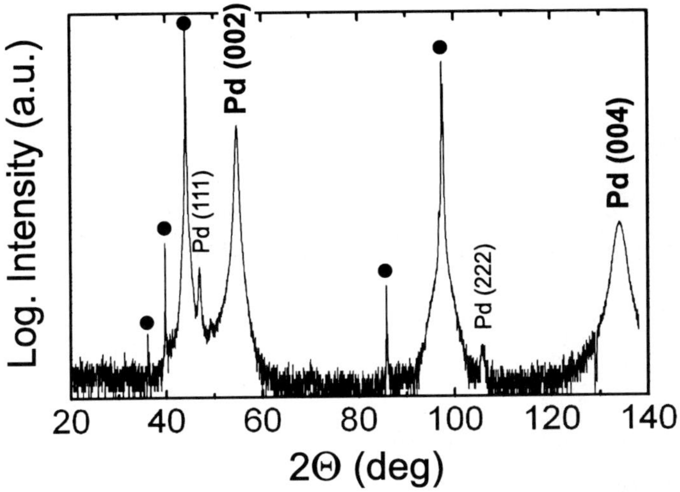

Figure 2. XRD Θ - 2Θ scan of a single Pd film (thickness t_{Pd} = 30nm) on a sapphire substrate pre-seeded with Fe. Bragg peaks labeled "•" originate from the sapphire substrate.

of the ML film. All images are taken at RT and represent an area of 50 × 50 nm^2. The solid lines indicate where the linescans for the height profiles are taken.

Fig. 3a shows the topography of the Pd buffer layer. The surface consists of rectangularly shaped hillocks with typical lengths and heights of $100nm \leq l_{hil,buff} \leq 200nm$ and $2nm \leq h_{hil,buff} \leq 5nm$, respectively. However, the surface roughness of the individual hillocks is typically $\sigma_{hil,buff} \simeq 1$ nm. In Fig. 3b we show an STM image of the surface of the 13th Fe layer within the ML stack. As a fingerprint of the buffer layer, the large hillocks can be seen to persist in the layer structure within the stack. In addition, we observe on top of these "buffer hillocks" the occurrence of smaller and rather elliptically or spherically shaped hillocks with diameters O $\simeq$ 20 nm. Fig. 3c shows the STM image of the surface of the terminating Pd layer. The topography of this layer is similar to that of the Fe layer shown in Fig. 3b, though the signature of the buffer layer is less pronounced.

Corresponding to the set of STM images in Fig. 3 we present in Fig. 4 a series of RHEED patterns taken at exactly the same stages of growth (see caption for details). The patterns in the left column of the figure [(a), (c), (e)] are obtained for the [011] direction of the Pd buffer being parallel to the incident electron beam. The patterns in the right column [(b), (d), (f)]

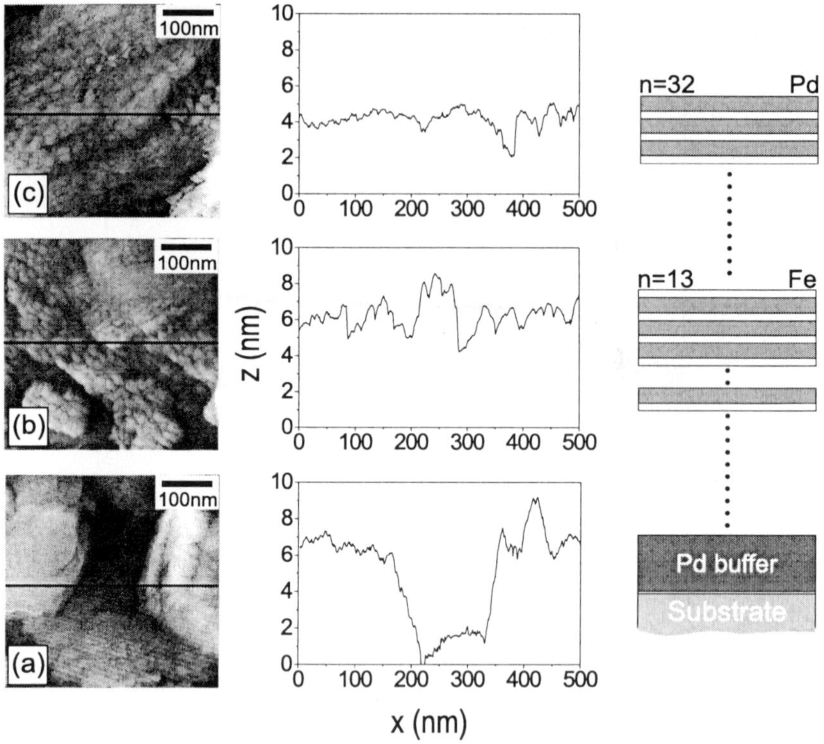

Figure 3. In-situ STM images of a $(Fe_{2.5nm}Pd_{4nm})_{32}$ multilayer (ML) on a $Pd_{30nm}(110)[100]$ | $Fe_{0.5nm}$ buffered $\alpha - Al_2O_3(11\bar{2}0)$ substrate: (a) Pd buffer layer, (b) 13th Fe layer of the ML stack, (c) 32nd (and terminating) Pd layer of the ML stack. The solid lines indicate where the linescans are taken to measure the height profiles, which are presented underneath the STM images.

are taken under conditions where the sample is rotated azimuthally by 45° with respect to this orientation. These two principal observation directions are referred to as θ_0 and θ_{45}, respectively. All RHEED patterns show clearly separated diffraction spots as expected in the case of volume diffraction and are thus in good agreement with the observation of hillocks in the STM images.

Figs. 4(a) and 4(b) show the RHEED patterns measured after the deposition of the Pd buffer. They are in good agreement with the results obtained from the single Pd film in fig. 1 and represent [110] and [010] orientations of fcc-Pd.

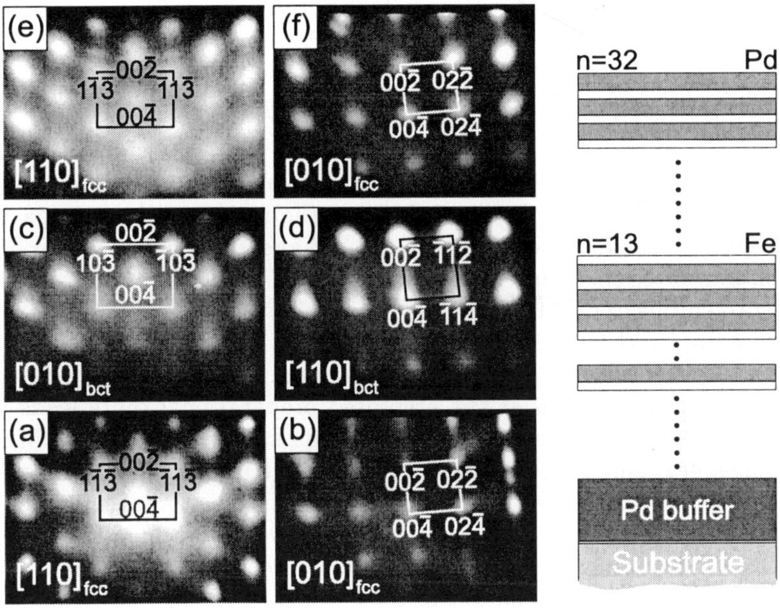

Figure 4. RHEED patterns taken at three different stages of growth of a $(Fe_{2.5nm}Pd_{4nm})_{32}$ multilayer (ML) on a $Pd_{30nm}(110)[100]$ buffer: (a)/(b) Pd buffer, (c)/(d) 13th Fe layer of the ML stack, (e)/(f) 32nd (and terminating) Pd layer of the ML stack. The azimuthal angles of the incident electron beam with respect to the substrate is the same in figures (a), (c), and (e). The RHEED patterns in figures (b), (d), and (f) are taken at azimuthal angles rotated by 45° with respect to those in figures (a), (c), and (e). The polygons are guides to the eye to visualize the symmetry of the diffraction pattern.

Figs. 4(c) and 4(d) show the RHEED patterns obtained from the surface of the 13th Fe layer within the ML stack. The diffraction patterns are similar to those of the Pd buffer [figs. 4(a) and 4(b)]. Whereas the in-plane lattice spacing of the Fe layer remains unchanged with respect to that of the Pd buffer, the vertical lattice spacings experience a significant reduction (i.e. the corresponding *reciprocal* lattice spacing in the RHEED patterns is *increased*). Thus the lattice of the Fe layer is tetragonally distorted. By means of continuously monitoring the growth of the Fe layer growth in equally prepared $(Fe_{4nm}/Pd_{4nm})_{32}$ ML films via in-situ RHEED we have shown that the tetragonal phase of an Fe layer with a thickness of $t_{Fe} = 2.5$nm is to be considered body center tetragonal (bct).[6] Accordingly, the RHEED patterns in fig. 4(c) and 4(d) represent diffraction along the [010] (θ_0) and [110] direction (θ_{45}) of bct-Fe. As a measure of the tetragonal distortion, the c/a-ratio is determined to be

$$\left(\frac{c}{a}\right)_{bct} = \frac{2d_{002}}{2d_{200}} = \frac{d_{200}^{-1}}{d_{002}^{-1}} = 1.12 \pm 0.02 \quad . \tag{1}$$

Here, c and a denote the out-of-plane and in-plane lattice constants of the bct crystal, respectively. d_{200}^{-1} and d_{002}^{-1} are the reciprocal lattice spacings as measured from the distances between the $10\bar{3}$ and $\bar{1}0\bar{3}$ reflections and between the $00\bar{2}$ and $00\bar{4}$ reflections in the RHEED pattern in fig. 4(c), respectively. We like to emphasize that within the experimental error, the determined c/a-ratio of $(c/a)_{bct}$=1.12 is in perfect agreement with the tetragonal distortion observed for according Fe layer thicknesses in the $(Fe_{4nm}/Pd_{4nm})_{32}$ ML films.[6] This finding thus shows that the presented preparational route leads to well reproduceable results regarding the orientation and structure of the ML films.

In figs. 4(e) and 4(f), we finally present the RHEED patterns of the 32nd Pd layer, which terminates the ML stack. As can be seen from a comparison with the results obtained from the Pd buffer, we obtain identical diffraction patterns representing the [011] (θ_0) and [001] directions (θ_{45}) of an fcc lattice. Neither the in-plane nor the out-of-plane lattice spacings are changed with respect to the underlying Pd buffer throughout the set of 32 bilayers of Fe and Pd. The epitaxial relationship between the individual Fe and Pd layers within the ML stack is Pd(001)[010]||bct-Fe(001)[110].

4 Conclusion

We have shown that Pd films, which are evaporated at elevated temperatures onto α-$Al_2O_3(11\bar{2}0)[0001]$ substrates (sapphire) pre-seeded with a thin Fe-layer, are strongly textured with their (001) plane lying parallel to the substrate surface. Utilizing such (001) textured Pd films as buffer layers allows for epitaxial growth of $(Fe/Pd)_n$ multilayers along the [001] direction in reproduceable quality. In contrast to other authors, we find that Fe grows in a tetragonally distorted phase (bct).

5 Acknowledgement

We like to thank Dr. J. Kästner for valuable discussions. Financial support from the Deutsche Forschungsgemeinschaft within the Sonderforschungsbereich 166 and the Graduate College "Structure and Dynamics of Heterogeneous Systems" is gratefully acknowledged.

References

1. H.C. Herper, E. Hoffmann and P. Entel, *Phys. Rev.* B, 1 (1999) and references therein.
2. A. Boufelfel, R.M. Emrick, and C.M. Falco, *Phys. Rev.* B **43**, 13152 (1991).
3. Ch. Müller, H. Mühlbauer and G. Dumpich, *Thin Solid Films* **310**, 81 (1997).
4. A. Schuhl, J.R. Childress, J.M. George, P. Galtier, O. Durand, A. Barthelemy and A. Fert, *J. Magn. Magn. Mat.* **121**, 275 (1993).
5. J.R. Childress, R. Kergoat, O. Durand, J.-M. George, P. Galtier, J. Miltat, and A. Schuhl, *J. Magn. Magn. Mat.* **130**, 13 (1994).
6. Ch. Müller, H. Mühlbauer, T. Steffl, B. Rellinghaus and G. Dumpich, submitted to *Phys. Rev.* B, (1999).

SUPPRESSION OF THE MARTENSITIC PHASE TRANSITION IN NI-RICH NI-AL THIN FILMS

CH. FELL, B. RELLINGHAUS, AND E.F. WASSERMANN
Experimentelle Tieftemperaturphysik, Gerhard-Mercator-Universität Duisburg, Lotharstr. 1, Duisburg, D-47048, Germany
E-mail: brell@ttphysik.uni-duisburg.de

We have prepared thin $\mathrm{Ni_xAl_{1-x}}$ alloy films in the concentration range $0.56 \leq x \leq 0.70$ by co-sputtering from element targets. Structural investigations utilizing TEM and XRD reveal that the films consist of ultra-fine grains with $d_{grain} \simeq 10$nm. Similar to bulk material the nano-grained Ni-Al samples undergo a structural phase change from (ordered) bcc to (ordered) fcc with increasing Ni content, which is, however, shifted towards higher Ni concentrations. From measurements of the electrical resistivity we find *no* indication for a martensitic phase transition for $x = 0.635$ and $x = 0.65$.

1 Introduction

The nucleation mechanism of martensitic phase transformations is an intensively debated problem.[1] In thin films and small particles, due to the reduced crystallite size, the energetic balance between chemical, elastic, and surface free energies is shifted with respect to those in bulk materials, and as a consequence, the martensitic phase transition occurs at reduced temperatures or is even completely suppressed.[2,3,4,5] Most investigations on micro- and nanocrystalline materials have been carried out in the Fe-Ni system, or have dealt with Fe or Co alloys. In the Ni-Al alloy system stoichiometric NiAl thin films play an important role in the field of magnetic recording, since they offer the possibility of providing a particular texture, which is advantageous for growing high coercivity magnetic material.[6] However, little is known on the structural properties of Ni-rich Ni-Al thin films, although in bulk form these alloys undergo a thermoelastic martensitic transformation.[7,8,9] Thus, the question arises as to whether the martensitic transformation is suppressed in Ni-rich Ni-Al films. In the present paper we report on the preparation of such Ni-rich Ni-Al alloy thin films by sputter deposition. We present results of investigations on the occurrence and the temperature range of stability of parent phase β-NiAl.

2 Experimental

The samples are prepared in a UHV preparation chamber by co-sputtering from two independently controlled sputter guns in Ar at room temperature (RT). The base pressure of the UHV chamber is 10^{-7}Pa and during deposition at a continuous gas flow of 0.6sccm/min it rises to $2.5 \cdot 10^{-1}$Pa. We use 99.995% Ni and 99.999% Al as primary target materials. 99.999% Ar is passed trough a gas purifying system capable to reduce both, O_2 and water contaminations to below 50ppb before being fed to the sputter guns immediately at the targets. The target to substrate distance is 130mm. We use glass, sapphire, and NaCl substrates, which are mounted on a substrate holder that rotates with a typical frequency of 1Hz. Deposition rate and layer thickness are monitored by quartz crystal oscillators, which are calibrated utilizing an optical interference method. The Ni concentration is varied between 56at% and 70at% ($0.56 \leq x \leq 0.70$). Accordingly, the deposition rate of RF-sputtered Ni is kept constant at $r_{Ni} \simeq 0.4$nm/s (RF power: $P_{RF} \simeq 270$W), while the deposition rate of DC-sputtered Al is varied in the range $0.26\text{nm/s} \leq r_{Al} \leq 0.42\text{nm/s}$ (DC power: $100\text{W} \leq P_{DC} \leq 170\text{W}$).

Structural investigations are carried out utilizing X-ray diffraction (XRD) and transmission electron microscopy (TEM). In order to prepare TEM samples, the films grown on NaCl subtrates are exposed to deionized water. After dissolution of the NaCl, the free NiAl films are placed onto Cu grids, which are then examined in a Philips CM12 scanning transmission electron microscope (STEM) (LaB_6 cathode, acceleration voltage = 120kV). The concentration of the alloy films is determined utilizing energy dispersive X-ray analysis (EDX) by means of an EDAX PW9900 system attached to the STEM.

The electrical resistivity is measured in 4-terminal geometry in the temperature range $4.2\text{K} \leq T \leq 800\text{K}$. The low-temperature ($4.2\text{K} \leq T \leq 300\text{K}$) experiments are carried out in a He^4 gas flow cryostat at normal pressure, whereas the high-temperature measurements ($300\text{K} \leq T \leq 800\text{K}$) are conducted in a tube oven in Ar atmosphere ($p_{Ar} \simeq 5 \cdot 10^3$Pa). Measuring currents are in the range $0.1\text{mA} \leq I \leq 1\text{mA}$.

3 Results and discussion

In order to characterize the as-deposited samples, we firstly focus on the structural investigations on a series of Ni_xAl_{1-x} thin films in the range $0.56 \leq x \leq 0.70$. All samples are prepared at RT and have a nominal thickness of $t = 100$nm. From TEM and XRD investigations the mean cystallite size is determined to be $d_{cryst} \leq 10$nm.

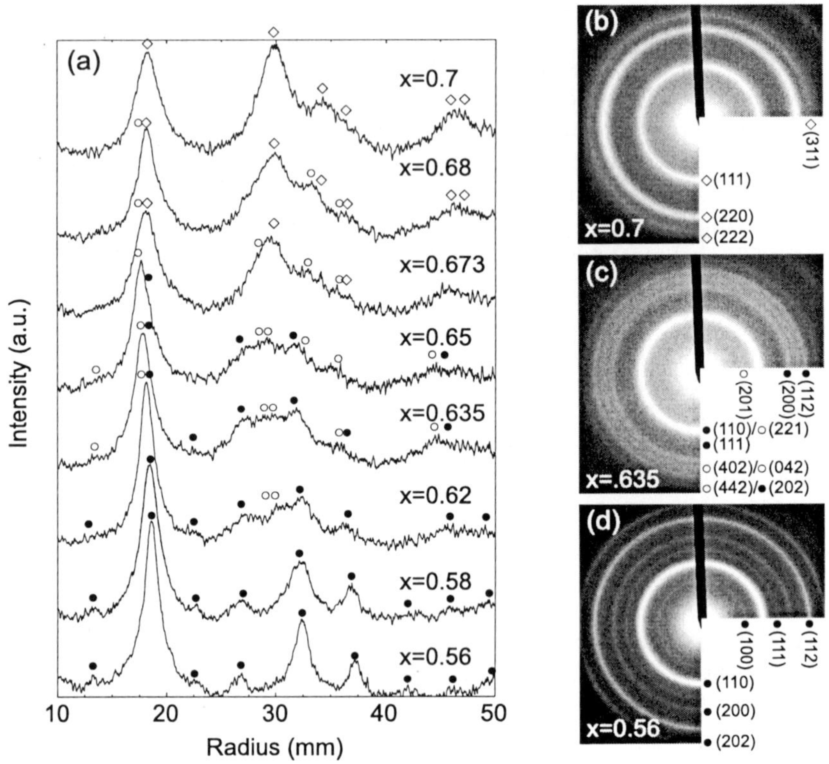

Figure 1. (a) Radial line scans through electron diffraction patterns of 8 Ni_xAl_{1-x} thin films. Background subtracted intensity is plotted vs. the radius of the diffraction patterns. The positions of diffraction fringes are labelled •, ○, and ◇ for β-NiAl, Ni_5Al_3, and Ni_3Al, respectively. Figures (b),(c), and (d) show diffraction patterns of the samples with $x = 0.70$, $x = 0.635$, and $x = 0.56$. The individual fringes are marked with their phase symbol and the according Miller indices.

In Fig. 1 we present the results of electron diffraction (ELD) experiments on 8 samples. Fig. 1(a) shows line scans taken radially from the center (radius, $R = 0$) to the periphery of the patterns. The positions of diffraction fringes are labelled according to the crystal structure they are associated with. In the concentration range under investigation we observe the occurence of the following phases: β-NiAl (•) has an ordered CsCl-type crystal structure [space group (SG): Pm3m], Ni_5Al_3 (○) has an orthorhombic unit cell (SG: Cmmm), and Ni_3Al (◇) exhibits an ordered $L1_2$ structure (SG: Pm3m). To illustrate

	Ni_xAl_{1-x}	
x	Phase (T = 300K)	$M_S^{bulk}(\beta\text{-NiAl})$
0.56	β-NiAl	0
0.58	β-NiAl	0
0.62	β-NiAl + Ni_5Al_3	250K
0.635	β-NiAl + Ni_5Al_3	450K
0.65	β-NiAl + Ni_5Al_3	660K
0.673	Ni_5Al_3 + Ni_3Al	-
0.68	Ni_5Al_3 + Ni_3Al	-
0.70	Ni_3Al	-

Table 1. Summary of the various structural phases present in our as-deposited sputtered Ni_xAl_{1-x} thin films at RT as determined from the analysis of electron diffraction patterns. For samples containing β-NiAl phase contributions the approximate martensite start temperatures, $M_S^{bulk}(\beta\text{-NiAl})$, of the according bulk alloys are also given (see Text for details).

the essential modifications in the ELD patterns with varying Ni concentration we show in Figs. 1(b)-(d) exemplarily the patterns obtained from samples with $x = 0.70, 0.635, 0.56$. The individual diffraction fringes are marked with the according Miller indices.

ELD patterns of samples with the lowest Ni contents ($x = 56, 58$) show a clear signature of the β-NiAl phase ($\bullet$). No evidence for the occurence of significant amounts of any other phase is found. In the range $0.62 \leq x \leq 0.65$, the films consist of both, β-NiAl and Ni_5Al_3 ($\circ$) phases, whereas for $x \geq 0.673$, the Ni_3Al phase ($\diamond$) evolves and becomes dominant for $x = 0.70$. These findings are summerized in Tab. 1. They are in principal agreement with the structural phase diagram of bulk Ni-Al.[9,10] However, the existence of significant β-NiAl phase contributions up to a Ni concentration of $x = 0.65$ is in contrast to the findings in bulk Ni-Al, where for $x \geq 0.62$, β-NiAl is known to undergo a thermoelastic martensitic phase transition from the CsCl-type B2 to a face-centered tetragonal (fct) $L1_0$ structure at temperatures above RT.[9,10,11] According to the bulk phase diagram the martensite start temperatures, M_S^{bulk}, for alloys with $x = 0.62, 0.635, 0.65$ are $M_S^{bulk} = 250K, 450K, 660K$, respectively.[10] As a consequence, the occurence of β-NiAl at RT is *not* expected at Ni concentrations $x > 0.62$.

The temperature dependence of the electrical resistivity has proved to be a sensitive physical property, in which the occurence of a martensitic phase

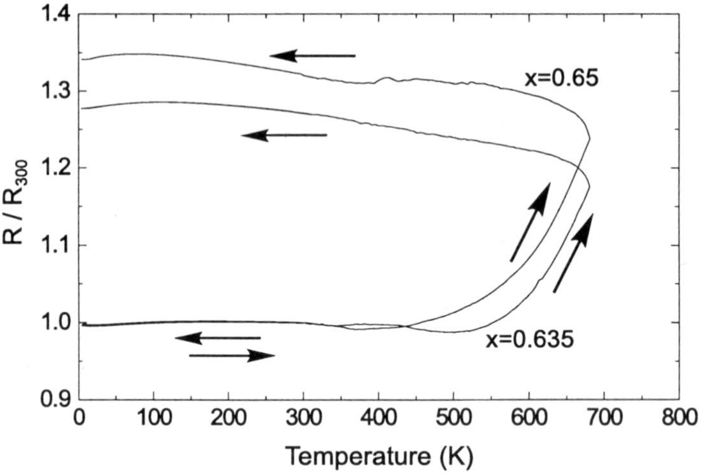

Figure 2. Temperature dependence of the electrical resistance, R, normalized to its value at T = 300K, R_{300}, for two sputtered Ni_xAl_{1-x} thin films with x = 0.635 and x = 0.65. Arrows indicate the chronological sequence of the measurements (see text for details).

transition shows up as a distinct and hysteretic increase of the resistivity with decreasing temperature.[11,12] Thus, to shed more light on the question as to whether or not the martensitic phase transition is suppressed in our sputtered Ni_xAl_{1-x} thin films, we present in Fig. 2 the results of resistivity measurements on the samples with x = 0.635 and x = 0.65. We plot the resistance normalized to its value at T = 300K, R/R_{300}, as function of the temperature. Starting at RT, both samples are cooled to T = 4.2K, then heated to T $\simeq$ 700K, and finally cooled to T = 4.2K, again.

During the first cooling cycle, and before reaching roughly T $\simeq$ 500K when being heated, the resistance of the films hardly changes with temperature. Specificially, in the sample with x = 0.635 there is *no* indication of a significant drop in R/R_{300} at T $\simeq$ 425K, i.e. where the martensitic phase transition occurs in the bulk material[12] (cf. Tab. 1). We like to point out that the resistance change associated with a martensitic tranformation in β-NiAl is a major effect and is expected to amount to some 30%.[11,12] In our Ni_xAl_{1-x} thin films with x = 0.635 and x = 0.65, however, the maximum changes in $R/R_{300}(T)$ below T $\simeq$ 500K are in the order of 1%. A comparably weak temperature dependence of the resistance is observed in additional low-temperature measurements on a sample with x = 0.62 (not shown here), though the martensite start temperature of the according bulk alloy is known to be $M_S^{bulk} \simeq 250K$.

As can be seen from Fig. 2, R/R_{300} increases strongly when the samples with $x = 0.635$ and $x = 0.65$ are heated to temperatures $500K \leq T \leq 700K$. From measurements on bulk Ni-Al alloys this resistivity increase is known to originate from the formation of Ni_5Al_3.[11] It sets in at lower temperatures in the sample with $x = 0.65$, and the effect is larger as compared to the film with the lower Ni concentration ($x = 0.635$), which is in good agreement with the findings in bulk Ni-Al.

A striking feature in $R/R_{300}(T)$ on finally cooling the samples down to $T = 4.2K$ is the anomalous increase of the resistance with decreasing temperature. We observe the occurence of a weak maximum at temperatures below roughly $T \simeq 100K$. A similar, though significantly smaller effect with its maximum lying at a somewhat larger temperatures is already observed in the initial cooling cycle. No hysteresis is found, and $R/R_{300}(T)$ is perfectly reversible below RT. The reason for this behavior is not yet understood. However, it is known that the occurence and growth of small precipitates leads to additional contribution in the electrical resistivity that increase with decreasing temperature.[13]

4 Conclusions

We have explored the structural properties of nano-grained Ni_xAl_{1-x} thin films in the concentration range $0.56 \leq x \leq 0.70$. Electron diffraction experiments on as-deposited samples at RT show the occurence of β-NiAl phase contributions in films with Ni contents as high as $x = 0.65$. These findings differ from the results on bulk Ni-Al alloys, which are known to undergo a martensitic phase transition from the B2 structure of β-NiAl to an fct $L1_0$ phase at temperatures above RT when $x > 0.62$. The ELD experiments thus indicate a suppression of the martensitic transformation in thin films. These results of our structural investigations are confirmed by measurements of the electrical resistivity. β-NiAl containing Ni_xAl_{1-x} films in the concentration range $0.62 \leq x \leq 0.65$ do *not* exhibit any features in the temperature dependence of the resistivity that indicate a martensitic transformation.

In order to understand the microphysical origin of the suppression of the martensitic transformation in Ni-rich Ni-Al thin films, furthergoing experiments including a systematic variation of the crystallite sizes and film thicknesses will be neccessary in the future.

5 Acknowledgements

We thank H. Zähres for conducting part of the TEM investigations and Prof. M. Acet for valuable discussions. Financial support from the Deutsche Forschungsgemeinschaft within the Sonderforschungsbereich 166 and the Graduate College "Structure and Dynamics of Heterogeneous Systems" is gratefully acknowledged.

References

1. For a review see: C. M. Wayman, *Mat. Sci. Forum* **56-58**, 1 (1990); and references therein.
2. R. E. Cech and D. Turnbull, *Transactions AIME* **206**, 124 (1956).
3. S. Kachi, Y. Bando, and S. Higuchi, *Japan. J. Appl. Phys.* **1** 307 (1962).
4. S. Kajiwara, S. Ohno, and K. Honma, *Phil. Mag.* **A63** no. 4, 625 (1991).
5. T. Tadaki, Y. Murai, A. Koreeda, Y. Nakata, and Y. Hirotsu, *Mat. Sci. Eng.* **A217/218**, 235 (1996).
6. L. Lee, D. E. Laughlin, and D. N. Lambeth, *J. Appl. Phys.* **79**, no. 8, 4902 (1996).
7. Y. K. Au and C. M. Wayman, *Scr. Metall.* **6**, 1209 (1972).
8. S. Chakravorty and C. M. Wayman, *Met. Trans.* **7A**, 555 (1976).
9. D. Schryvers, *J. de Physique IV, Supplement on J. de Physique III* **5**, C2-225 (1995).
10. P. Nash, M. F. Singleton, and J. L. Murray, in *Phase Diagrams of Binaray Nickel Alloys*, ed. by P. Nash (Materials Park, Ohio, USA, 1991), p. 3.
11. J. Lašek, T. Chráska, P. Křečk, and P. Bartuška, *Scr. Mater.* **37**, no. 6, 897 (1997); and references therein.
12. T. Chráska, J. Lašek, and P. Chráska, *Mat. Sci. Eng.* **A244**, 263 (1998).
13. F. Richter, W. Bendick, W. Pepperhoff, *Z. Metallkde.* **65**, 42 (1974).

MICROSCOPIC AND MESOSCOPIC MODELS OF FILM GROWTH BY ENERGETIC CLUSTER IMPACT (ECI)

JOHANNES NORDIEK, MICHAEL MOSELER*, OLIVER RATTUNDE AND HELLMUT HABERLAND

Freiburger Materialforschungszentrum, Stefan-Meier-Str. 21, 79104 Freiburg, Germany

E-mail: Johannes.Nordiek@physik.uni-freiburg.de

Microscopic and mesoscopic models have been applied in order to describe the deposition of thin films onto solid surfaces by Energetic Cluster Impact (ECI). On the *atomic* scale, Molecular Dynamic (MD) Simulations of the impact of a single cluster onto the surface have been carried out. These simulations reveal, that the local slope of the surface is diminished by a downhill particle current induced by the impact. Introduction of this downhill current into a *continuum* description of the development of the film profile leads to the Edwards-Wilkonson-Equation. The parameters of this equation can be estimated from the MD-results. The analytic solution of the Edwards-Wilkinson-Equation gives the development in time of the power spectrum and the roughness of the film. The results are in good agreement with Atomic-Force-Microscopy-measurements (AFM) on copper films deposited by ECI.

1 Introduction

A widely used method for the production of thin films is the deposition of atoms or molecules onto a solid surface. These particles may come from a vapor or from a beam, but in either case they are deposited onto random positions of the surface. Therefore, as long as this *Random Deposition* is the dominating process, the films will inevitably become rough.[1] Only if one supplies additional energy, redistribution processes like surface diffusion can be activated, and thus the roughness is reduced. In conventional deposition techniques, the most common method for supplying this activation energy consists in heating the substrate to several hundred Kelvins above room temperature.

Many materials, however, are not resistant to such high temperatures. Therefore, instead of heating the substrate, sometimes the incoming particles are accelerated to high kinetic energies, in order to activate other kinds of lateral mass transport. Unfortunately, these energetic ions can penetrate the solid, thus causing radiation damage or sputtering.[2] In contrast, energetic cluster ions consisting of $10^3 - 10^4$ atoms can deposit a large amount of energy

*PRESENT ADDRESS: SCHOOL OF PHYSICS, GEORGIA INSTITUTE OF TECHNOLOGY, 837 STATE STREET, ATLANTA, GEORGIA 30332, USA

at the surface without penetrating the substrate. This is due to the fact that the energy per atom is fairly low, even if the cluster is accelerated to a very high kinetic energy. Therefore, Energetic Cluster Impact (ECI) has been introduced as a method for producing high quality thin films of a great variety of materials.[3]

Molecular Dynamic simulations have been used in order to study the impact of a single cluster onto a solid surface.[4,5] If the impact energy is not too high, these simulations deliver quite reliable results. For higher energies, however, one must take into account, that the pressure wave induced by the impact will be reflected at the boundaries of the necessarily finite simulational cell and thus can significantly affect the results. In order to overcome this problem, we have designed a new kind of boundary conditions, which enable us to embed the simulational cell into a much larger continuum mechanical zone thus reducing the intensity of the reflected wave significantly.[6] In section 2 we will briefly explain how these boundary conditions work, and we will present an application of them to the energetic impact of a copper cluster onto a tilted copper surface.

The film formation by ECI consists of the consecutive impact of millions of clusters onto the surface. This process, of course, can not be simulated by means of Molecular Dynamics. Therefore, we have developed a mesoscopic model, using a continuum equation for the development of the film profile.[7] From the MD-simulations we deduced a downhill particle current as the main smoothing mechanism in the ECI-process, which in the continuum limit leads us to a stochastic differential equation: the Edwards-Wilkinson-equation. Its derivation as well as its solution and consequences for the roughness of the ECI-films are discussed in section 3.

Finally, in section 4 we summarize our results and give a short outlook at possible extensions of the model.

2 MD Simulation of Cluster Impact

2.1 Non-Reflecting Boundary Conditions

The main idea of our method is that only a rather small region surrounding the point of impact of the cluster will suffer substantial displacements of the atoms. In this region the trajectories of all atoms have to be calculated explicitly, and non-linear contributions of the interatomic potential have to be taken into account. Contrary to this, in the surrounding of this primary zone only slight displacements of atoms occur, which can be described by assuming harmonic interactions. Therefore, in this secondary zone a continuum

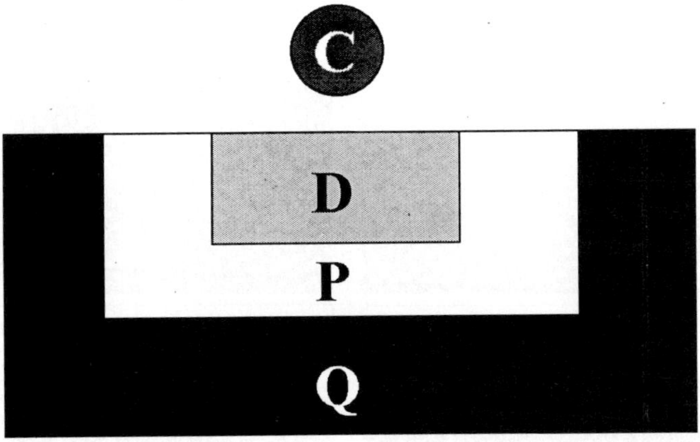

Figure 1. The simulational system: C: Cluster; D: Impact zone (MD); P: Crystalline zone (MD); Q: Crystalline zone (FD)

description by the equations of the linear theory of elasticity is appropriate. Numerically they can be handled by a Finite-Difference-scheme(FD). Thus, we apply in our simulations the model sketched in fig.1: The cluster C and the impact zone D as well as the crystalline zone P with nonlinear forces are treated by Molecular Dynamics. In the surrounding secondary zone Q we use Finite Differences. The discretization is such, that one fcc-cell consisting of four atoms is represented in Q by one point, which has four times the mass of a single atom. The strength of the harmonic forces in the secondary zone is adjusted to the elastic constants of the respective material. Of course, due to this coarse graining disturbances with very short wavelengths can not intrude from the MD-zone into the secondary zone, and thus would still be reflected. Therefore additional damping forces are introduced, acting on the atoms at the boundary between the MD and the continuum zone. So, while the low frequency disturbances are transmitted from the primary to the secondary zone, the high-frequency contributions are damped out at the boundary. A thorough mathematical foundation of this method, and the determination of the damping parameters have been given by Moseler et al.[6] Comparing results gained with these new boundary conditions to fully molecular dynamic calculations have shown the reliability of this scheme.

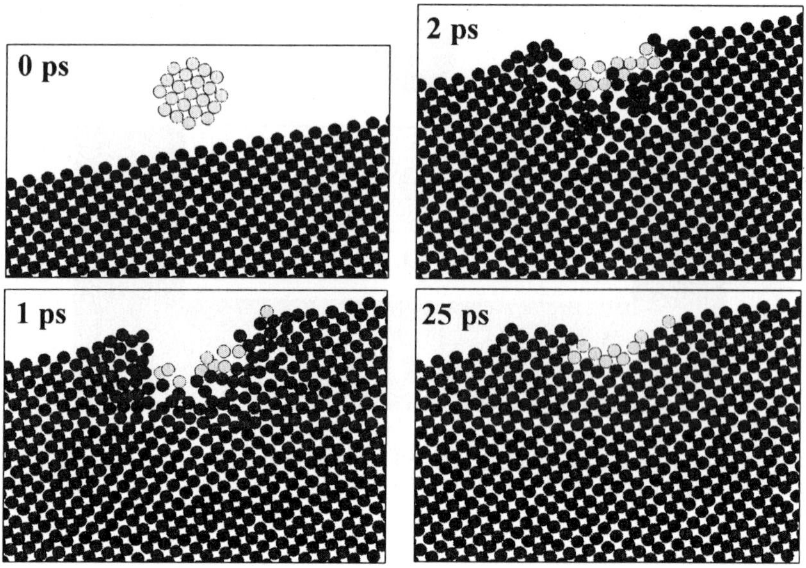

Figure 2. A 87-atom copper cluster hits with 900 eV onto a tilted copper surface. The particle displacements locally diminish the slope.

2.2 Cluster Impact onto Sloped Surface

With this new type of boundary conditions we are able to simulate the energetic impact of clusters onto solid surfaces. For example, we have adopted the scheme to a 87-atom copper cluster hitting with 900 eV of kinetic energy onto a sloped copper surface. The interaction of the copper atoms in the MD-zone has been modeled by an Embedded Atom Method potential (EAM) using the parameterization for copper by Oh and Johnson.[8] About 28.000 atoms have been included in the MD-zone. The total system (consisting of the MD and the FD-zone) represented more than 10^6 atoms.

Snapshots of the atomic coordinates are given in fig.2. They reveal the main smoothing mechanism in the ECI-process: The impacting cluster preferentially moves substrate atoms downhill and thereby diminishes locally the slope of the surface. Let u_i be the displacement of atom i along the direction of the slope. A measure of the total downhill movement of atoms is then given

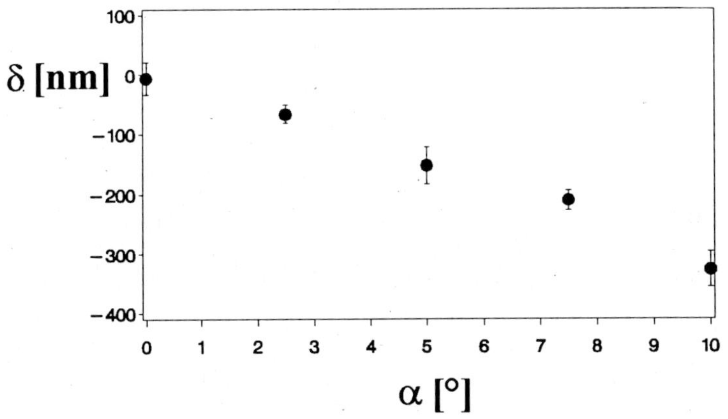

Figure 3. The displacement sum δ for a copper 1000 cluster with 10keV of kinetic energy for various angles of impact α. See eq.(1) for a definition of δ.

by the displacement sum

$$\delta = \sum_i u_i. \tag{1}$$

Of course, δ depends on the angle α of the initial slope. As an example, fig.3 shows the function $\delta(\alpha)$ for a Cu_{1000} cluster and 10 keV of kinetic energy. For small angles, it is roughly linear.

3 The Mesoscopic Model

3.1 The Edwards-Wilkinson-Equation

The mesoscopic model of film formation disregards the atomic structure of matter. Additionally, it assumes that the deposited film has no overhangs, such that its profile can be described by a single valued function $h(\vec{x}, t)$ of the two-dimensional surface-vector $\vec{x}$ and the time t. Temporal changes of h have two different origins: On the one hand, h increases due to the deposition of further clusters. Because of the random nature of the deposition process and because of the absence of temporal or spatial correlations of the incoming clusters, the random deposition can be modeled by a Gaussian White Noise, $\eta(\vec{x}, t)$ with mean and covariance determined by the parameters of the cluster beam. If the frame of reference is moving with the growing film, the moments are:

$$< \eta(\vec{x},t) > = 0 \tag{2}$$

$$< \eta(\vec{x},t)\eta(\vec{x}',t') > = r(\Omega N_{cl})^2 \delta(\vec{x}' - \vec{x})\delta(t' - t). \tag{3}$$

Here, r is the rate of deposited clusters. The volume of each cluster is given by the product of the atomic volume Ω and the number N_{cl} of atoms per cluster.

On the other hand, there exists a lateral current of atoms originating from the downhill particle movement. Of course, it depends linearly on the displacement sum δ, and thus, according to the MD results, is proportional to the slope of the surface (fig.3). Moreover, it is proportional to the rate of deposited clusters and the atomic volume:

$$\vec{j}_h = -cr\Omega\nabla h(\vec{x},t) \tag{4}$$

with $c = \delta/\alpha$. The contribution of the downhill movement to the development of the film profile ist given by the divergence of this current.

Adding these two contributions, we end up with the *Edwards-Wilkinson-Equation* for the temporal development of the film profile:

$$\frac{\partial h}{\partial t} = \nu\nabla^2 h + \eta, \tag{5}$$

where

$$\nu := cr\Omega \tag{6}$$

is called the smoothing parameter.

3.2 Power Spectrum and Roughness of the Films

On account of its linear character, the Edwards-Wilkinson-Equation can be easily solved analytically.[9] For the Power Spectrum of h on an initially flat surface one obtains:[10]

$$\widetilde{C}(k,t) = \frac{D}{\nu L^2 k^2}\left(1 - e^{-2\nu k^2 t}\right) \tag{7}$$

with

$$D := \frac{rN_c^2\Omega^2}{2} \tag{8}$$

Experimental data for the case of 2000-atom copper clusters deposited with 10 keV can be fitted to this formula. The same values of the fit parameters ν and D have been used for different mean film thicknesses $s = r\Omega t$.

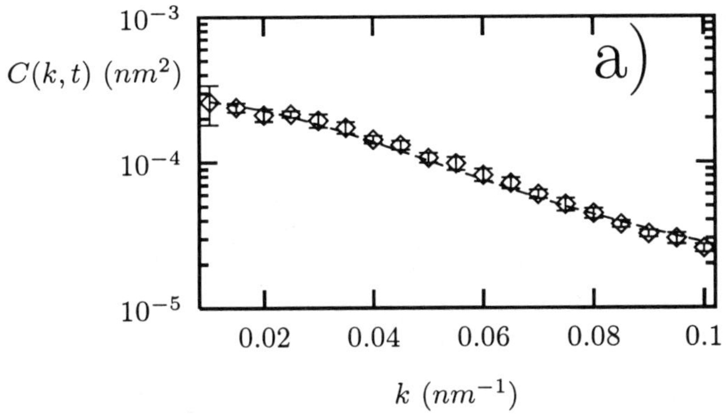

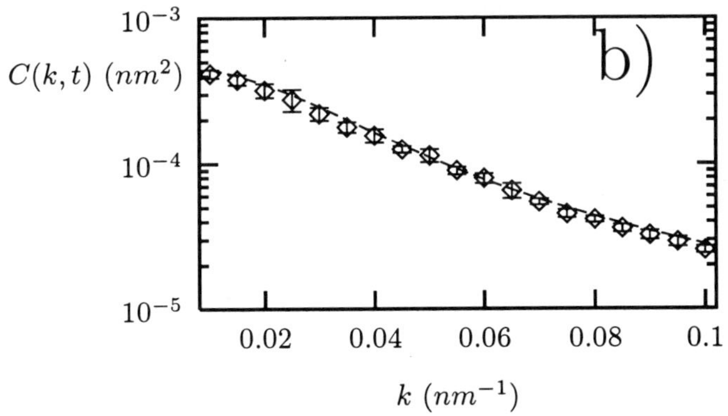

Figure 4. The power spectrum of copper ECI-films for different values of the mean film thickness. The dots are experimental data obtained by AFM. The lines are fits of eq.(7)to these data. The same values of the fit parameters ($\nu = 7 \pm 1$nm , $D = 12.8 \pm 0.5$nm^3) have been used for the two curves.

Comparisons of these fits to the actual data are shown in fig.4. The fitted values for ν agree rather well with those obtained from the MD simulations according to equation (6), which nicely confirms the validity of our model. There is a trend of the simulations to overestimate the downhill current. This is probably due to the fact, that the reduction of the downhill current by the polycrystallinity of the substrate is neglected in the simulations.

Integration of the Power Spectrum over all k-space (leaving out $k > k_0$ where the continuum-approximation loses its validity) yields the surface roughness. Two limiting cases are especially interesting: For

$$s << s_0 := \frac{L^2}{2\pi\nu^2}. \tag{9}$$

one has

$$R^2 = \frac{D}{4\pi\nu}\left[\ln(2\nu k_0 s) + C\right]. \tag{10}$$

For $s >> s_0$, on the other hand, the Roughness saturates to the value

$$R^2_{sat} = \frac{D}{2\pi\nu}\ln\left(\frac{Lk_0}{\pi}\right). \tag{11}$$

Saturation would not take place without the smoothing mechanism.

4 Extensions of the Model

There are several possible extensions to the Mesoscopic Model:

- Experimentally, one has not only clusters of one size, but a whole distribution $f(N)$ of cluster sizes in the beam. In this case the covariance of the source term in the Edwards-Wilkinson-Equation becomes

$$D = \frac{r\Omega^2}{2} < N^2 >= \frac{r\Omega^2}{2}\left(< N >^2 + \sigma_N^2\right). \tag{12}$$

  The strength of the downhill current, on the other hand, depends only on the first moment of the cluster size distribution. Therefore the roughness of the surface increases with the width of $f(N)$. In order to get the film as smooth as possible, it would be best to have only clusters of one size in the beam.

- Instead of starting the deposition on a flat surface ($R_0 = 0$), one can choose a finite roughness as the initial condition. If R_0 is larger then the saturation value, ECI is an effective means of smoothing an initially rough surface (fig.5). Experiments are in progress, in order to confirm this prediction.

- In contrast to metal clusters, noble gas clusters will evaporate after the impact with a surface. Therefore, they do not yield a source term in the Edwards-Wilkinson-Equation. On the other hand, the downhill current continues to exist. Thus, the stochastic character of the film-development vanishes and we end up with the deterministic equation

$$\frac{\partial h(\vec{x},t)}{\partial t} = \nu\nabla^2 h(\vec{x},t). \tag{13}$$

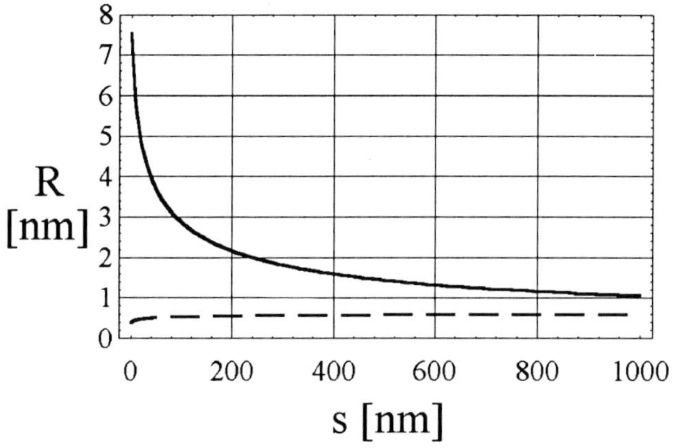

Figure 5. Roughening of an initially flat surface (dashed line) and smoothing of an initially rough surface (continuous line) by cluster deposition.

Now, the roughness of a surface diminishes in time according to the formula:

$$R^2(t) = \int_{\pi/L}^{k_0} dk \frac{L^2}{2\pi} k e^{-2\nu k^2 t} \widetilde{C}(k,0). \tag{14}$$

Thus, a beam of energetic noble gas clusters should be an efficient method of smoothing surfaces without depositing further material.

References

1. A.L. Barabasi and H.E. Stanley, *Fractal Concepts in Surface Growth* (Cambridge University Press, Cambridge, 1995).
2. *Atomic and Ionic Collisions in Solids and at Surfaces*, ed. R. Smith (Cambridge University Press, Cambridge, 1997).
3. H. Haberland et al., *J. Vac. Sci. and Technol.* **A12**, 2925 (1994).
4. C. Cleveland and U. Landman, *Science* **257**, 355 (1992).
5. H. Haberland et al., *Phys. Rev.* **B51**, 11061 (1995).
6. M. Moseler et al., *Phys. Rev.* **B56**, 15439 (1997).
7. M. Moseler et al., *Comp. Mat. Sci.* **10**, 452 (1998).
8. D.J. Oh and R.A. Johnson, *J. Mater. Res.* **3**, 471 (1988).

9. S.F. Edwards and D.R. Wilkinson, *Proc. Roy. Soc. Lond.* **A381**, 17 (1982).
10. J. Villain, *J. Phys. I* **1**, 19 (1991).

REFLECTION HIGH-ENERGY ELECTRON DIFFRACTION AND ^{119}SN MÖSSBAUER INVESTIGATION OF EPITAXIAL α-SN FILMS

B. ROLDAN CUENYA, M. DOI*, O. MARKS AND W. KEUNE
Laboratorium für Angewandte Physik, Gerhard-Mercator Universität, D-47048 Duisburg, Germany
E-mail: roldan@uni-duisburg.de

K. MIBU
Institute for Chemical Research, Kyoto University, Uji, Kyoto-fu 611, Japan
E-mail: mibu@scl.kyoto-u.ac.jp

Structure and growth of epitaxial Sn films on InSb(001), Cu(001) and fcc-Fe/Cu(001) substrates were investigated by reflection high-energy electron diffraction (RHEED) and ^{119}Sn Mössbauer spectroscopy. The Sn films grow epitaxially in the α-Sn(001) phase up to 1100 Å thickness on InSb, up to 5.5 ML on Cu(001), and up to 2.7 ML on fcc-Fe/Cu(001). Various surface reconstructions as a function of Sn coverage have been observed. The in-plane lattice parameter of the α-Sn overlayer was studied as a function of coverage.

1 Introduction

Gray tin (α-Sn, a=6.489 Å) is a non-polar semiconductor with diamond structure and a band gap nearly equal to zero (0.08 eV at 300 K).[1] Although α-Sn is unstable above 12.3°C (transformation α-Sn $\rightarrow$ tetragonal metallic β-Sn, a=5.83 Å, c=3.18 Å) it can be stabilized by heteroepitaxy at room temperature (RT) on an appropriate substrate. It is well known that α-Sn epitaxy is feasible on InSb because of the close lattice matching.[1–6] InSb (a=6.4798Å) is a polar semiconductor with a wider band gap (0.17 eV at 300 K). Compared to InSb, α-Sn has a very high electron mobility.[7] The existence of a finite thickness-dependent band gap [8] (by confinement of carriers in a one-dimensional square-well potential) in α-Sn films and formation of a two-dimensional electron gas (2DEG)[3] has been reported in the literature. These results suggest that α-Sn films have a potential for microwave, infrared and other device applications.[9] For instance, a heterostructure of InSb/α-Sn/InSb, where the electrons in the α-Sn layer could be two dimensionally confined, might have potential for infrared laser emission.[1] CdTe is another semiconducting substrate for heteroepitaxial growth of metastable α-Sn.[2,10–12]

To the best of our knowledge, there are no literature reports on epitaxial growth of α-Sn films on metallic substrates. This motivated us to attempt growth of α-Sn on clean and fcc-Fe coated Cu(001) substrates (fcc-Fe: a=3.59 Å; Cu: a=3.615 Å), although the lattice misfit with α-Sn is rather large (9.6 % and 10.2 % for fcc-Fe and Cu, respectively).

Moreover, the ^{119}Sn Mössbauer effect is a useful method to distinguish α-Sn and β-Sn via the isomer shift (chemical shift).[13, 14] Therefore, ^{119}Sn Mössbauer spectroscopy was applied to characterize epitaxial α-Sn layers on InSb(001).

2 Experimental procedure

We used an MBE system (molecular beam epitaxy) equipped with a RHEED system to prepare epitaxial Sn films with a maximum thickness of 1100 Å on InSb(001), 8 Å on Cu(001) and 20 Å on fcc-Fe/Cu(001). We used a commercial InSb(001) wafer and a Cu(001) single crystal with a polished surface. Just before being loaded into the chamber, the substrates were rinsed in ethanol. Auger electron spectra (AES) of the InSb and Cu substrate surfaces prior to in situ cleaning showed considerable C and O contamination. The substrates were cleaned by sputtering with low energy argon ion bombardment (500 eV) at a current density of 1 $\mu A/cm^2$ (P=5.5×10^{-5} mbar) and at a temperature of 200°C, 1 hour for InSb, and 400°C, 45 min for Cu; annealing followed at 350°C (InSb) and 600 °C (Cu) for about 3 hours. After this treatment, no surface impurities were detected by AES, and RHEED and LEED images of the substrates revealed flat surfaces, with reconstructions (4x2) for InSb and (1x1) for Cu.

We deposited Sn on the different substrates with a deposition pressure of 5.0×10^{-10} mbar (base pressure 6.0×10^{-11} mbar) nominally at room temperature (RT) and with a deposition rate between 0.025-0.075 Å/s for Sn and 0.05 Å/s for Fe. The real substrate temperature during deposition was somewhat higher than RT, since we measured 40 °C at the sample holder. The evaporation rate and film thickness were monitored with a calibrated quartz-crystal microbalance and were estimated from the bulk density of β-Sn.

Surface structures of the α-Sn films grown on the different substrates have been studied by RHEED along different azimuthal directions. RHEED during growth was performed with an angle of incidence of $\sim$ 3 ° and an electron energy of 12 keV (40 μA). The film thickness was also studied by RHEED intensity oscillations. Since the varying surface reconstructions of the growing Sn/InSb(001), Sn/Cu(001) and Sn/fcc-Fe(001)/Cu(001) films lead to irregular intensity oscillations of the specularly reflected RHEED beam, which conceal the real Sn film growth, we measured the intensity oscillations of the diffusively scattered electrons, located along the (0,0) rod of the specular beam position between the specular beam and the shadow edge, for monitoring the growth mode. The separation of the RHEED spots in reciprocal space is a measure for the in-plane lattice parameter in real space. The in-plane lattice spacing was determined from the measured distance between (1,1) and (-1,-1) RHEED reflections along the $[100]_{fcc}$ azimuth and between the (1,0) and (-1,0) reflections for the $[110]_{fcc}$ azimuth. The RHEED patterns appearing on the phosphorous screen were recorded during growth every few seconds by a CCD camera connected to a computer for data storage and later analysis. This method allows the real-time monitoring of film growth and in-plane lattice spacing.

By least-squares fitting the measured intensity profile of the RHEED spots with a single Lorentzian line, the position of the RHEED reflection was obtained with high precision. Consequently, the in-plane lattice spacing during deposition of Sn was determined with an uncertainty of only 0.005 Å. The RHEED probing depth, over which the lattice spacing is averaged, is about 1 to 2 ML.

3 Results and discussion

3.1 α-Sn/InSb(001)

Fig.1 shows typical RHEED patterns of the clean and Sn covered InSb(001) substrate along the [110] azimuth. For the clean InSb(001) surface (Fig.1(a)) streaks of the 4-fold surface reconstruction are observed, in addition to the fundamental reflections. The former are typical for a flat, well ordered InSb(001) surface with a (4x2) surface reconstruction.[2] The intensity profile measured along the horizontal direction in k-space (Fig 1(a), right-hand side) also demonstrates the 4-fold reconstructions. All the intensity profiles were measured during growth at the same location along the streaked spots in reciprocal space, where negligible contribution from Kikuchi lines exists. After deposition of 0.8 monolayers (ML) of Sn, the 4-fold surface reconstructions disappear and a (1x1) pattern is observed (Fig 1(b)). This is supported by the corresponding intensity profile, which shows only the fundamental reflections. At a Sn coverage of 2 ML (Fig 1(c)), 2-fold surface reconstructions appear which are typical for a clean, well-ordered α-Sn(001) surface with (2x2) reconstructions.[2] We followed the (2x2) pattern up to 9 ML Sn coverage.

In Fig 2(a) we observe RHEED-intensity oscillations. The minima correspond approximately to completely closed layers. The distance between neighboring minima is equivalent to 1 ML α-Sn coverage (1ML=a/4 for α-

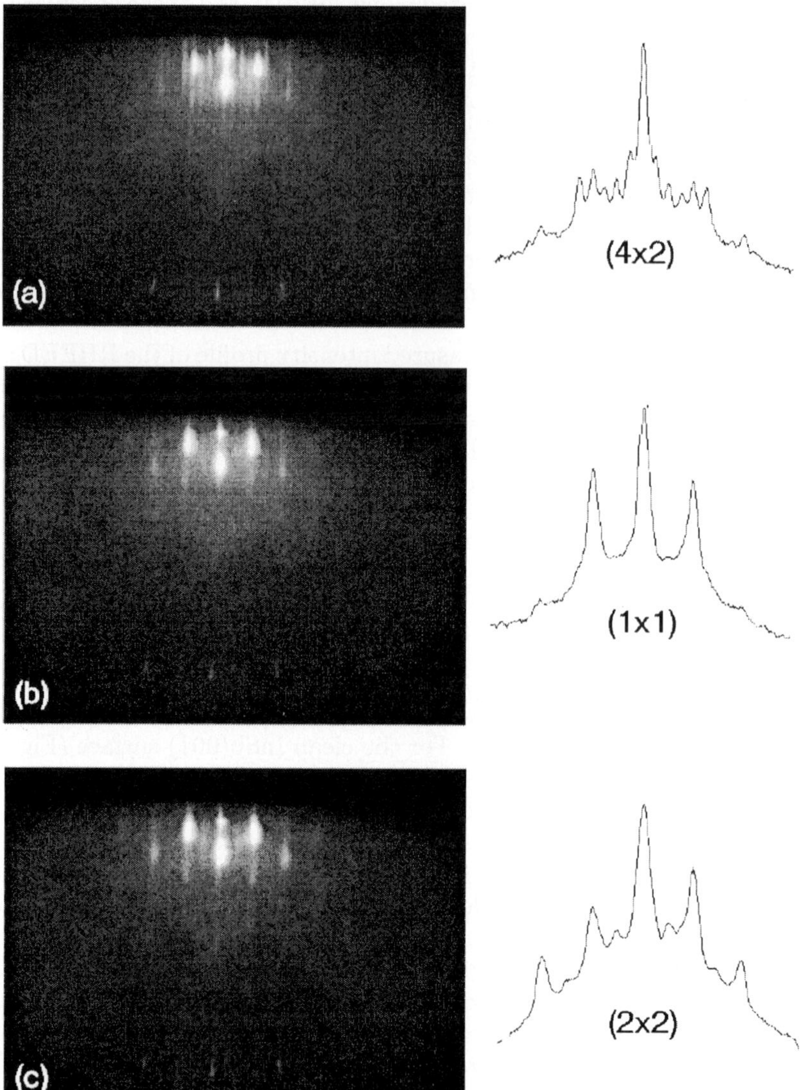

Figure 1. RHEED patterns of (a) clean InSb(001) substrate; covered by (b) 0.8 ML α-Sn and (c) 2 ML α-Sn ($[110]_{fcc}$ azimuth).

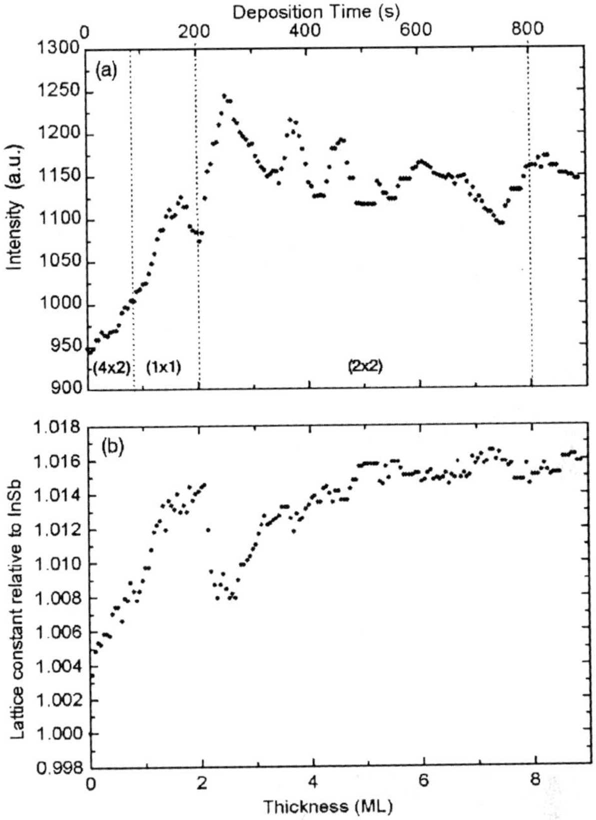

Figure 2. (a) RHEED intensity near the (0,0) streak and (b) in-plane lattice parameter of α-Sn (relative to that of the InSb(001) surface), both as a function of α-Sn coverage ($[110]_{fcc}$ azimuth).

Sn), as indicated by the calibrated quartz-crystal oscillator. The thickness scale in Fig 2 was obtained from the average distance between minima. We conclude that for a certain thickness range α-Sn grows nearly layer-by-layer. Also the RHEED patterns observed up to 8 ML indicated 2D growth.

The in-plane lattice parameter of the α-Sn film shows non-monotonous behavior as a function of coverage (Fig 2(b)). Irregularities of the lattice-parameter-versus-coverage behavior are observed at coverages where the type of surface reconstructions changes. Above 5 ML coverage the in-plane lattice spacing remains 1.5 % larger than that of the InSb substrate surface, and is even 1.4 % larger than that of bulk α-Sn.

In order to provide an independent proof of the α-Sn-film structure, we show a ^{119}Sn Mössbauer spectrum (conversion electron Mössbauer spectrum, CEMS) of an 1100 Å thick α-Sn film on InSb (Fig 3(a)). ^{119}Sn in a $CaSnO_3$ matrix was used as the Mössbauer source. The typical single line of single-phase α-Sn is observed. For comparison, we also measured the ^{119}Sn Mössbauer spectrum in transmission for a bulk β-Sn foil (Fig 3(b)) and for bulk $CaSnO_3$ as a reference (Fig 3(c)). All spectra were fitted with a Lorentzian single line.

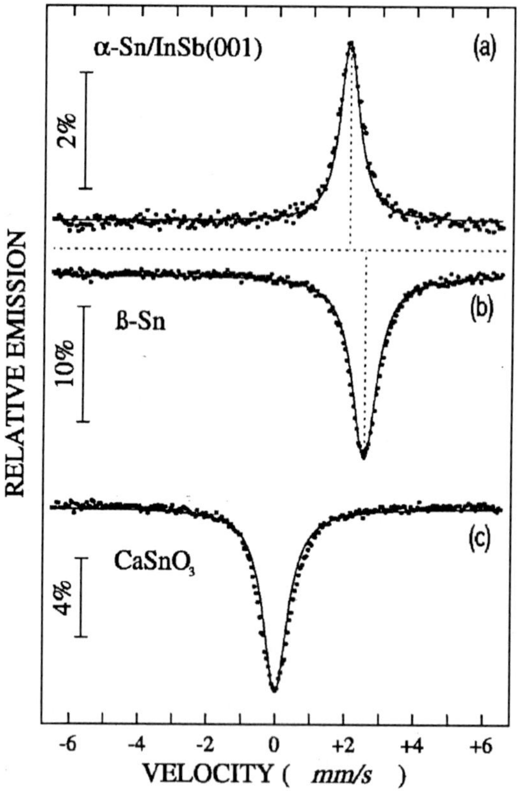

Figure 3. Mössbauer spectra at RT of (a) 1100 Å α-Sn on InSb(001) measured by CEMS; (b) β-Sn foil (c) $CaSnO_3$ powder (reference). (b) and (c) measured in transmission.

The Mössbauer parameters obtained are given in Table 1, together with literature values for bulk α-Sn and β-Sn.[13, 14] The isomer shift (chemical shift) of our α-Sn film is in very good agreement with that of the literature for α-Sn, and is clearly different from that of β-Sn.

Table 1. Mössbauer parameters of α-Sn/InSb and β-Sn in comparison to the corresponding values from the literature.

sample	isomer shift δ (mm/s) (T=RT)	linewidth Γ (mm/s) (T=RT)
α-Sn film on InSb, 1100 Å	$+ 2.049 \pm 0.004$	$+0.680 \pm 0.008$
α-Sn (bulk)[13, 14]	$+ 2.03 \pm 0.02$	
β-Sn (foil)	$+ 2.520 \pm 0.002$	$+0.854 \pm 0.004$
β-Sn (bulk)[13, 14]	$+ 2.56 \pm 0.01$	

3.2 α-Sn/Cu(001)

Fig.4 displays RHEED patterns of the clean and Sn covered Cu(001) substrate along the [110] azimuth. Fig.4(a) shows the typical p(1x1) pattern of a clean Cu(001) surface directly after the cleaning and annealing procedure. After the deposition of 1 ML Sn (Fig 4(b)), the streaks of a (2x2) surface reconstruction in addition to the fundamental reflections are visible along the [110] azimuth and also along [100] (not shown). At a coverage of 1.3 ML (Fig 4(c)), the 2-fold surface reconstruction disappears, and a (3x6) pattern is observed. After deposition of 1.7 ML Sn, a 4-fold surface reconstruction appears (c(4x4)),(Fig 4(d)) and a (4x4) pattern is retained at least up to 2.7 ML. Above 2.7 ML the intensity of the RHEED-reflections becomes very weak and the pattern disappears.

In Fig.5 (a), there is a RHEED intensity oscillation with a minimum near 1 ML coverage, indicating a smooth surface of the first monolayer. As expected, the surface is rough at 1.5 ML coverage. Above 1.5 ML coverage the intensity drops rapidly with increasing coverage, since the RHEED pattern generally becomes faint.

The in-plane lattice parameter (Fig.5(b)) remains constant and similar to the Cu bulk value. Half of the lattice parameter of bulk α-Sn is 3.244 Å, which is closer to that of Cu (3.615 Å) than half of the lattice parameter of β-Sn (2.915 Å). Therefore it is reasonable to assume the presence of the α-Sn phase in our films on Cu(001).

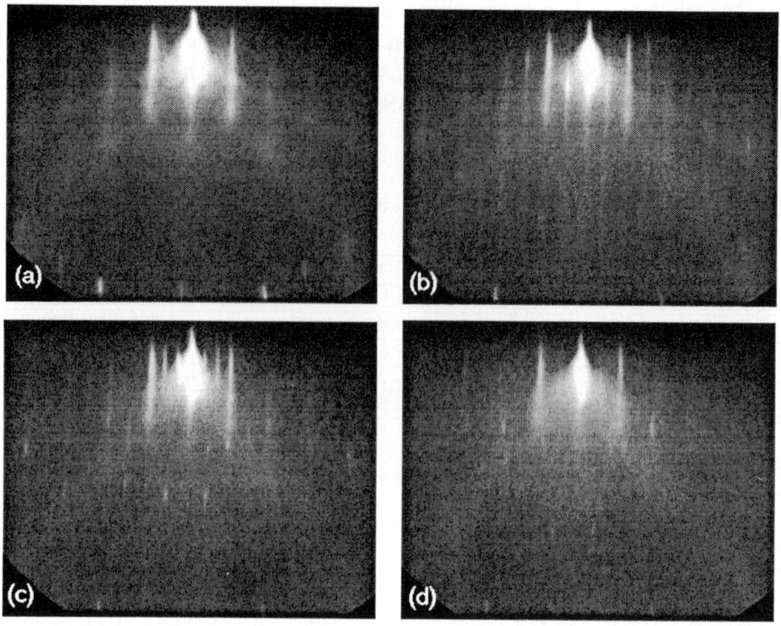

Figure 4. RHEED patterns along the $[110]_{fcc}$ azimuth of (a) clean Cu(001) covered by (b) 1 ML Sn, (c) 1.3 ML Sn, and (d) 1.7 ML Sn.

3.3 Sn/fcc-Fe/Cu(001)

Fcc (γ-)Fe is a well-known metastable system, that can be stabilized at room temperature up to a thickness of $\sim$ 10 ML by epitaxial growth.[15, 16] This is possible because of the similarity of the lattice parameters of Cu (3.615 Å) and γ-Fe (3.588 Å)).

By means of RHEED we studied the metastable growth of α-Sn on an 8 ML Fe film epitaxially grown on a Cu(001) substrate.

Fig.6(a) shows the typical (1x1) RHEED pattern of an fcc-Fe surface along the $[100]_{fcc}$ azimuth. After deposition of 1 ML Sn (Fig.6(b)), a 2-fold surface reconstruction appears, and (according to RHEED along the other azimuth $[110]_{fcc}$) a c(2x2) pattern is observed. This is typical for 1 ML of α-Sn(001)(compare with Fig. 4(b)). At and above 2.3 ML, the streaks of the 2-fold surface reconstruction disappear (Fig.6(c)), and above that thickness the intensity of the RHEED (1,1) and (-1,-1) reflections becomes weak and gradually disappears. This demonstrates that epitaxial growth above 2.3 ML

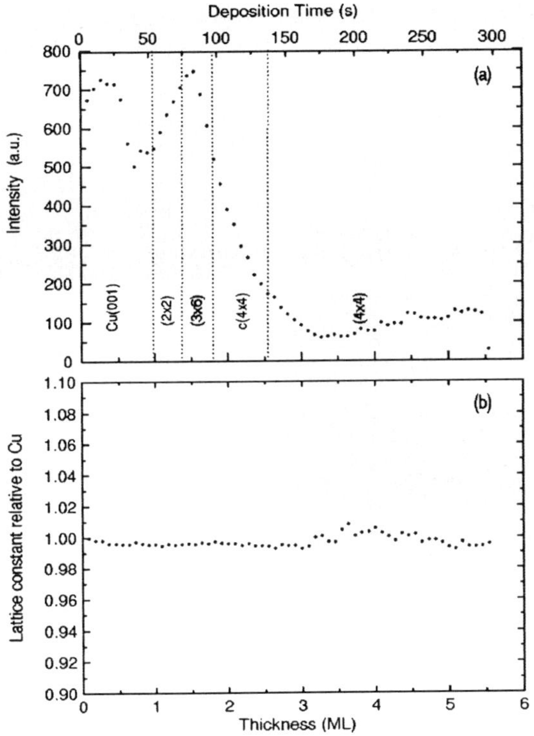

Figure 5. RHEED intensity near the (0,0) streak, and (b) in-plane lattice parameter of α-Sn (relative to that of the Cu(001) surface), both as a function of the α-Sn coverage ($[110]_{fcc}$ azimuth).

becomes increasingly worse.

Fig.7(a) displays RHEED intensity oscillations of diffusely scattered electrons near (0,0) with minima at 2, 3 and maybe 4 ML coverage, indicating a smooth surface of the 2nd, 3th and (maybe) 4th monolayer, and nearly layer-by-layer growth. Above 4 ML coverage the intensity gradually drops with increasing coverage, and neither layer-by-layer growth nor epitaxy is observed. This could be related to the α-Sn $\rightarrow$ β-Sn transition above 4 ML coverage.

Fig.7(b) shows that the in-plane α-Sn lattice parameter drops slowly with the coverage. Caused by the weak RHEED intensity, we only can calculate the lattice spacing for the first 3 ML. Similar to the case of α-Sn on Cu(001),

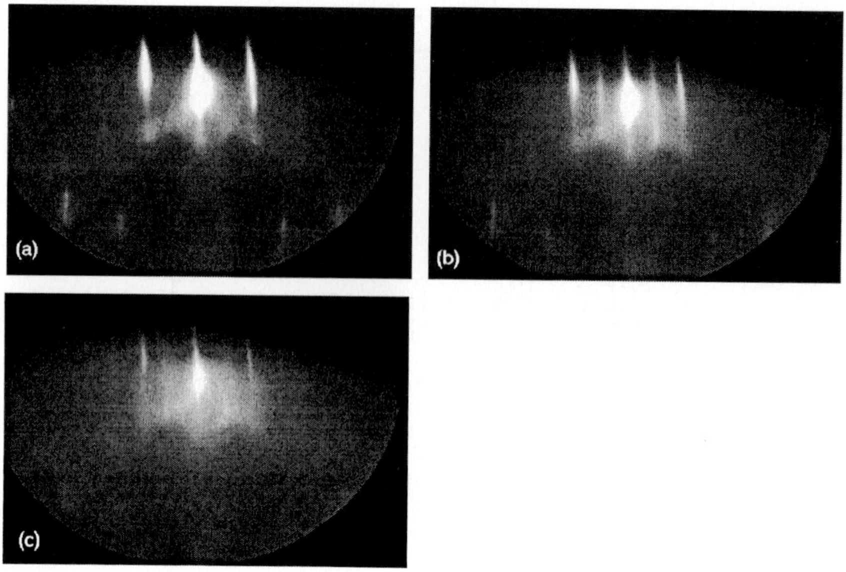

Figure 6. RHEED patterns along $[100]_{fcc}$ azimuth of a) 8 ML fcc-Fe/Cu(001) covered by (b) 1 ML Sn, (c) 2.3 ML Sn.

half of the lattice parameter of bulk α-Sn (3.244 Å) is closer to that of γ-Fe (3.588 Å) than half of the lattice parameter of β-Sn (2.915 Å). This is the reason why we can assume the presence of the α-Sn phase in our films on fcc-Fe/Cu(001) below 4 ML.

4 Conclusions

In this work epitaxial growth of Sn films on semiconducting InSb(001) and metallic Cu(001) and fcc-Fe(001)/Cu(001) substrates was studied mainly by RHEED. In accordance with the literature, we observed growth of α-Sn(001) films on InSb(001) up to 1100 Å thickness. As an independent proof, ^{119}Sn Mössbauer spectroscopy (CEMS) demonstrated the growth of the pure α-Sn phase for 1100 Å Sn on InSb. Epitaxial α-Sn films can be grown up to 5.5 ML on Cu(001) and up to 2.7 ML on fcc-Fe/Cu(001). Various surface reconstructions as a function of Sn coverage have been found in the RHEED pattern. With increasing Sn coverage, (1x1) followed by (2x2) reconstructions were observed on InSb(001), (2x2) followed by (3x6), c(4x4) and (4x4) on

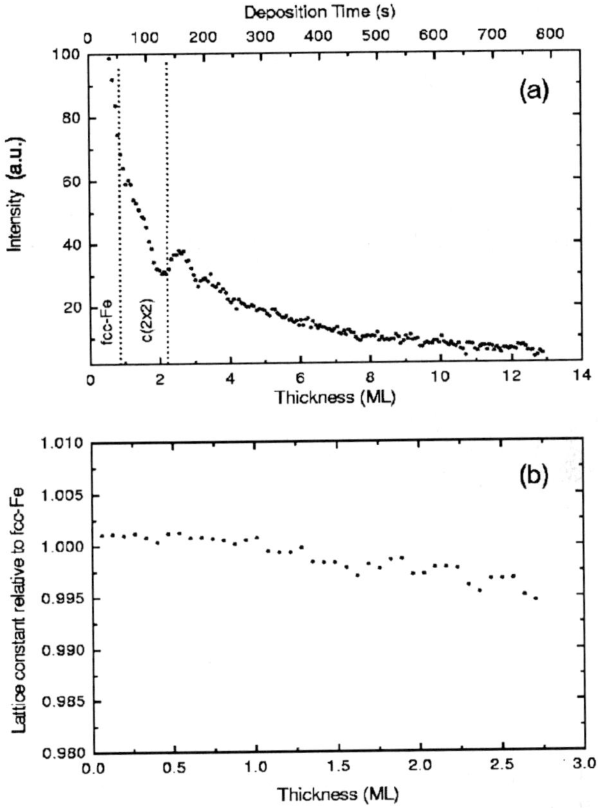

Figure 7. (a)RHEED intensity near the (0,0) streak, and (b) in-plane lattice parameter of α-Sn (relative to that of the fcc-Fe(001) surface), both as a function of the α-Sn coverage ($[100]_{fcc}$ azimuth).

Cu(001), and c(2x2) on fcc-Fe/Cu(001). The in-plane lattice parameter of α-Sn(001) above 5 ML coverage on InSb was found to be 1.5% enhanced relative to that of the substrate. By contrast, the in-plane lattice parameter of α-Sn on Cu(001) was observed to be equal to that of the Cu(001) surface up to 5.5 ML coverage. In the case of fcc-Fe/Cu(001), the in-plane lattice parameter of α-Sn decreases by 0.5% at 2.7 ML coverage as compared to that of fcc-Fe/Cu(001).

Acknowledgements

We are grateful for the valuable technical assistence by U. von Hörsten. This work was supported by the Deutsche Forschungsgemeinschaft (GRK 277/1-97).

* permanent address: Department of Materials Science and Engineering, Nagoya University, Furo-cho, Chikusa-ku Nagoya, 464-01, Japan

E-mail: doi@numse.nagoya-u.ac.jp

References

1. H. Omi, H. Saito, T. Osaka, *Phys. Rev. Lett.* **72**, 2596 (1994).
2. R. F. C. Farrow, D. S. Robertson, G. M. Williams, A. G. Cullis, G. R. Jones, I. M. Young, P. N. J. Dennis, *J. Crystal Growth* **54**, 507 (1981).
3. W. T. Yuen, W. K. Liu, S. N. Holmes, R. A. Stradling, *Semicond. Sci. Technol.* **4**, 819 (1989).
4. W. T. Yuen, W. K. Liu, B. A. Joyce, R. A. Stradling, *Semicond. Sci. Technol.* **5**, 373 (1990).
5. W. T. Yuen, W. K. Liu, R. A. Stradling, B. A. Joyce, *J. Crystal Growth* **111**, 943 (1991).
6. B. F. Mason, B. R. Williams, *Surf. Sci.* **262**, 169 (1992).
7. R. Dingle, *Festkörperprobleme XV* (Friedr. Vieweg u. Sohn Verlagsgesellschaft mbH, Braunschweig, 1975), p. 21.
8. L. W. Tu, G. K. Wong, J. B. Ketterson, *Appl. Phys. Lett.* **55**, 1327 (1989).
9. T. Osaka, H. Omi, K. Yamamoto, A. Ohtake, *Phys. Rev.* B **50**, 7567 (1994).
10. L. W. Tu, G. K. Wong, J. B. Ketterson, *Appl. Phys. Lett.* **54**, 1010 (1989).
11. L. W. Tu, G. K. Wong, S. N. Song, Z. Zhao, J. B. Ketterson, *Appl. Phys. Lett.* **55**, 2643 (1989).
12. M. Kimata, T. Suzuki, K. Saino, K. Kawamura, A. Hobbs, *J. Vac. Sci. Technol.* B **12**, 1184 (1994).
13. A. Svane, N. E. Christensen, C. O. Rodriguez, M. Methfessel, *Phys. Rev.* B **55**, 12572 (1997).
14. *Mössbauer Effect Data Index (1974)*, edited by J. G. Stevens and V. E. Stevens (IFI/Plenum, New York, 1975), p. 153.
15. W. A. Jesser, J. W. Matthews, *Philos. Mag.* **15**, 1097 (1967).
16. W. A. Jesser, J. W. Matthews, *Philos. Mag.* **17**, 595 (1968).

KINETICS AND DYNAMICS OF HETEROGENEOUS MAGNETIC STRUCTURES

THERMAL ACTIVATION AND HIGH FREQUENCY MAGNETISATION PROCESSES IN MAGNETIC MEDIA

J. D. HANNAY AND R. W. CHANTRELL

Computational Magnetism Group, School of Electronic Engineering and Computer Systems, University of Wales Bangor, Dean Street, Bangor, LL57 1UT, UK

E-mail: hannay@sees.bangor.ac.uk, roy@sees.bangor.ac.uk

A computational model of thermally activated magnetisation reversal in magnetic materials is described. The dynamic magnetic properties of the material are simulated using the Langevin formalism which is essentially the Landau-Lifshitz-Gilbert equation of motion augmented by a random thermal field whose properties are determined by the fluctuation-dissipation theorem. The model can include anisotropic crystalline, magnetostatic (dipole) and exchange interactions. The simulation is used to calculate pulsed-field magnetisation and remanence curves which are obtained by applying a field pulse to a system in the remanent magnetisation state. These simulations show an increased non-local damping of the system when interactions are included which potentially explains the difference in effective damping constants between ferromagnetic resonance (FMR) and magnetisation reversal experiments. Application of the Langevin equation leads to a magnetic response in the form of correlated magnetisation fluctuations (spin waves).

1 Introduction

Due to modern demands, the operation frequencies and storage densities of today's recording disk media are rapidly increasing. These materials are comprised of fine, magnetically well decoupled grains with an irregular lattice thin-film structure in order to provide sufficient signal-to-noise ratio (SNR) [1,2]. However, the reduced grain size will increase the effect of thermal fluctuations upon the magnetic state of the system, therefore any realistic model must include finite temperature effects.

High speed switching in longitudinal thin-film media is currently a matter of considerable importance, because of the potential limitations on data rates that this phenomenon provides. It is imperative, for our rapidly growing system requirements, that we characterise and understand the mechanisms of the precessional response of the magnetisation, which governs the upper limits of switching speed. Therefore, to conduct meaningful studies of thin-films we utilise a dynamical approach which is able to model thermally activated high-speed magnetisation reversal using the stochastic Langevin equation of motion. Our model assumes a randomly placed set of grains with a log-normal particle size distribution and also an in plane random distribution of easy axes which has been shown to give a more realistic underlying physical structure [3]. The effect of this underlying randomness is generally to give rise to a reduction in correlation lengths. The model includes short-range intergranular exchange coupling as well as long-

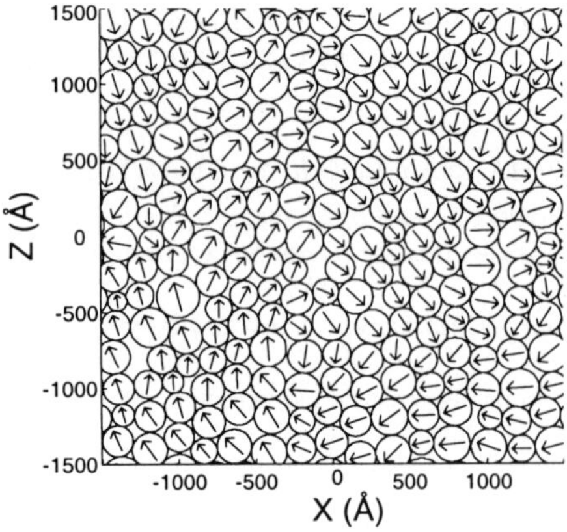

Figure 1. A sample of strongly exchange coupled thin-film material with an irregular structure in its coercive state.

range magnetostatic (dipole) interactions to allow investigation of interaction effects on the high-speed reversal process. We apply the model to simulate pulsed-field magnetisation studies, which are currently the main experimental method of high-speed switching investigation.

2 Computational Model

The thin film medium is simulated by an irregular lattice structure of spherical grains as shown in figure 1. The granular array uses periodic boundary conditions in the x and z directions thus eliminating any boundary effects, and has a finite spread into the out of plane y direction. The grains are assumed to be single domain and are represented by single spins obeying coherent reversal rotations.

The total magnetic energy of the ith grain includes the Zeeman energy from the applied field, uniaxial crystalline anisotropy energy, inter-granular exchange coupling energy and the magnetostatic energy. Let the magnetic moment of grain i be $\boldsymbol{\mu}_i = \mathbf{M}_i V_i$, where $\mathbf{M}_i$ and V_i are the magnetisation vector and the volume of the ith grain respectively. The total energy for grain i is then given by

$$E_{tot,i} = -\boldsymbol{\mu}_i \cdot \mathbf{H}_{app} + KV_i \sin^2\theta_i - \frac{2A^*}{\overline{V}\,\overline{D}^2 M_s^2} \cdot \sum_{\{j \neq i : S_{ij} < R_{ex}\}} \frac{\boldsymbol{\mu}_i \cdot \boldsymbol{\mu}_j}{d_{ij}^2} - \sum_{\{j \neq i : r_{ij} < R_{mag}\}} \frac{3(\boldsymbol{\mu}_i \cdot \mathbf{r}_{ij})(\boldsymbol{\mu}_j \cdot \mathbf{r}_{ij})}{r_{ij}^5} - \frac{\boldsymbol{\mu}_i \cdot \boldsymbol{\mu}_j}{r_{ij}^3} \tag{1}$$

where K is the uniaxial anisotropy energy constant, θ_i is the angle between the easy axis orientation and the magnetisation direction, A^* is the effective exchange energy constant, M_s the saturation magnetisation and $\mathbf{r}_{ij}$ is the vector which connects the centres of grains i and j. $\overline{V}$ is the mean grain volume of the system, $\overline{D}$ the mean grain diameter, d_{ij} is the centre to centre distance between grains i and j divided by the mean grain diameter and S_{ij} is the distance between the surfaces of grains i and j.

The exchange energy is analogous to that from spin-spin exchange and here we introduce a phenomenological parameter R_{ex} which specifies that only grains within this value are included in the exchange energy calculation. In this study, R_{ex} has a value of approximately $0.06\,\overline{D}$.

The magnetostatic interaction energy for one grain i is approximated as a sum of dipolar contributions from other grains within a centre to centre cut off radius R_{mag} which extends to $8\,\overline{D}$.

The reduced effective field acting on the ith grain is given by

$$\mathbf{h}_{eff,i} = -\frac{1}{H_K} \cdot \frac{\partial E_{tot,i}}{\partial \mu_i} \tag{2}$$

where $H_K = 2K/M_s$ is the magnetocrystalline anisotropy field. If we combine equations (1) and (2) we have

$$\begin{aligned} \mathbf{h}_{eff,i} &= \mathbf{h}_{app} + (\hat{\mathbf{m}}_i \cdot \hat{\mathbf{e}}_i)\hat{\mathbf{e}}_i + C^* \sum_{\{j \neq i: S_{ij} < R_{ex}\}} \frac{v_j \hat{\mathbf{m}}_j}{d_{ij}^2} \\ &+ \frac{M_s^2 \overline{V}}{2K} \sum_{\{j \neq i: r_{ij} < R_{mag}\}} \frac{v_j}{r_{ij}^3} [3(\hat{\mathbf{m}}_j \cdot \hat{\mathbf{r}}_{ij})\hat{\mathbf{r}}_{ij} - \hat{\mathbf{m}}_j] \end{aligned} \tag{3}$$

where $\mathbf{h}_{app}$ is the reduced externally applied field vector, $\hat{\mathbf{m}}_i = \mathbf{M}_i / M_s$ is the magnetisation unit vector of grain i, $\hat{\mathbf{e}}_i$ is the easy axis unit vector of grain i, v_j is the reduced volume of grain j normalised with respect to the mean grain volume and $C^* = A^* / (\overline{D}^2 K)$ is the reduced effective exchange coupling parameter.

The time evolution for the magnetisation vector of each grain is governed by the phenomenological stochastic Gilbert equation [4]. This particular form for the equation of motion is used in preference to that of Landau and Lifshitz [5] because it can be directly derived from a Lagrangian function and a Rayleigh dissipation function [4]. However, the Gilbert equation is usually cast in the Landau-Lifshitz form to aid simplicity, resulting in the so called stochastic Landau-Lifshitz-Gilbert (LLG) equation which, by using a reduced time variable $\tau = t|\gamma|H_k/(1+\alpha^2)$, can be written as

$$\frac{d\hat{\mathbf{m}}_i}{d\tau} = -\hat{\mathbf{m}}_i \times \left(\mathbf{h}_{eff,i} + \mathbf{h}_{th,i}(\tau)\right) - \alpha\, \hat{\mathbf{m}}_i \times \left[\hat{\mathbf{m}}_i \times \left(\mathbf{h}_{eff,i} + \mathbf{h}_{th,i}(\tau)\right)\right] \tag{4}$$

where the magneto-mechanical or gyromagnetic ratio $\gamma = -1.76 \times 10^7$ rad Oe^{-1}s^{-1} and α represents the dissipation or damping constant.

Equation (4) is the Langevin equation of this particular system because the reduced effective field that is acting on each grain i has been augmented by a fluctuation field $\mathbf{h}_{th,i}(\tau)$. This stochastic field accounts for the interaction between the simple magnetic spin and the many degrees of freedom associated with the surrounding complex dissipative system which causes random fluctuations of the magnetisation. The resultant relationship between cause and effect is expressed in the fluctuation-dissipation theorem [6]. In order to define the statistical properties of $\mathbf{h}_{th,i}(\tau)$, assumptions following those made in Brownian motion [7] (also known as the Wiener process) theory are made [8]. Therefore we assume, by virtue of the 'Central Limit Theorem', that the random field has the form of Gaussian white noise and has the properties

$$\mathbf{h}_{th,i}(\tau) = \hat{\mathbf{x}}\, h^1{}_{th,i}(\tau) + \hat{\mathbf{y}}\, h^2{}_{th,i}(\tau) + \hat{\mathbf{z}}\, h^3{}_{th,i}(\tau) \tag{5}$$

$$\left\langle h^x{}_{th,i}(\tau) \right\rangle = 0 \tag{6}$$

$$\left\langle h^x{}_{th,i}(\tau)\, h^y{}_{th,i}(\tau') \right\rangle = \frac{k_B T \alpha}{K V_i (1+\alpha^2) \Delta\tau} \cdot \delta_{xy}\, \delta(\tau - \tau') \tag{7}$$

where k_B is Boltzmann's constant, T the temperature, $\Delta\tau$ is the reduced time integration step, δ_{xy}. is the Kronecker delta and $\delta(\tau-\tau')$ is the Dirac delta function. Equation (6) states that the ensemble mean of $\mathbf{h}_{th,i}(\tau)$ is zero. Equation (7) states that different components of $\mathbf{h}_{th,i}(\tau)$ are uncorrelated and identical components are uncorrelated for $\tau' \neq \tau$. Finally, it is also assumed that the stochastic random field is different for each grain i.

3 Studies of Magnetic Thin-Films

The intrinsic material parameters used for this study, unless otherwise stated are as follows: K = 4.2×10^6 erg/cm^3, M_s = 1400 emu/cm^3, C^* is in the range of 0.0 - 0.4 and the average grain diameter is 200Å with all simulations done at a temperature of 300 °K.

We first consider the behaviour of a model Stoner and Wohlfarth (SW) system consisting of isolated grains without magnetostatic or exchange coupling in order to demonstrate a precessional effect which is sensitive to the damping parameter α. The remanent magnetisation of the system without interactions for short field pulses below 0.5 ns is simulated. Each magnetic moment is initially oriented parallel to the positive z-direction and then allowed to reach its equilibrium orientation in the absence of an external field, before applying a pulsed field of specified duration. Figure 2 shows some rather interesting results for different values of the damping constant in a reversing field equal to H_K. If we firstly study the curve for α = 0.05 we observe some unusual peaks and troughs in the remanence. We can see that more moments have switched direction for a pulse width slightly less than 0.05 ns

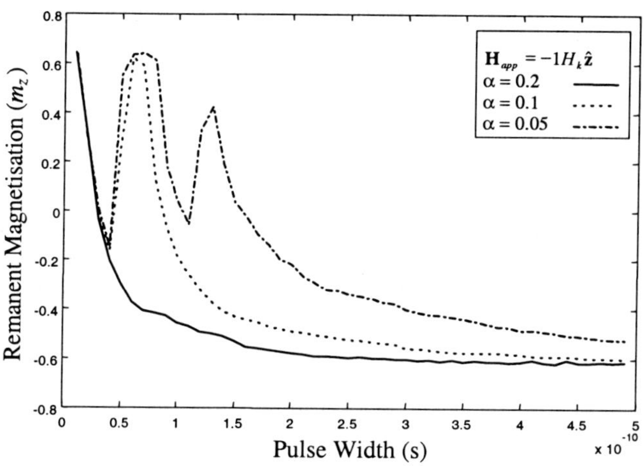

Figure 2. Remanent magnetisation against pulse width for different damping values with an applied field of $-1H_K$.

than for a longer pulse approximately 0.05 – 0.2 ns wide. This phenomenon is due to the pulse width possessing a similar value to the time period of the gyromagnetic precessional motion. Hence for a short pulse the precession can cross over the hard plane and if the field is removed before the precession has a chance to cross back over the hard plane again, it will switch into the opposite easy direction. However for a longer pulse the magnetisation will cross back over the hard plane and return to its original easy direction. If we increase the damping constant then we effectively reduce the time period of the precessional motion and therefore the magnetisation has time to react properly to the applied field. We will come back to this result when interpreting the dynamics of an interacting system, the results for which follow.

The model is now used with magnetostatic and exchange interactions included in the local field calculations of each grain and we investigate the effects of exchange coupling strength on the dynamic coercivity which is defined as the reverse field strength required for the dynamic magnetisation of the system to reach zero. The damping parameter $\alpha = 0.1$ is used for this study.

Figure 3 agrees with previous work [9] showing that the field required to produce a relatively small reduction in switching time dramatically increases for field pulses shorter than 0.1 ns. We also note that even at these very short time scales the rate of change of dynamic coercivity decreases as the inter-granular exchange coupling strength is reduced. It is reasonable to expect that this trend will expand into larger time scales which agrees qualitatively with experimental data [10].

We now present calculations of the remanence curves for the interacting system of grains. As before each grain starts oriented along the positive z-axis and is

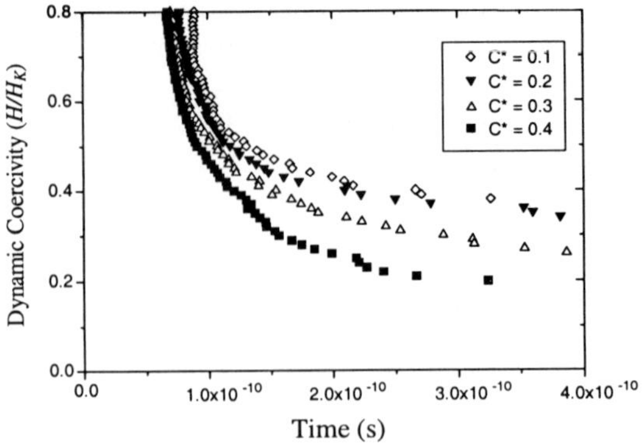

Figure 3. Dynamic coercivity versus switching time for samples with varying C* of 0.1 – 0.4.

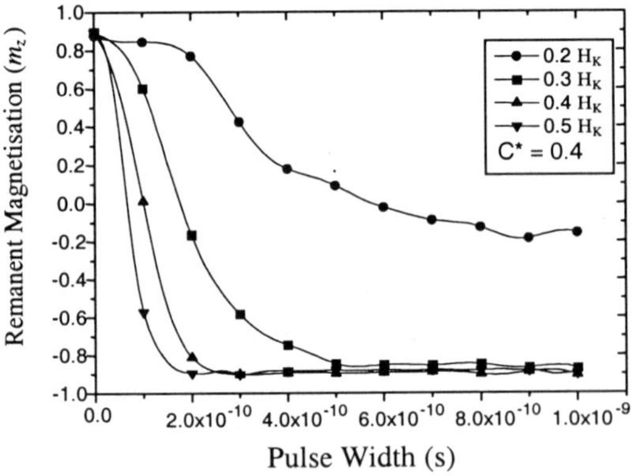

Figure 4. Remanent magnetisation curve versus pulse width for various applied fields with C* = 0.4.

allowed to reach its equilibrium position without applying an external field source Once equilibrium is reached we apply a reverse pulsed field to the system for a specified time and then let the magnetisation reach equilibrium again to obtain the remanent state after a specified pulse width.

Figure 4 shows the simulated remanence curves with different magnitudes of applied field for a highly exchange coupled system of grains. We note that over the timescale of <1.5ns there are no extrema such as are seen in figure 2 for the completely non-interacting system. A possible interpretation of this is in terms of an effective increase in the damping constant α since the effect of increased α in the completely non-interacting case is to damp out the extrema. Thus, interactions between grains would be considered as providing an extra damping mechanism. This of course has implications for the magnetisation reversal process at high frequencies, and potentially explains the difference in effective damping constants between FMR and magnetisation reversal experiments.

Doyle and co-workers have recently demonstrated such an effect for γ-Fe_2O_3 particulate recording media [11]. In this work, FMR experiments suggested $\alpha < 0.1$, however sensible switching times could only be achieved via the deterministic single spin Landau-Lifshitz-Gilbert formalism with critical damping ($\alpha = 1$). FMR measurements are made close to saturation and therefore would not be expected to exhibit complete damping effects which might be expected to be present only when non-uniform local magnetisation structures develop during reversal. These modes of reversal would contribute to the increased damping observed experimentally.

4 Correlated Magnetisation Fluctuations

The studies outlined so far are based on the direct numerical solution of the Langevin equation with random thermal perturbations. However, it might be expected that a coupled system of interacting spins would respond in the form of correlated magnetisation fluctuations, in an analogous manner to the formation of magnetisation ripple in response to random changes in the local anisotropy direction. Here we present a simple demonstration of magnetisation fluctuations in a strongly exchange coupled chain of spheres, with physical parameters identical to those outlined in section 3.

We demonstrate the existence of correlated magnetisation fluctuations which represent the collective response of a coupled system to the random thermal perturbations. The simulation is carried out starting with all spins aligned along the chain axis (zero temperature solution). We then introduce the thermal perturbations for a given temperature and allow the system to achieve thermal equilibrium using the Langevin equation approach. The simulation is then continued for a further period, during which the off-axis magnetisation components are recorded. We then perform a two-dimensional Fourier transform of this data in space and time which gives the characteristic frequencies associated with each mode. Typical results are shown in Figure 5 for three values of the wavenumber k.

The linear chain is a simple model system in which to investigate the magnetisation fluctuations in detail. In particular the simulated spin waves should be expected to exhibit the classical spin wave dispersion relation. This is achieved by plotting, for all values of k, the frequency of the peak amplitude against their associated wavenumber (Figure 6).

It can be seen that there is a spread of frequencies associated with each value of k with a definite peak. Also, the dispersion relation has the $\omega \propto 1 - \cos(ka)$ behaviour expected for exchange induced spin waves. This represents an unambiguous demonstration of thermally excited spin waves resulting from correlated magnetisation fluctuations.

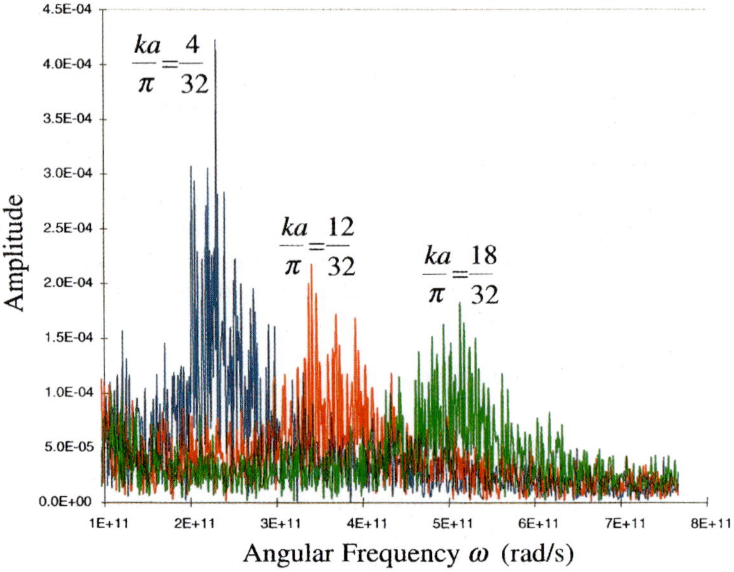

Figure 5. Frequency spectrum showing three excited spin wave modes for comparison.

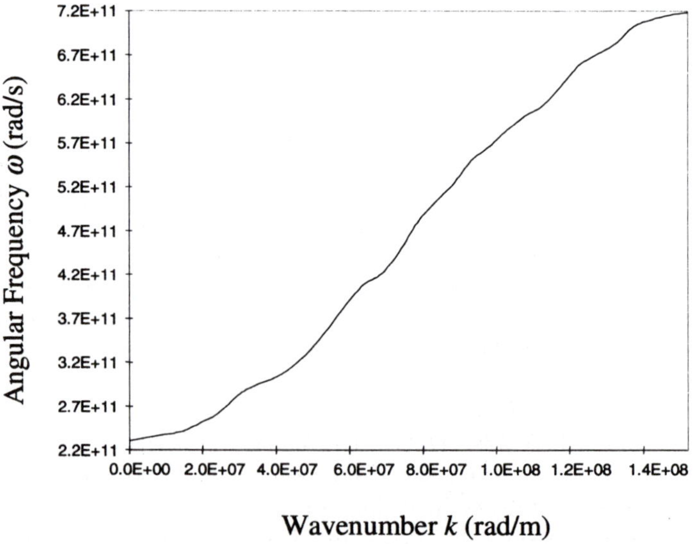

Figure 6. Dispersion relation for the simulated spin wave modes in a strongly exchange coupled chain.

5 Conclusions

A dynamic model of thermally activated high speed switching in thin-film media has been described. We have shown that in the absence of interactions there is a gyromagnetic precessional effect upon the switching mechanism. This phenomenon is very sensitive to the dissipation term and can be damped out. We have found that the inclusion of exchange coupling leads to more rapid changes of coercivity as write frequencies increase. The addition of magnetostatic and exchange interactions to the system speeds up the overall switching process and the precessional effects appear to be damped out. This suggests an interpretation in terms of an effective increase in the damping parameter of the system due to inter-granular exchange and magnetostatic coupling hence we obtain results which compare favourably with experiment using a relatively low value of α as FMR experiments suggest. Given the implications in terms of high speed switching it would appear worthwhile to extend the comparison of α via switching and resonance experiments to longitudinal media. We have also discussed some of the recent work which has been carried out in the simulation of thermally assisted magnetisation reversal, and its relation to correlated magnetisation fluctuations.

Acknowledgements

The financial support of the UK EPSRC is gratefully acknowledged. J. D. Hannay also wishes to thank Seagate (Northern Ireland) for their support of a CASE studentship.

References

1. T. Yogi, T. Ching, T. A. Nguyen, K. C. Ju, G. L. Gorman, and G. Castillo, *IEEE Trans. Magn.* **26**, 2271 (1990).
2. T. Yogi, T. A. Nguyen, S. E. Lambert, G. L. Gorman, and G. Castillo, *IEEE Trans. Magn.* **26**, 1578 (1990).
3. N. S. Walmsley, A. Hart, D. A. Parker, C. Dean, and R. W. Chantrell, *J. Magn. Magn. Mater.* **170**, 81 (1997).
4. T. L. Gilbert, *Phys. Rev.* **100**, 1243 (1955).
5. L. D. Landau and E. M. Lifshitz, *Physik. Z. Sowjetunion* **8**, 153 (1935).
6. H. B. Callen and T. A. Welton, *Phys. Rev.* **83**, 34 (1951).
7. M. C. Wang and G. E. Uhlenbeck, *Rev. Mod. Phys.* **17**, 323 (1945).
8. W. F. Brown, *Phys. Rev.* **130**, 1677 (1963).
9. R. W. Chantrell and J. D. Hannay, *J. Magn. Soc. Jpn.* **21**, 283 (1997).
10. H. J. Richter and R. Y. Ranjan, *J. Magn. Magn. Mater.* **193**, 213 (1999).
11. W. D. Doyle, S. M. Stinnett, C. Dawson, and L. He, *J. Magn. Soc. Jpn.* **22**, 91 (1998).

STOCHASTIC MAGNETIZATION DYNAMICS IN MAGNETIC NANOSTRUCTURES: FROM NÉEL-BROWN TO SOLITON-ANTISOLITON CREATION

HANS-BENJAMIN BRAUN

Paul Scherrer Institut, CH-5232 Villigen, Switzerland

1 Introduction

The understanding of magnetization dynamics in magnetic nanostructures has profound practical implications. Most of our information is stored magnetically. The ongoing quest for increasing magnetic storage density requires the use of ever decreasing volume elements for the storage of a single bit. Since the energy barrier preventing the magnetization from reversal is at best proportional to the particle volume, thermal decay puts the ultimate limit on the maximally attainable storage density.

The successful use of magnetic storage near this "super-paramagnetic limit" therefore requires a detailed understanding of the reversal mechanism in magnetic nanostructures. The first leap in our understanding of super-paramagnetism has been made by Néel[1] and Brown[2] who computed the switching probability if the magnetization in a small particle is subject of thermal noise. Their prediction was based on the assumption that the magnetization remains uniform throughout the reversal process.

It is only during the past few years that these predictions could be tested experimentally. At the same time it has been realized that the approach to the super-paramagnetic limit does not only depend on the particle volume but also depends in a delicate way on the particle geometry. As is discussed in Wernsdorfer's contribution to these proceedings, recent experiments have confirmed that for sufficiently small particles, the Arrhenius exponent scales indeed with their volume, a strong indication that the magnetization dynamics is indeed uniform and that the Néel-Brown mechanism is operative. On the other hand, experiments on a single nanowire[3] revealed an energy barrier that is a factor of 200 below the Néel-Brown value, clearly challenging the hypothesis of uniform magnetization reversal. Similar conclusions have also be reached from observations on arrays of nanoparticles.[4]

Most theoretical attempts to generalize the Néel-Brown theory for elongated samples focussed on the linear stability of the metastable state. For cylindrical samples, linear instability modes have been found such as curling or buckling. While this approach[5] is suited to describe an instability at zero temperature, it cannot give information about the reversal mechanism at finite temperature. Moreover, this method is not able to capture possible topological defects that obstruct a particular reversal path.

The understanding of stochastic dynamics in nanostructures thus requires an intrinsically nonlinear approach that is capable of dealing with the nucleation problem at finite temperatures. Given the complexity of the problem, i.e. equations of motion for the magnetization that are nonlinear, nonlocal and stochastic, it is not too surprising to see that a general solution for realistic particle sizes is beyond reach — even for numerical simulations at present.

It is thus of particular importance to develop and study simplified models with the goal of capturing the essential physics of a nanoparticle. A rather significant role is taken by numerical simulations which can extend analytical results, test the validity of approximations and provide the crucial link between simplified models and the often rather complex aspects of even the cleanest experiments. An important step in this direction has been made by Hinzke, Nowak and Usadel described in their contribution to this volume.[6] They consider an anisotropic spin chain, a system that has been proposed as a model for a nanowire with transverse dimensions smaller than the exchange length. Using Monte-Carlo and Langevin simulations they find agreement with analytical expressions for the nucleation rate.[7]

In the light of this recent development, I present in the following a self-contained review of analytical results for the nucleation rate. Analogies are stressed between the simple case of uniform reversal and the more involved situation of nonuniform nucleation. Finally, barrier energies are given for the cross-over regime between Néel-Brown and soliton-antisoliton nucleation.

The exciting new developments in experimental techniques[3,4,8] and numerical approaches[6] give us hope that a thorough understanding of the magnetization dynamics in nanostructures lies within reach.

2 Energy of nanoparticles

We are interested in magnetic phenomena on a length scale considerably larger than the lattice spacing. In this case the total energy of a ferromagnetic sample of volume V is given by

$$E = E_m + \frac{1}{M_0^2} \int_V d^3r \Big\{ A[(\nabla M_x)^2 + (\nabla M_y)^2 + (\nabla M_z)^2] + \\ - K_{\mathrm{e,cryst}} M_z^2 + K_{\mathrm{h,cryst}} M_x^2 - M_0^2 \mathbf{M} \cdot \mathbf{H}_{\mathrm{ext}} \Big\}, \tag{1}$$

where $\mathbf{M}(\mathbf{r})$ is the magnetization of constant magnitude M_0. The first term describes the exchange interaction with exchange constant A. The remaining terms are crystalline anisotropies of easy-axis and hard-axis (easy-plane) type with anisotropy constants $K_{\mathrm{e,h,cryst}} > 0$, and the last term is a Zeeman term due to an external field. The dipolar or magnetostatic energy is nonlocal and can be expressed as

$$E_m = -\frac{1}{2}\int_V d^3r \int_V d^3r' \sum_{i,j} M_i(\mathbf{r})M_j(\mathbf{r}')\, \frac{\partial^2}{\partial r_i \partial r_j}\, \frac{1}{|\mathbf{r}-\mathbf{r}'|}, \tag{2}$$

where the integrations extend over the sample volume. The magnetization obeys the dynamics of the Landau-Lifshitz equation

$$(1+\alpha^2)\partial_t \mathbf{M} = -\gamma \mathbf{M}\times\mathbf{H}_{\mathrm{eff}} - \frac{\alpha\gamma}{M_0}\mathbf{M}\times[\mathbf{M}\times\mathbf{H}_{\mathrm{eff}}] \tag{3}$$

with the effective field $\mathbf{H}_{\mathrm{eff}} = -\delta E/\delta\mathbf{M}$ and gyromagnetic ratio $\gamma > 0$. The last term describes a Gilbert damping term with damping constant $\alpha > 0$. The Landau-Lifshitz equation is not yet consistent with the fluctuation dissipation theorem as it only includes damping but no random forces. The latter can be included via a random field term, i.e. by replacing

$$\mathbf{H}_{\mathrm{eff}} \to \mathbf{H}_{\mathrm{eff}} + \boldsymbol{\zeta}, \tag{4}$$

where $\boldsymbol{\zeta}$ is a random magnetic field with correlations $\langle\zeta_i(\mathbf{x},t)\zeta_j(0,0)\rangle = g_{ij}D_0\delta_{ij}\delta(t)\delta(\mathbf{x})$ with $D_0 = 2\alpha k_B T/\gamma M_0$, and g_{ij} the metric tensor associated with the particular choice of coordinates. The resulting equations of motion with the energy (1) are nonlocal, nonlinear and stochastic — an impossible task to treat in full generality.

We are here mainly concerned with nanoparticles where one or more dimensions are comparable or smaller to a domain wall width. This allows us to consider low dimensional realizations of the above equations, since the formations of domains is suppressed on sufficiently short length scales: A twist of the magnetization by an angle π over a length scale L costs an exchange energy density $(A/M_0^2)(\partial_x\mathbf{M})^2 \simeq \pi^2/L^2$. On the other hand, the potential gain in magnetostatic energy density is at most of the order πM_0^2. The creation of domain walls or nonuniform structures is therefore no longer favorable for sample dimensions such that

$$L < L_{\mathrm{mag}} = \pi\sqrt{A/\pi M_0^2}. \tag{5}$$

We shall now discuss two limiting cases, so called single-domain particles and nanowires where the magnetization can vary along the wire. In the latter case we shall see that a uniform remanent state does *not* imply that the magnetization dynamics is uniform. Indeed, we shall find nucleus configurations that are localized , either at the sample end or in the interior of the particle.

3 Thermal fluctuations in single domain particles, Néel-Brown theory

We first discuss the situation where all dimensions of the particle satisfy the inequality (5), and the particle is in a "single-domain" state with a magnetization that is uniform at all times. The energy of such a single-domain particle of ellipsoidal shape then takes the following form

$$E = V\Big\{-K_e m_z^2 + K_h m_x^2 - HM_0 m_z\Big\}, \tag{6}$$

where $\mathbf{m} = \mathbf{M}/M_0$ is the magnetization unit vector and V the particle volume. The effective anisotropy constants contain now *both* demagnetizing and crystalline anisotropies

$$\begin{aligned} K_e &= K_{\mathrm{e,cryst}} - 2\pi M_0^2(N_x - N_y), \\ K_h &= K_{\mathrm{h,cryst}} + 2\pi M_0^2(N_z - N_y). \end{aligned} \tag{7}$$

with demagnetizing factors defined as

$$N_i = -\frac{1}{4\pi}\frac{\partial^2}{\partial x_i^2}\int_V d^3r'\frac{1}{|\mathbf{r}-\mathbf{r}'|}, \tag{8}$$

For simplicity we have assumed that the crystalline anisotropies are aligned with the sample axes. Note, however, that with an appropriate redefinition of the N's, the energy retains the form (23) for non-ellipsoidal samples if they have mirror symmetric cross-section.

An important consequence of the energy (23) with (7) is the fact that the orientation of the remanent magnetization is not necessarily determined by magnetostatic shape effects as is commonly believed. If the crystalline anisotropies exceed $2\pi M_0^2$, the remanent state can be dominated by the crystalline anisotropy. This has recently been demonstrated in ultra-thin nanostructures[8] where a remanent magnetization has been observed that is oriented perpendicular to the shape induced anisotropy along the nanowire.

The magnetization obeys the dynamics of the Landau-Lifshitz equation with the spatially uniform effective field $\mathbf{H}_{\mathrm{eff}} = -\partial E/\partial \mathbf{m}$. It is convenient to pass to spherical coordinates in which the Landau-Lifshitz equations can be expressed as follows (cf. appendix)

$$\Omega_0^{-1}\sin\theta\,\partial_t\phi = \frac{\partial(E/E_0)}{\partial\theta} - \frac{\alpha}{\sin\theta}\frac{\partial(E/E_0)}{\partial\phi} + \eta_0^{-1}[-\eta_\theta + \alpha\eta_\phi]$$

$$\Omega_0^{-1}\partial_t\theta = -\frac{1}{\sin\theta}\frac{\partial(E/E_0)}{\partial\phi} - \alpha\frac{\partial(E/E_0)}{\partial\theta} + \eta_0^{-1}[\eta_\phi + \alpha\eta_\theta]. \tag{9}$$

with the energy

$$E/E_0 = -\tfrac{1}{2}\sin^2\theta\cos^2\phi + Q^{-1}\cos^2\theta - h[\sin\theta\cos\phi + 1] \tag{10}$$

Here we have used the following abbreviations

$$E_0 = 2K_eV, \quad Q^{-1} = \frac{K_h}{K_e}, \quad \eta_0 = \frac{2K_e}{M_0}, \quad h = \frac{H}{\eta_0}, \quad \Omega_0 = \frac{2\gamma K_e}{M_0(1+\alpha^2)}. \tag{11}$$

The Gaussian random noise fields have zero mean and obey the correlation

$$\langle\eta_i(0)\eta_j(t)\rangle = (D_0/V)\delta_{ij}\delta(t), \tag{12}$$

where $D_0 = 2\alpha k_B T/\gamma M_0$ controls the magnitude of the fluctuating field.

We are now interested in the switching probability of a magnetization prepared in the metastable state antiparallel to an external field as shown in Fig. 1. The magnetization will have to pass over a saddle point which is the extremum of the energy (23). For non-vanishing hard-axis anisotropy the saddle point can be obtained from $\partial E/\partial\phi = 0$ within the easy plane $\theta = \pi/2$. The saddle point lies at $\cos\phi_s = -h$ and has energy

$$E_s = K_eV(1-h)^2. \tag{13}$$

In order to compute the transition rate within the framework of the Kramers-Langer theory[12] we need the quadratic expansion of the energy around the saddle point and metastable state, respectively.

$$E/E_0 \simeq \begin{cases} E_s/E_0 + \frac{1}{2}(h^{s\varphi}\varphi^2 + h^{sp}p^2), & \text{saddle} \\ \frac{1}{2}(h^{m\varphi}\varphi^2 + h^{mp}p^2), & \text{metastable state} \end{cases} \tag{14}$$

with polar fluctuations defined via $\phi = \phi_s + \varphi$ and out of easy-plane fluctuations $\theta = \pi/2 - p$. The coefficients are given by

$$h^{s\varphi} = -(1-h^2), \qquad h^{sp} = Q^{-1},$$
$$h^{m\varphi} = 1-h, \qquad h^{mp} = 1-h+Q^{-1}.$$

The negative $h^{s\varphi}$ is associated with the unstable mode near the saddle point. Using this expansion, the stochastic Landau-Lifshitz equations of motion take in the vicinity of the saddle point the following form

$$\begin{aligned}\Omega_0^{-1}\partial_t\varphi &= -h^{sp}p - \alpha\, h^{s\varphi}\varphi + \eta_0^{-1}(\eta_p + \alpha\eta_\varphi), \\ \Omega_0^{-1}\partial_t p &= h^{s\varphi}\varphi - \alpha\, h^{sp}p + \eta_0^{-1}(-\eta_\varphi + \alpha\eta_p).\end{aligned} \tag{15}$$

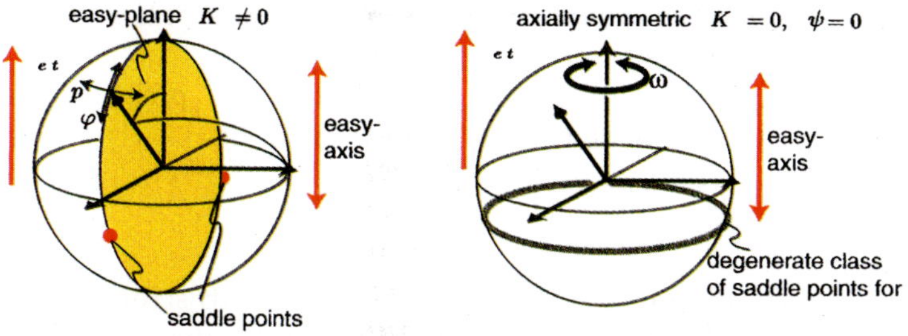

Figure 1. Definitions of the various angles in the single-domain particle. The angles φ and p denote fluctuations around metastable or saddle configurations. In the absence of an easy-plane (hard-axis) anisotropy there is a continuously degenerate "class" of saddle points which gives rise to a $1/\sqrt{T}$ temperature dependence of the prefactor in Eqn (23).

This can also be written [a] compactly $\Omega_0^{-1}\partial_t\psi_i = -\sum_j M_{ij}[h_j\psi_j - \eta_j/\eta_0]$, $i = (\varphi, p)$ with $M_{11} = M_{22} = \alpha$ and $M_{12} = -M_{21} = 1$. The (linearized) noise terms are defined in the same way as the magnetization components angles, hence $\eta_p \simeq -\eta_\theta$. They obey the same correlations as in (12).

In order to compute the switching rate, it is now convenient to pass from the above Langevin equation to a Fokker-Planck equation, as discussed in the appendix. The latter describes the dynamics of the probability density $\varrho(\{\psi\}) = \langle\prod_n \delta(\psi_n - \psi_n(t))\rangle$ where $\psi_n(t)$ are stochastic trajectories which solve the above Langevin equations, and $\langle..\rangle$ denotes the average with respect to the noise distribution $\exp\{-(V/2D_0)\int dt \sum_n \eta_n^2\}$. The "Fokker-Planck" equation has the form of a continuity equation

$$\Omega_0^{-1}\partial_t\rho + \sum_i \frac{\partial J_i}{\partial\psi_i} = 0 \quad . \tag{16}$$

with the probability current

$$J_i = -\sum_j M_{ij}\Big[h^{sj}\psi_j + \frac{k_BT}{2K_eV}\frac{\partial}{\partial\psi_j}\Big]\rho. \tag{17}$$

The rate is obtained by integrating this probability current near a surface

[a]Note that the metric in the easy plane $\theta = \pi/2$ is Euclidean and thus we do not have to distinguish between upper and lower indices of ψ_i.

through the saddle point. For the somewhat technical details of the integration procedure we refer to Ref. [10,7].

The probability per unit time for a magnetization prepared in the metastable state to switch into the stable state is given by

$$\Gamma = \frac{\lambda_+}{2\pi} \left(\frac{h^{m\varphi}}{|h^{s\varphi}|} \right)^{1/2} \left(\frac{h^{mp}}{h^{sp}} \right)^{1/2} e^{-\beta E_s}, \tag{18}$$

where $\beta = 1/k_B T$. This formula is valid in the moderate to strongly damped regime. The frequency λ_+ measures how fast the magnetization is dynamically driven away from the saddle point. More precisely, it is obtained by inserting $(\varphi_+(t), p_+(t)) = e^{\lambda_+ t}(\varphi_+, p_+)$ into the non-stochastic part of eqn (15), i.e.

$$\lambda_+ \psi_i^+ = -\Omega_0 \sum_j M_{ij} h_j \psi_j^+. \tag{19}$$

The rate Γ consists of the familiar Arrhenius factor and a prefactor that is associated with the fluctuations around metastable state and saddle point. It has the form of renormalized partition function $Z(S)/Z(M)$ in Gaussian approximation with the unstable direction turned into a stable one.

Inserting all definitions, we obtain for the nucleation rate in the moderately damped regime

$$\Gamma = \frac{\lambda_+}{\pi} \left(\frac{1 + Q[1-h]}{1+h} \right)^{1/2} e^{-\beta E_s}, \tag{20}$$

with the escape frequency

$$\lambda_+ = \Omega_0 Q^{-1}(-\alpha[1-q] + \sqrt{4q + \alpha^2[1+q]^2}), \tag{21}$$

where $q = Q|h^{s\varphi}| = Q(1-h^2)$. Note that the damping constant α exclusively enters the escape frequency. The result (20) includes a factor of 2 due to the existence of the two equivalent saddle points, $\pm\phi_s$.

This result is only valid as long as the fluctuations around the saddle point are small. It fails if one of the directions at the saddle point becomes "flat", i.e. if one of the h^{si}'s becomes zero. This is the case for vanishing hard-axis anisotropy K_h since the resulting energy has full rotational invariance around the z-axis. If one integrates the probability current across a surface near the saddle point, the corresponding Gaussian integrations diverge and have to be replaced by the finite integrations over the azimuthal angle ω (see, e.g. Ref. [10]). In the limit $h^{sp} \to 0$ we must cure the potentially divergent Gaussian integration as follows

$$\sqrt{\frac{2\pi}{\beta E_0 h^{sp}}} \to \int dp = \int_0^{2\pi} d\omega \, \sin\phi_s = 2\pi \sin\phi_s \tag{22}$$

The escape frequency becomes now simply $\lambda_+ = \alpha\Omega_0|h^{s\varphi}|$, and $h^{mp} = h^{m\varphi}$. Inserting all this into (37) we obtain for the switching rate for an axially symmetric situation

$$\Gamma = \frac{\alpha}{1+\alpha^2}\frac{2\gamma K_e}{M_0}\left(\frac{\beta V K_e}{\pi}\right)^{1/2}(1+h)(1-h)^2\,e^{-\beta E_s}. \tag{23}$$

This is the celebrated result by Néel and Brown.[1,2] One should notice that in contrast to the biaxial case, the prefactor now carries a temperature dependence as a result of the continuous symmetry. We shall encounter a similar connection between zero modes and temperature dependence for spin chains below.

4 Nucleation in nanowires

So far we have considered the reversal process of a sample where all dimensions satisfy the inequality (5), or equivalently, a single large spin. However for nanowires, recent experiments[3] revealed an energy barrier that is at variance with the Néel-Brown value.

In order to see whether (local) nucleation is possible in such nanowires, we have to consider a model which takes into account the magnetization variation along the sample. There will of course be nonlocal contributions to the energy due to magnetostatic interactions. However, if the particle diameter is smaller than the domain wall width, one can describe the resulting energy with anisotropies that are renormalized by magnetostatic interactions. The neglected nonlocal terms are small in the squared ratio of wire diameter vs. wall width.[11] We describe the nanowire by the following energy

$$\mathcal{E} = \mathcal{A}\int dx\Big\{A(\partial_x\mathbf{m})^2 - K_e m_x^2 + K_h m_z^2 - M_0\mathbf{H}\cdot\mathbf{m}\Big\}. \tag{24}$$

with effective anisotropy constants that contain local magnetostatic interactions as in (7). The external field is pointing within the easy-plane $\theta = \pi/2$, i.e., $\mathbf{H} = H(\cos\psi, \sin\psi, 0)$. Within the easy-plane, the energy can be rewritten as follows

$$E = 2K_e\mathcal{A}\int dx\left\{\frac{A}{2K_e}(\partial_x\phi)^2 + U(\phi,\psi)\right\}, \tag{25}$$

with the local potential energy felt by a single spin

$$U(\phi,\psi) = \frac{1}{2}\sin^2\phi - h\cos(\phi-\psi). \tag{26}$$

The integrated Euler-Lagrange equations in the easy-plane read as follows:

$$\frac{A}{2K_e}(\partial_x\phi)^2 - U(\phi,\psi) = c, \tag{27}$$

where c is an integration constant. Being separable, this equation can easily be formally integrated

$$\int \frac{d\phi}{\sqrt{c+U(\phi,\psi)}} = \frac{\sqrt{2}}{\delta_0}\int dx, \tag{28}$$

with $\delta_0 = \sqrt{A/K_e}$ is the domain wall width.

We now consider the case of a field applied along the easy-axis, i.e. $\psi = 0$. The desired nucleus configuration should merge asymptotically smoothly into the metastable state $\phi = \pi$. This fixes $c = -h$. The above integral may then be performed and we obtain the nucleus as a soliton-antisoliton pair

$$\phi_{s\bar{s}} = 2\arctan\left(\frac{\sinh R}{\cosh(\xi/\delta)}\right) = \phi_K(\frac{\xi}{\delta} - R) + \phi_K(-\frac{\xi}{\delta} - R). \tag{29}$$

The intrinsic (dimensionless) width, $\delta := \coth R$, depends on the field $h = \mathrm{sech}^2 R$, and $\xi = x/\delta_0$. The second representation shows that the nucleus consists of two solitons $\phi_K(w) = 2\arctan e^w$ located at $\xi = \pm R\delta$. The corresponding saddle point energy per area is given by

$$\mathcal{E}_s = 2A\int dx(\partial_x\phi_{s\bar{s}})^2 = 8\sqrt{AK}[\tanh R - R\,\mathrm{sech}^2 R]. \tag{30}$$

It tends towards the energy of two isolated domain walls for $h \to 0$ and decreases with decreasing separation of the soliton-antisoliton pair.

In order to compute the rate, we have to expand the energy near the saddle point and the metastable state, similar to the single-domain case above. However, the coefficients in the expansion of the energy around the saddle point,

$$E = E_s + 2K_e\mathcal{A}\int dx\,\{\tfrac{1}{2}\varphi\,\mathcal{H}^{s\varphi}\,\varphi + \tfrac{1}{2}p\,\mathcal{H}^{sp}\,p\}, \tag{31}$$

are no longer constants but Schrödinger-type operators

$$\mathcal{H}^{s\varphi} = -\partial_\xi^2 + V^-, \qquad \mathcal{H}^{sp} = -\partial_\xi^2 + V^+ + Q^{-1} \tag{32}$$

with the potentials

$$V^\pm = \delta^{-2}[1-2\,\mathrm{sech}^2(\frac{\xi}{\delta}-R)-2\,\mathrm{sech}^2(\frac{\xi}{\delta}+R)\pm 2\,\mathrm{sech}(\frac{\xi}{\delta}+R)\mathrm{sech}(\frac{\xi}{\delta}-R)]. \tag{33}$$

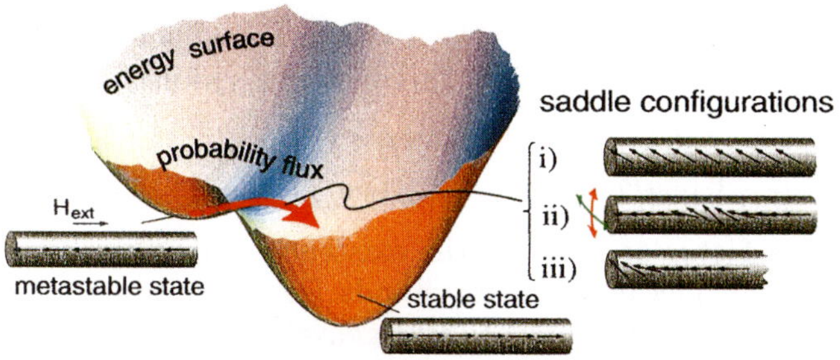

Figure 2. A schematic view of the (infinitely dimensional) energy surface near the saddle point. Saddle point configurations are shown which indicate i) uniform (Néel-Brown) nucleation for short sample lengths $L < L_{\text{crit}}$. For larger sample lengths one may have ii) soliton-antisoliton nucleation in the interior of a sample (or a toroidal sample with periodic boundary conditions) for $L > L_{\text{crit}}$ or iii) nucleation from the sample end for $L > L_{\text{crit}}/2$. The arrows indicate fluctuations φ, p around a magnetization configuration and the dynamics near the saddle point. The escape frequency λ_+ is the positive eigenvalue corresponding to the mode of fastest decay of the nucleus.

Correspondingly, fluctuations near the metastable state are described by the "free" operators

$$\mathcal{H}^{m\varphi} = -\partial_\xi^2 + \delta^{-2}, \qquad \mathcal{H}^{mp} = -\partial_\xi^2 + \delta^{-2} + Q^{-1}. \tag{34}$$

Note that far away from the nucleus, the two sets of operators are identical. Fluctuations around the saddle point reflects two essential properties of the saddle point.

Translational symmetry implies that $\phi_{s\bar{s}}$ is energetically degenerate with $\phi_{s\bar{s}}(x - x_0)$. This degeneracy must show up as a zero mode in the excitation spectrum , i.e., $\phi'_{s\bar{s}}$ is eigenfunction of $\mathcal{H}^{s\varphi}$ with eigenvalue zero. Due to the opposite winding sense of the two solitons forming the $s\bar{s}$-pair, this zero mode has one node. From this we infer the existence of a "ground state" with negative energy and the nucleus is indeed unstable. Fluctuations in azimuthal direction always have a positive eigenvalue except for $K_h = 0$. In this latter case there is a continuous symmetry around the easy-axis.

Similarly to the case of a single domain particle, one has to study the stochastic dynamics near the saddle point

$$\Omega_0^{-1}\partial_t\psi_i(x,t) = -\sum_j M_{ij}[\mathcal{H}_j\psi_j(x,t) - \eta_j(x,t)\eta_0^{-1}] \tag{35}$$

with the following noise correlation

$$\langle \eta_i(x,t)\, \eta_j(0,0)\rangle = (D_0/\mathcal{A})\delta_{ij}\delta(x)\delta(t). \tag{36}$$

The corresponding Fokker-Planck equation is derived in the appendix. The rate is obtained by integrating the probability flux near the saddle point[7,10]

$$\Gamma = \frac{\lambda_+}{\pi}\frac{L}{\delta_0}\sqrt{\frac{\beta\mathcal{A}\mathcal{E}_s}{2\pi}}\left(\frac{\det\mathcal{H}^{m\varphi}}{|\det'\mathcal{H}^{s\varphi}|}\right)^{1/2}\left(\frac{\det\mathcal{H}^{mp}}{\det\mathcal{H}^{sp}}\right)^{1/2} e^{-\beta\mathcal{A}\mathcal{E}_s}, \tag{37}$$

where we have included a factor of 2 due to the existence of two equivalent nucleus configurations $\pm\phi_s$. The determinant stands for the product of all eigenvalues, the prime denoting omission of the eigenvalue zero. λ_+ is the escape frequency as before, now satisfying the coupled Schrödinger equation

$$\lambda_+\psi_i^+ = -\Omega_0 \sum_j M_{ij}\mathcal{H}_j\psi_j^+ \tag{38}$$

with $\mathcal{H}_j = \mathcal{H}^{sj}$. The eigenvalue λ_+ can be evaluated analytically in some limiting cases. Somewhat surprisingly, this approximate expression agrees with numerical results[7] in a a large parameter range and is possibly even exact. It is given by Eqn (21) but with $q = Q|E_0^{s\varphi}|$ where the ground state energy $E_0^{s\varphi}$ of $\mathcal{H}^{s\varphi}$ can be evaluated numerically. In the limits of large and small fields we have

$$E_0^{s\varphi} = -3R^2, \quad R\to 0; \qquad E_0^{s\varphi} = -8e^{-2R}, \quad R\to\infty. \tag{39}$$

The origin of the unusual, L-dependent term in Eqn (37) can be understood as follows: Taking the naive, unprimed product of all eigenvalues, $|\det\mathcal{H}^{s\varphi}|$ would lead to a divergence of the prefactor since translational invariance causes one eigenvalue to be zero. It is helpful to recall that the determinant originates from an integration over the probability flux resulting in a Gaussian integration over the energy (31). We thus can replace the potentially dangerous Gaussian integration by a direct integration over the center of mass coordinate of the nucleus. More precisely, infinitesimal translations are described by

$$\mathcal{T}_{dX}\circ\phi_s - \phi_s = dX\,\phi_s'(x) = d\varphi_1\,\chi_1^{s\varphi}(x) \tag{40}$$

where $\chi_1^{s\varphi}$ is the first excited state of $\mathcal{H}^{s\varphi}$. Thus $\chi_1^{s\varphi}\propto\phi_s'(x)$ and

$$\int d\varphi_1 = [\delta_0\int dx\,(\phi_s')^2]^{1/2}\int\frac{dX}{\delta_0} = 2\frac{L}{\delta_0}[\tanh R - R\,\mathrm{sech}^2 R]^{1/2} \tag{41}$$

where the sqrt-term plays the role of the Jacobian of the transformation. Note that the rhs equals $(\mathcal{E}_s/2\sqrt{AK_e})^{1/2}L/\delta_0$. The result of the defective Gaussian integration thus has to be replaced as follows

$$\sqrt{2\pi/\beta\mathcal{A}E_1^{s\varphi}} \to \sqrt{\mathcal{E}_s}L/\delta_0. \tag{42}$$

Some of the remaining terms also can be further evaluated: The exact result[9] for the determinant in φ-direction is

$$\frac{\det \mathcal{H}^{m\varphi}}{|\det' \mathcal{H}^{s\varphi}|} \equiv \frac{\det[-\partial_\xi^2 + \delta^{-2}]}{|\det'[-\partial_\xi^2 + V^-]|} = \frac{8\tanh^3 R \sinh^2 R}{\tanh R - R\,\mathrm{sech}^2 R}. \tag{43}$$

The p-determinants can be evaluated in some limiting cases.[7] Here we use the fact that for large $Q^{-1}\delta^2$ the p-fluctuations are irrelevant, i.e., $\det \mathcal{H}^{mp}/\det \mathcal{H}^{sp} = 1$.

Combining all results, we find that for large $Q^{-1}\delta^2$ or equivalently $Q(1-h)$ small, the switching rate is given by,

$$\Gamma = \frac{4}{\pi^{3/2}}\lambda_+ \frac{L}{\delta_0}\sqrt{\sigma_0 \beta \mathcal{A}}\ \tanh^{3/2} R\ \sinh R\, e^{-\beta\mathcal{A}\mathcal{E}_s}, \tag{44}$$

where $\sigma_0 = 2\sqrt{AK_e}$. Using above expression for λ_+ with $q = Q|E_0^{s\varphi}|$ and numerical results on $E_0^{s\varphi}$, the rate Γ may be computed for small Q for all fields. Simplified results are obtained in the following limits: For $\alpha^2 > Q|E_0^{s\varphi}$ we have $\lambda_+ = 2\Omega_0(\alpha+\alpha^{-1})|E_0^{s\varphi}|$, while for $\alpha^2 < Q|E_0^{s\varphi}|$ we have $\lambda_+/\Omega_0 = \sqrt{Q|E_0^{s\varphi}|}$. For fields close to the coercivity, i.e. small $Q(1-h)$, we may use (39) to obtain

$$\Gamma = \frac{\gamma K_e}{M_0}\frac{L}{\delta_0}\sqrt{\sigma_0\beta\mathcal{A}} \times \begin{cases} \frac{48}{\pi^{3/2}}\frac{1}{\alpha}\,(1-h)^{9/4}\,e^{-\beta\mathcal{A}\mathcal{E}_s}, & \alpha \gg \sqrt{Q(1-h)}, \\ \frac{8\sqrt{3}}{\pi^{3/2}}\frac{\sqrt{Q}}{(1+\alpha^2)}\,(1-h)^{5/4}\,e^{-\beta\mathcal{A}\mathcal{E}_s}, & \alpha \ll \sqrt{Q(1-h)}. \end{cases} \tag{45}$$

Recall that $1-h = R^2$ for $h \to 1$. The first equation has a α^{-1} dependence, a characteristic of rates in the overdamped limit. The second expression represents the essentially damping independent rate in the moderately damped regime. Both results are in accordance with general rate theory. For small fields we can use $E_0 = -8e^{-2R} = -2h$ and the switching probability per unit time is

$$\Gamma = \frac{\gamma K_e}{M_0}\frac{L}{\delta_0}\sqrt{\sigma_0\beta\mathcal{A}} \times \begin{cases} \frac{32}{\pi^{3/2}}\frac{1}{\alpha}\,\sqrt{h}\,e^{-\beta\mathcal{A}\mathcal{E}_s}, & \alpha \gg \sqrt{Qh}, \\ \frac{8\sqrt{2}}{\pi^{3/2}}\frac{1}{(1+\alpha^2)}\sqrt{Q}\,e^{-\beta\mathcal{A}\mathcal{E}_s}, & \alpha \ll \sqrt{Qh}, \end{cases} \tag{46}$$

and Q small. In the overdamped limit, the slow decay of the nucleus leads to a $\sqrt{h}$-suppression of the prefactor. In the moderately damped limit, the faster decay frequency leads to a field-independent prefactor. We shall encounter the same situation below for complete rotational invariance around the easy-axis.

4.1 Rates in the case of rotational degeneracy

We now discuss the case of vanishing hard-axis anisotropy, i.e. $K_h = 0$. We make now use of the fact that the ground state of $\mathcal{H}^{sp}$, χ_0^{sp}, corresponds to a rotation of the nucleus,

$$\mathcal{R}^x_{d\alpha} \circ \mathbf{n}_s - \mathbf{n}_s = d\omega \; \sin\phi_s(x)\, \hat{\mathbf{z}} = dp_0 \; \chi_0^{sp}(x)\, \hat{\mathbf{z}}, \tag{47}$$

where ω denotes the rotation angle around the x-axis. Thus

$$\int dp_0 = [\int \frac{dx}{\delta_0} \; \sin^2\phi_s]^{1/2} \int_0^{2\pi} d\omega = 4\pi\sqrt{\rho_+}, \tag{48}$$

where $\rho_\pm = 4[\tanh R \pm R\mathrm{sech}^2 R]$. Correspondingly, the Gaussian integration has to be replaced by

$$\sqrt{2\pi/\beta\mathcal{A}E_0^{sp}} \rightarrow 4\pi\sqrt{\rho_+} \tag{49}$$

with $\rho_+ = 4[\tanh R + R\mathrm{sech}^2 R]$. For $Q^{-1} = 0$ we have $\lambda_+ = \alpha|E_0^{s\varphi}|$. In this limit, $\mathcal{H}^{sp} = -\partial_\xi^2 + V^+$ also possesses a zero mode, i.e. $\mathcal{H}^{sp}\chi_0^{sp} = 0$, and the ratio of the determinants can be computed exactly:

$$\left.\frac{\det\mathcal{H}^{mp}}{\det'\mathcal{H}^{sp}}\right|_{Q^{-1}=0} \equiv \frac{\det[-\partial_\xi^2 + \delta^{-2}]}{\det'[-\partial_\xi^2 + V^+]} = \frac{8\tanh^3 R\cosh^2 R}{\tanh R + R\;\mathrm{sech}^2 R} \tag{50}$$

where the prime denotes omission of the zero mode. In the case of rotational degeneracy we thus obtain for the switching rate (37)

$$\begin{aligned}\Gamma &= \frac{\lambda_+}{\pi}\frac{L}{\delta_0}\sigma_0\beta\mathcal{A}\sqrt{\rho_+\rho_-}\left(\frac{\det\mathcal{H}^{m\varphi}}{|\det'\mathcal{H}^{s\varphi}|}\right)^{1/2}\left(\frac{\det\mathcal{H}^{mp}}{\det'\mathcal{H}^{sp}}\right)^{1/2} e^{-\beta\mathcal{A}\mathcal{E}_s} \\ &= (32/\pi)\alpha\Omega_0|E_0^{s\varphi}|\;\beta K_e V\;\tanh^3 R\sinh 2R\; e^{-\beta\mathcal{A}\mathcal{E}_s}\end{aligned} \tag{51}$$

where $V = L\mathcal{A}$ is the particle volume. Here we have used that for vanishing hard-axis anisotropy the unstable mode is exactly given by $(\varphi_+, p_+) = (\chi_0^{s\varphi}, \chi_0^{sp})$ and hence $\lambda_+ = \alpha\Omega_0|E_0^{s\varphi}|$. In particular, we obtain from (39)

$$\Gamma = \frac{\alpha}{1+\alpha^2}\frac{\gamma K_e}{\pi M_0}\beta K_e V \times \begin{cases} 2^7 e^{-\beta\mathcal{A}\mathcal{E}_s}, & h \rightarrow 0, \\ 3\cdot 2^6(1-h)^3\; e^{-\beta\mathcal{A}\mathcal{E}_s}, & h \rightarrow 1. \end{cases} \tag{52}$$

There are two remarkable points regarding this result: First, the rotational symmetry around the x-axis has now led to a factor $\sqrt{\beta}$ in addition to the already existing $\sqrt{\beta}$ factor of the zero mode. Its origin is quite analogous to the rotationally invariant case (23) of the Néel-Brown theory. Secondly, the α dependence now interpolates continuously between the high damping limit and the low damping limit where $\Gamma \propto \alpha$ vanishes linearly. This is remarkable since this is a characteristic of underdamped theories, while the present derivation relies on moderate damping. A possible reason for this coincidence could be the following: In the absence of a hard-axis anisotropy, the dynamics is entirely governed in the variable ϕ which measures essentially the easy-axis energy. A Fokker-Planck equation in ϕ is therefore closely related to one in energy whose integration yields the rate in the underdamped limit.

We thus reach an important conclusion from this result: *Each continuous symmetry that is broken by the saddle point contributes a factor* $1/\sqrt{T}$ *to the prefactor*. Comparing the corresponding expression from uniform reversal and nonuniform nucleation we also conclude that the exponent ϵ in the prefactor $(1-h)^\epsilon$ becomes larger with increasing number of continuous symmetries.

As an example, we may conclude from these observations that owing to the lack of broken symmetries, nucleation at the sample end leads to a temperature independent prefactor for $K_h \neq 0$. In the absence of a hard-axis anisotropy the broken rotational invariance will lead to a $1/\sqrt{T}$ dependence of the prefactor. Note, however, that this complete symmetry is only present for fields pointing along the easy-axis.

5 Cross-over between Néel-Brown and soliton-antisoliton mechanism

So far we have considered the energy barriers in two opposite limits, grains with $L < L_{\text{mag}}$ or nanowires arbitrary sample lengths. But at what length scales does the cross-over between the two regimes occur ?

In order to answer this question, we study eqn(27) with $\psi = 0$ but now with the boundary conditions $\partial_x\phi|_{0,L} = 0$. For infinite L and $c = -h$ we obtain the soliton-antisoliton nucleus, whereas for smaller values of c we obtain the generalization of $s\bar{s}$ pairs to a chain of finite length. These generalizations of the $s\bar{s}$- nucleus to samples of finite length are given by

$$\phi^L_{s\bar{s}} = 2\arctan\Big(\frac{w_1}{\text{dn}(\frac{x}{\delta}, k)}\Big), \tag{53}$$

where dn is a Jacobian elliptic function. Here, the elliptic modulus k is related to the sample length as follows:

$$L = 2K(k)\delta. \tag{54}$$

The elliptic modulus k controls the nonlinearity of the problem. For $k \to 0$, $\phi^L_{\text{s}\bar{\text{s}}}$ describes a harmonic excursion from the uniform Néel-Brown barrier, for $k \to 1$ we obtain the nonlinear $s\bar{s}$-pair. Note that $\phi(0) = 2\arctan w_1$ and

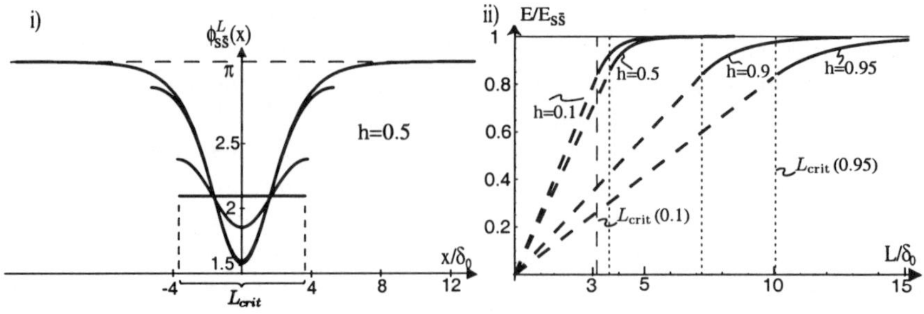

Figure 3. i) Critical nucleus (53) for samples of various length with periodic boundary conditions for $h = 0.5$. Below sample lengths of L_{crit}, $\phi^L_{s\bar{s}}$ is constant and takes the Néel-Brown value. ii) Saddle point energy (58) as a function of sample length, normalized to the soliton-antisoliton value. For $L < L_{\text{crit}}$ the barrier is given by the Néel-Brown expression (dashed line). For $L > L_{\text{crit}}$ there is a rapid approach to the soliton-antisoliton energy. Nucleation from the sample end is obtained by restricting above configuration to the interval $x > 0$. Correspondingly the barrier energy is halved and the above value has to be replaced by $L_{\text{crit}}/\delta_0 = \pi/\sqrt{1-h^2}$.

$\phi(L/2) = 2\arctan(w_1/\sqrt{1-k^2})$. More precisely, both k and the intrinsic width δ are a function of external field and integration constant c,

$$\delta_0/\delta = w_2\sqrt{\tfrac{1}{2}|c+h|}, \qquad k = \sqrt{1 - w_1^2/w_2^2}, \tag{55}$$

via

$$w_{1,2}^2 = [-(1+c) \pm \sqrt{1+h^2+2c}]/(h+c). \tag{56}$$

The plus-sign refers to w_1^2 and $0 < w_1^2 < w_2^2$, and $-\frac{1}{2}(1+h^2) < c < -h$. Symmetries allow us to restrict ourselves to the cases $w_1, w_2 > 0$.

We can now compute the energy of this configuration relative to the metastable state $\phi = \pi$. Using the Euler-Lagrange equations we have

$$\mathcal{E} - \mathcal{E}_m = \int_{-L/2}^{L/2} dx \Big\{ 2A(\partial_x\phi)^2 - 2K_e(c+h) \Big\} \tag{57}$$

This leads to the following result for the energy barrier

$$\mathcal{E}[\phi^L_{s\bar{s}}] - \mathcal{E}_m = 8\sqrt{AK_e}\frac{\delta_0}{\delta}\Big\{E(k) + (1-\frac{k^2}{\alpha^2})K(k) + (\alpha^2 + \frac{k^2}{\alpha^2} - 2)\,\Pi(\alpha^2,k)\Big\} + \\ -2K_eL(c+h) \tag{58}$$

where we defined $\alpha^2 = k^2/(1+w_1^2)$, and E, K, Π are elliptic integrals. In order to obtain the barrier energy as a function of the sample length L one has to express both L and $\mathcal{E}$ by the parameter c. Examples are shown in Fig. 3

As expected, this barrier energy reduces to the Néel-Brown and soliton-antisoliton values: For $c \to -h$ one reaches the soliton-antisoliton limit. This implies $w_1^2 = h/(1-h)$, $\alpha^2 = 1-h$ and $|w_2|\sqrt{\frac{1}{2}|c+h|} = \sqrt{1-h}$, and thus $\delta_0/\delta = \sqrt{1-h}$ for $w_2^2 \to \infty$. For $k=1$ the second and third term in curly brackets reduce to $(\alpha - \alpha^{-1}) \ln \sqrt{(1+\alpha)/(1-\alpha)}$ and one recovers the energy of the soliton-antisoliton pair.

Approaching the uniform case c will tend towards $c = -\frac{1}{2}(1+h^2)$, and $k \to 0$. This leads to $w_1^2 = w_2^2 = (1+h)/(1-h)$ and $\sqrt{\frac{1}{2}|c+h|} = \frac{1}{2}(1-h)$. In this case, we have $K = E = \Pi = \pi/2$, the entire curly bracket in (11) reduces to zero, and we are left with

$$\mathcal{E} - \mathcal{E}_m = K_e L [1-h]^2 \tag{59}$$

which is just the Néel-Brown barrier for uniform nucleation.

For $L < L_{\text{crit}}$, there are no longer nonuniform solutions since the minimal value that satisfies Eqn (54) is

$$L_{\text{crit}} = 2K\delta|_{k\to 0} = 2\pi\delta_0/\sqrt{1-h^2}. \tag{60}$$

For $L < L_{\text{crit}}$ only the Néel-Brown solution exists.

We conclude that the cross-over occurs in a narrow range of sample lengths and has almost the character of a first order phase transition, cf. Fig. 3. The sharpness of this transition region justifies our rough estimate of the criterion (5). Remarkably, this criterion is even *exact* if we use the value $K_h = \pi M_0^2$, the appropriate value for a cylindrical sample: For nucleation from sample end the critical sample length is half the value of (60) which agrees with (5) for vanishing field.

6 Outlook and conclusions

Let us conclude with a few remarks on experiments and nucleation from the sample end. In the limit of particle diameters that are small compared to the wall width, the demagnetizing induced hard-axis is small and we may use all of above results simply multiplying all energy barriers by a factor 1/2 and replacing L by $2L$ in (54). We then obtain an energy barrier for nucleation at the sample end $E = (8/3)\mathcal{A}_{\text{end}}\sqrt{AK_e}[1-h]^{3/2}$ for $h \to 1$. This is within a factor of three from the observed value.[3,11] Note that the prefactors are affected in a more subtle way since the translational zero mode does of course no longer exist. In fact only even parity wave-functions contribute to the determinants.

At the moment there are not sufficient experimental data to decide whether the above theory is a good description of the phenomena in nanowires, or whether further modifications are needed. On the other hand, the recent progress in numerical simulations of the stochastic Landau-Lifshitz equations together with experimental progress give us hope that after decades of research, we are about to understand the secrets of nanomagnets.

We did not at all touch here any aspects of quantum effects. For two recent reviews emphasizing the close similarities to the topics discussed here I refer to Ref. [13].

Appendix

In order to make contact between the approach used in the main text and the numerical simulations by Hinzke and Nowak presented elsewhere in this volume, we discuss here the reduction of the stochastic Landau-Lifshitz equations to a Fokker-Planck equation. In this appendix we show that the approach of Hinzke and Nowak of adding the noise term directly to both precessional and damping terms in the Landau-Lifshitz equation is equivalent to adding diagonal noise terms to the linearized Langevin equation as in Ref. [7]. The Landau-Lifshitz equation in the presence of random noise can be written as

$$(1+\alpha^2)\frac{1}{\gamma}\,\partial_t \mathbf{m} = -\mathbf{m}\times\mathbf{H} - \alpha\mathbf{m}\times[\mathbf{m}\times\mathbf{H}] \tag{61}$$

where $\mathbf{H} = \mathbf{H}_{\text{eff}} + \boldsymbol{\zeta}$ is the sum of reversible and stochastic fields and $\mathbf{m} = \mathbf{M}/M_0$ is the magnetization unit vector. We have $\mathbf{H}_{\text{eff}} = -\frac{1}{M_0}(\delta E/\delta\Theta^i)\mathbf{g}^i$ and the stochastic field $\boldsymbol{\zeta}$ is characterized by the correlation function

$$\langle \zeta_i(\mathbf{x},t)\zeta_j(0,0)\rangle = D_0 g_{ij}\delta(\mathbf{x})\delta(t) \tag{62}$$

where g_{ij} is the metric tensor associated with the chosen coordinates Θ^i. In cartesian coordinates we have of course $g_{ij} = \delta_{ij}$. Below, we shall see that the fluctuation dissipation theorem implies that $D_0 = 2\alpha k_B T/\gamma M_0$.

The Landau-Lifshitz equations conserve the magnitude of the magnetization and only two components of the noise field are relevant. For this reason it is convenient to pass to a standard polar frame with the radial unit vector pointing along the magnetization, $\mathbf{m} = \mathbf{g}_1$, i.e., the only nonvanishing component is $\Theta^1 = 1$. The increment is $d\mathbf{m} = d\Theta^i\mathbf{g}_i$ with $d\Theta^i = (0, d\theta, d\phi)$. Note that on the unit sphere the metric is $g_{ij} = \text{diag}(1, 1, \sin^2\theta)$. We have

$$\mathbf{m}\times\mathbf{H} = \Theta_i H_j \epsilon^{ijk}\mathbf{g}_k \tag{63}$$

$$\mathbf{m}\times[\mathbf{m}\times\mathbf{H}] = \Theta^l\Theta_i H_j\epsilon^{ijk}\epsilon_{lkr}\mathbf{g}^r = -H_\theta\, g_\theta - \frac{H_\phi}{\sin^2\theta}\, g_\phi \tag{64}$$

and $\epsilon_{ijk} = 1/\epsilon^{ijk} = \pm\sqrt{\det \mathrm{g_{ij}}}$ for cyclic and anticyclic permutations, respectively. The Landau-Lifshitz equations of motion (61) then read

$$\frac{1+\alpha^2}{\gamma}\sin\theta\partial_t\phi = -H_\theta + \alpha\frac{H_\phi}{\sin\theta} - \eta_\theta + \alpha\eta_\phi$$
$$\frac{1+\alpha^2}{\gamma}\partial_t\theta = \frac{H_\phi}{\sin\theta} + \alpha H_\theta + \eta_\phi + \alpha\eta_\theta \tag{65}$$

where $H_i(\mathbf{x}) = -\frac{1}{M_0}\frac{\delta E}{\delta\Theta^i(\mathbf{x})}$. Here we have also introduced the normalized, polar components of $\boldsymbol{\zeta}$, i.e. $\eta_i = \zeta_i/\sqrt{g_{ii}}$ as a stochastic field. It is characterized by the correlation functions

$$\langle\eta_i(\mathbf{x},t)\;\eta_j(0,0)\rangle = D_0\delta_{ij}\delta(\mathbf{x})\delta(t) \tag{66}$$

with $i,j = \theta,\phi$. In quasi 1D systems, the functional derivative has to be replaced by $\frac{\delta(E/\mathcal{A})}{\delta\Theta^i(x)}$ and $\delta(\mathbf{x}) \to \delta(x)/\mathcal{A}$ in the noise correlation functions , while for uniform situations one has $\frac{\partial(E/V)}{\partial\Theta^i}$ and $\delta(\mathbf{x}) \to 1/V$. [Recall that V, $\mathcal{A}$ are the sample volume and cross sectional area, respectively.]

In order to evaluate the nucleation rate we need the dynamics of the magnetization field near the saddle point. We now specify to the case where the magnetization only depends on one coordinate. Near the saddle point with $\theta = \pi/2$, the stochastic Landau Lifshitz equation has the form of a Langevin equation

$$\tfrac{1+\alpha^2}{\gamma}\partial_t\psi_i(x,t) = -\sum_j M_{ij}[\tfrac{2K_e}{M_0}\mathcal{H}_j\psi_j(x,t) - \eta_j(x,t)] \tag{67}$$

with $\langle\eta_i(x,t)\eta_j(0,0)\rangle = D\delta_{ij}\delta(x)\delta(t)$, $D = D_0/\mathcal{A}$. Here we introduced a compact notation with $M_{11} = M_{22} = \alpha$ and $M_{12} = -M_{21} = 1$. $\mathcal{H}_j$ is given by (32) and $\psi_j = (\varphi, p)$. The computation of the rate is based on a Fokker-Planck equation for the probability density

$$\rho[\{\psi_i\}] = \langle\prod_{n,x}\delta(\psi_n(x) - \psi_n(x,t))\rangle \tag{68}$$

where the product over x indicates the functional nature of the δ-function and the $\psi_n(x,t)$'s are trajectories satisfying the Langevin equation. The expectation value is taken with respect to the Gaussian white noise distribution $\exp\{-\frac{1}{2D}\int dx\;dt\sum\eta_i^2(x,t)\}$. For the dynamics of the probability distribution we have

$$\tfrac{1+\alpha^2}{\gamma}\partial_t\rho = \tfrac{1+\alpha^2}{\gamma}\langle\int dy\sum_i\partial_t\psi_i(y,t)\frac{\delta}{\delta\psi_i(y,t)}\prod_{n,x}\delta(\psi_n(x) - \psi_n(x,t))\rangle$$

$$= \int dy \sum_i \frac{\delta}{\delta\psi_i(y)} \sum_j M_{ij} \Big\{ \frac{2K_e}{M_0} \mathcal{H}_j \psi_j(y)\rho - \langle \eta_j(y,t) \prod_{n,x} \delta(\psi_n(x) - \psi_n(x,t)) \rangle \Big\} \tag{69}$$

We still have to evaluate the expectation value of the noise.

$$\begin{aligned} \langle \eta_j(y,t) \prod_{n,x} \delta(\psi_n(x) - \psi_n(x,t)) \rangle &= D \langle \frac{\delta}{\delta\eta_j(y,t)} \prod_{n,x} \delta(\psi_n(x) - \psi_n(x,t)) \rangle \\ &= -D \int dz \sum_k \frac{\delta}{\delta\psi_k(z)} \langle \frac{\delta\psi_k(z,t)}{\delta\eta_j(y,t)} \prod_{n,x} \delta(\psi_n(x) - \psi_n(x,t)) \rangle \\ &= -\frac{D}{2} \frac{\gamma}{1+\alpha^2} \sum_k M_{kj} \frac{\delta\rho}{\delta\psi_k(y)} \end{aligned} \tag{70}$$

In the first step, we have use the explicit Gaussian form of the noise correlation function with partial integration. In the last step, we have used the integrated version of the Langevin equation. The factor 1/2 arises from a regularization of the noise correlation function. Note that such a straightforward integration of the Langevin equation is only possible for linear stochastic equations, otherwise one has to distinguish between Stratonovich and Ito definitions.

Collecting everything, we arrive at the Fokker-Planck equation

$$\frac{1+\alpha^2}{\gamma} \partial_t \rho + \int dx \sum_i \frac{\delta J_i[\psi]}{\delta\psi_i(x)} = 0 \tag{71}$$

with the probability current

$$J_i[\psi] = -\sum_j M_{ij} \Big\{ \frac{2K_e}{M_0} \mathcal{H}_j \psi_j(y) + \frac{D\gamma}{2\alpha} \frac{\delta}{\delta\psi_j(y)} \Big\} \rho, \tag{72}$$

where we have used the antisymmetry of M in the second term. The noise amplitude is now determined by the fact that the net probability current has to vanish for the equilibrium distribution $\rho_{\text{eq}} = \exp\{-\beta \mathcal{A} K_e \int dx \sum_i \psi_i \mathcal{H}_i \psi_i\}$. From this we obtain

$$D = 2\alpha k_B T / \gamma M_0 \mathcal{A}. \tag{73}$$

Proceeding in the same way as above, we may now convince ourselves that a noise which couples only to the diagonal term[7] in the Langevin equation

$$\frac{1+\alpha^2}{\gamma} \partial_t \psi_i(x,t) = -\frac{2K_e}{M_0} \sum_j M_{ij} \mathcal{H}_j \psi_j(x,t) + \zeta_i(x,t), \tag{74}$$

with $\langle \zeta_i(x,t) \zeta_j(0,0) \rangle = D(1+\alpha^2) \delta_{ij} \delta(x) \delta(t)$ leads to the same Fokker-Planck equation (71),(72) as the Langevin equation (67). With appropriate rescaling, it is thus immaterial whether stochastic fields are included in the Landau-Lifshitz equation in both damping and precessional terms or only in the precessional term.

References

1. L. Néel, *Ann. Géophys.* **5**, 99 (1949).
2. W.F. Brown, *Phys. Rev.* **130**, 1677 (1963); *IEEE Trans. Magn.* **MAG-15**, 1196 (1979).
3. W. Wernsdorfer, B. Doudin, D. Mailly, K. Hasselbach, A. Benoit, J. Meier, J.-Ph. Ansermet, and B. Barbara, *Phys. Rev. Lett.* **77**, 1873 (1997).
4. A. Kent, S. von Molnár, S. Gider, and D.D. Awschalom, *J. Appl. Phys.* **76**, 6656 (1994).
5. A. Aharoni in *Introduction to the Theory of Ferromagnetism* (Oxford, University Press, 1997).
6. D. Hinzke, U. Nowak and K.D. Usadel, *these proceedings.*
7. H.B. Braun, *Phys. Rev. Lett.* **71**, 3557 (1993); *Phys. Rev.* B **50**, 16501 (1994).
8. C. Stamm, F. Marty, A. Vaterlaus, V. Weich, S. Egger, U. Maier, U. Ramsperger, H. Fuhrmann, and D. Pescia, *Science* **282**, 449 (1998).
9. H.B. Braun, *Phys. Rev.* B, 16485 (1994).
10. H.B. Braun, *J. Appl. Phys.* **76**, 6310 (1994).
11. H.B. Braun, *J. Appl. Phys.* **85**, 6172 (1999).
12. J.S. Langer, *Ann. Phys.* **54**, 258 (1969).
13. H.B. Braun in *Complementarity between Neutron and Synchrotron X-Ray Scattering*, ed. A. Furrer (World Scientific, Singapore, 1998); and in *Mesoscopic Quantum Effects*, ed. J.R. Friedman and S. Han (Nova, New York, 1999).

MAGNETIC ANISOTROPY PROPERTIES OF TRANSITION METALS: AB-INITIO VERSUS SEMI-EMPIRICAL METHODS?

ŠTĚPÁN PICK

Jaroslav Heyrovský Institute of Physical Chemistry, Academy of Sciences of the Czech Republic, Dolejškova 3, CZ-182 23 Prague 8, Czech Republic

HUGUES DREYSSÉ

Institut de Physique et Chimie des Matériaux de Strasbourg, UMR CNRS 46, 23 rue du Loess, BP 20 CR, F-67037 Strasbourg Cedex, France

In this contribution we review the different strategies used to determine the magnetic anisotropy properties. For bulk transition metals partial satisfactory results have been obtained within ab-initio schemes. For low-dimensional system, semi-empirical tight-binding Hamiltonian provides an efficient framework to get the differences of energy associated with different orientations of the magnetization, and to describe the orbital magnetism. Different "technical" difficulties encountered in both approaches are discussed, as the "irregular" behavior of the Magnetic Anisotropy Energy, in the **k**-space and in the real-space. In a near future, real-space ab-initio and semi-empirical methods should gain a larger interest. Another effort should be done on a fully ab-initio determination of the orbital moments.

1 Introduction

The magnetic (or magnetocrystalline) anisotropy energy (MAE) in bulk transition-metal (TM) ferromagnetic crystals attains extremely small values from the point of view of the computational solid-state physics: $\sim 4.10^{-6} - 5.10^{-5}$ (eV/atom) for Fe, Co, and Ni. Magnetic phenomena are of collective character and MAE can thus decide upon energy changes that are by no means negligible. For example, it is one of the main factors controlling the form of domain walls in magnets. The calculation on the first-principles level represents considerable problem and the results can depend on the details of calculation.[1] Various factors can be blamed for it. The density functional employed by theoreticians represents a simplification which can become yet more serious in future applications to nonhomogeneous magnetic structures.[2] The spin-orbit interaction, which is believed to be the cause of MAE[3], is introduced into the Schrödinger equation by a perturbative manipulation with the Dirac Hamiltonian. Naturally, the direct use of the Dirac operator is possible.[1,4] However, it's application without full quantization of fields has been questioned by field theoreticians[5] and is still object of discussion.[6] Nevertheless, presently it seems that an important difficulty to be solved lies in the numerical convergency (see Ref.[7] and below; for recent development we refer the reader to Refs.[1,8,9]). A crude guess of the MAE magnitude in cubic TM crystals follows from a simple reasoning when one accepts that spin-orbit coupling (SOC) is responsible for it. (We do not consider other mechanisms influencing the preferred magnetization direction. The most important is the interaction of magnetic dipoles responsible for the *shape* anisotropy. At interface with considerable misfit between the metallic radii of constituent atoms *elastic* or *magnetoelastic* interaction can be important as well[10,11,12,13].) One finds that in the 2nd order perturbation expansion in SOC the anisotropy can arise from terms $< L_i n_i .. L_j n_j .. >$, where $\mathbf{L} = \{L_i\}$ is the orbital moment operator, $\mathbf{n} = \{n_i\}$ is unit vector in the magnetization direction, .. stands for terms possessing the crystal-lattice symmetry, and $< ... >$ is mean value calculated for the unperturbed ground state. For crystals with cubic symmetry, however, the symmetry allows the substitution $n_i n_j \rightarrow (1/3)\delta_{ij}$ and the anisotropy can appear only in the terms of higher order (quartic) in SOC. Since the SOC parameter ξ is 5-10 meV for 3d TM's, the arguments based on the 4-th order perturbation theory yield a rough idea on the possible range of MAE values consistent with the data presented above. For surfaces and films the symmetry admits two distinct quadratic invariants $n_x^2 + n_y^2$ and n_z^2 (the z-axis being along the film normal) and a considerable enhancement of MAE can be anticipated.[14] Actually, in TM films the MAE is of order meV, i.e. $\sim \xi^2$.[15,16,17] Hypothetically, MAE and magnetic orbital momentum can be enhanced in appropriate multicomponent or antiferromagnetic superstructures.[18] This is because d-bands get narrow and contribute by small energy denominators in the perturbation theory. It is mainly the perpendicular anisotropy that attracts attention because of application in recording media. Nevertheless, only ultrathin films or other nanostructures are of interest from this point of view. In massive ferromagnetic films, starting from thickness of several layers, the dipole-dipole interactions overweight and force the in-plane magnetization.

2 Ab-initio schemes

Theoretical speculations on MAE origin date more than half century back.[3,19] A model of surface anisotropy appeared only later[14] and first band-theoretical calculations for TM surfaces or films were published in seventies.[20,21,22] Yet later realistic studies of TM films on the first-principles level were performed by Gay and Richter.[23] The latter paper shows that the MAE value depends critically on the number of **k**-points in the 2-dim Brillouin zone (BZ) employed in the calculation: as confirmed also by other groups[15,24] one needs to use about 1000 points or more to get a reasonably converged values. (For comparison, in the bulk MAE calculation[7] more than 500 000 **k**-points were used.) In nineties, TM-film MAE has been studied successfully by the first-principles local-density-functional methods by a number of authors as documented by few selected examples.[15,16,17,25,26] We mention also the numerical investigation.[27] The first-principle studies employ often the force theorem[28] that allows one not to recalculate the self-consistent potentials when the magnetization direction is changed. Numerical tests[15] confirm correctness of this approximation. Below, we are going to mention some other simplifications.

The small value of MAE represents itself a serious numerical problem. However, the main complication in most calculational schemes can be shown explicitly.[16] As we have seen, MAE is expected to be of order ξ^2 for uniaxial TM systems. On the other hand, the (quasi)degenerate energy eigenvalues in the vicinity of intersecting (or touching) bands are perturbed by $O(\xi)$ by SOC. When the band crossing occurs near the Fermi level E_F, the contribution to the *numerical* BZ integrals can be highly overestimated. As a function of E_F (or d-band filling) MAE might display irregular oscillations. The "brute-force" remedy consists in employing a large number of BZ points. It is well possible but is highly demanding and needs additional effort the check the convergency. The modification of the **kp**-method suggested recently[29] for extrapolation of band structures has not been yet tested for MAE calculations. In this respect, several approximations deserve attention:

1) *The state-tracking method.*[16,30] The main idea of this method is to reject the above "dangerous" contributions. The structure of energy-band crossing in two-dimensional structures is known.[31] Contribution from random crossings to energy integrals is $O(\xi^3)$ and it can be neglected. However, intersections along high-symmetry BZ directions and band degeneracy in high-symmetry points can contribute[32] by $O(\xi^{5/2})$ or $O(\xi^2)$, respectively. Hence, some points in BZ can turn as dangerous, although the practical application of the method[16] is encouraging. Note, that a similar idea was used in the tight-binding study.[33]

2) *The torque method.*[34] Starting from the phenomenological expression for MAE angular dependence in films (θ is the angle between the film magnetization and normal, respectively)

$$E(\theta) = K_2 \sin^2\theta + K_4 \sin^4\theta, \tag{1}$$

one finds that

$$E(90^o) - E(0^o) = dE(\theta)/d(\theta),\ \theta = 45^o. \tag{2}$$

The derivative in eq. (2) can be evaluated in microscopical theory by using the Hellmann-Feynman theorem.[34] The numerical tests of the torque method are favorable and recently the same conclusion has been made also for a semi-empirical tight-binding model.[35] The result shows that one needs to evaluate the expectation value of the operator[35]

$$0.5\xi(L_x\sigma_z - L_z\sigma_x) \tag{3}$$

in the *exact* ground state at $\theta = 45^o$. Above, σ_i are Pauli matrices and it is supposed that the change of θ is equivalent to rotation with the rotation axis y. The expression for the torque term within the Dirac-equation formalism is similar and is given in ref.[4].

3) *Scaling method.*[9] Were the SOC parameter large, the evaluation of large energy differences would be more convenient. The method starts from an exaggerated ξ value and then supposes that MAE for 3-dimensional crystals scales as ξ^4. For films, naturally, ξ^2 scaling should be assumed. Validity of the scaling assumption in the range of ξ employed must be checked by numerical tests.

3 Semi-empirical tight-binding Hamiltonian

The preceding discussion shows that for a number of important systems the MAE calculation can be performed on first-principles level. Nevertheless, experimentalists have discovered a number of nanostructures with non-trivial bearing on MAE such as film steps,[36,37,38] surface strips,[39] alloys[40] and adsorbates.[41] In similar situations also the orbital magnetism can reach anomalous values.[42] TM clusters deserve also attention.[43] Here, semi-empirical models might be also of help.

Attempt to get *realistic* results for TM thin-film MAE from a semi-empirical model was initiated by Bruno[33] and continued in refs.[32,44]. Typical TB Hamiltonian reads

$$H = H^o + \sum_q(-0.5\lambda_q\sigma_Z + \xi\vec{L}\cdot\vec{S}). \tag{4}$$

Above, H^o is the paramagnetic Hamiltonian, the second term leads to magnetic moment orientation along the axis Z, λ_q equals to the exchange splitting at the site q, and the third term is the SOC operator with the SOC parameter ξ. The sum extends over lattice sites q. It is convenient to identify the magnetization direction with axis Z which is independent of the coordinate system used when describing the lattice geometry and electron orbitals. As a consequence only the SOC operator depends explicitly on the magnetization direction. Strictly speaking, the Coulomb integrals in H^o and λ should be

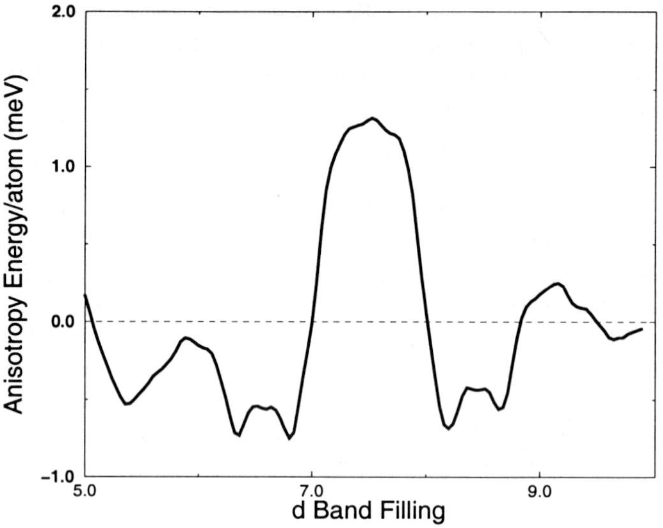

Figure 1. Magnetic anisotropy energy per atom versus the d-band filling for a free-standing (001) monolayer [?] Negative values correspond to favorable perpendicular direction of the magnetization. For the occupation $0 < N_d < 5$, the curve is practically symmetric with respect to the point N_d=5.

recalculated self-consistently when the direction Z changes. Another possibility - to neglect these corrections - is justified by a TB analogue[35,45,46] of the force theorem. In this case the MAE is expressed as the change of the band energy

$$E_{band} = \int^{E_F} (E - E_F) d\rho(E), \tag{5}$$

where $\rho(E)$ is density of electronic states. We shall commence by describing two general trends derived within TB models.

1) *Moment analysis.*[44] As it is well known, the n-th moment of the Hamiltonian is defined[47] as $\mu_n = TrH^n$, with the trace evaluated e.g. for all d-electron orbitals in the film elementary cell. It can be shown that the moments depend on the magnetization direction only staring from $n = 6$. (To avoid a misunderstanding, let us stress that the decisive terms in μ_6 are of order ξ^2.)

The so-called *oscillation theorem*[46,48] asserts that, as the atomic d-electron occupation changes in the range $0 \leq N_d \leq 10$, MAE changes its sign four times at least. There is evidence that the there are really four well-resolved sign changes in this range, apart from additional zero crossings when MAE is very small,[16,32,49] see also Figs. 1 and 2. An *a posteriori* analysis allows to relate the prominent features on the MAE curve to the film bend-structure features.[16,17] General rules having predictive power, however, are not known. 2) *MAE versus orbital magnetism.*[33,44] The common origin of orbital magnetism and MAE is SOC. If the hopping matrix elements between unlike spins in the SOC operator can be neglected, one gets from the 2nd-order perturbation formulas

$$\Delta E = \pm(\xi/4\mu_B)\Delta L, \tag{6}$$

where ΔE and ΔL are changes of the electronic energy and of the orbital moment, respectively, caused by the change of the magnetization orientation. The relation between MAE and orbital moment has been studied by a number of authors.[25,33,44,50,51] An important correlation between the two quantities exists. However, the above formula is not fully reliable even for strong ferromagnets.

In studies[22,33] considerable attention has been paid to crystal-filed splitting parametrization as a possible effect controlling MAE. It has been argued later[27,32,44] that the crystal-field splitting has appeared partly as an artifact of the model[33] and is most likely yet smaller when one goes from free-standing films to films grown on an appropriate substrate. The latter effect is important. For example, inclusion of substrate changes the preferred magnetization orientation of a hexagonal free-standing Co monolayer from in-plane into perpendicular one.[17] This happens despite of the fact that the corresponding change of the local density of the cobalt electronic states might seem quite negligible. The calculations[32,44] used the real-space recursion scheme method[47] to study ultrathin TM films. The effort has continued in ref.[52]. In principle, the recursion scheme enables one to study also more complicated systems lacking the periodicity. First steps in this direction have been done.[53] There are reasons[1,32] to believe that MAE is rather "local" (i.e. it depends critically on interactions within several atomic coordination spheres). The recursion method is especially suitable in such situations. In Figs. 1 and 2 we present results of such calculations for a simple model of free-standing (001) and (111) ferromagnetic films.[32] Fig. 2 documents the influence of model parameters on the result.

The first recursion-scheme studies[44] suffered from oscillations on the MAE curve similarly as in **k**-space calculations mentioned above. It appears, however, that the reason is different. In the recursion-scheme calculations, the partial local densities of electronic states $\rho_I(E)$ associated with an appropriately chosen basis of spin-orbitals $\{\phi_I\}$ are constructed in the form of continued fractions.[47] If the orbitals are build to respect the symmetry of the unperturbed Hamiltonian ($\xi = 0$), the unperturbed density ρ_I is nonzero on a subset Σ_I of the whole energy spectrum $\Sigma = \bigcup \Sigma_I$. For $\xi \neq 0$, ρ_I acquires nonzero values also on some $\Sigma_J, J \neq I$, where it is, however, much smaller than the density on Σ_I. Construction of such uneven densities if not effective in the recursion scheme and the convergency is extremely poor (150 levels of the continued fraction were used in[32] to get converged results). When new *mixed* basis is used with spin-orbitals that represent a "mix" of the symmetry-adapted orbitals, the new densities are much more uniformly spread on Σ. In

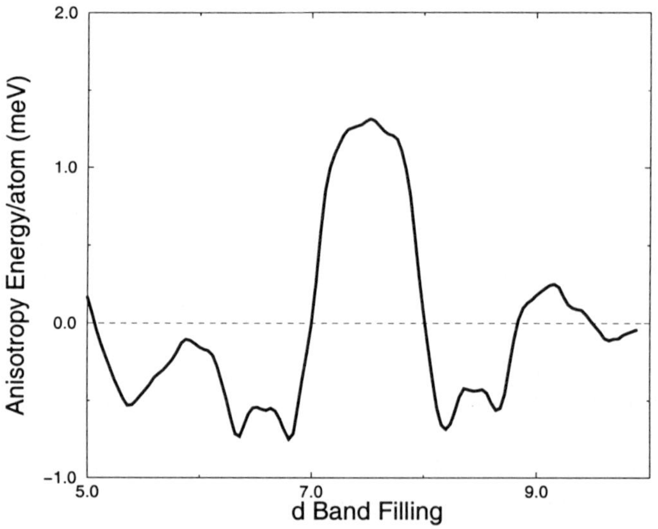

Figure 2. Magnetic anisotropy energy per atom versus the d-band filling for a free-standing (111) monolayer [?] Negative values correspond to favorable perpendicular direction of the magnetization. Full (long-dashed) line corresponds to 2.9 eV (1.9 eV) exchange splitting. The short-dashed line is calculated for exchange splitting 1.9 eV and non-zero crystal-field splitting 0.4 eV of d-orbital energies. The recursion scheme with the use of the *mixed basis* (see the text and ref. [?] was employed.

this way an extremely well convergent scheme was obtained; particularly, the converged MAE results were reached with only 25 levels of the continued fraction.[32] Recently, it has been shown[35] that the torque method (see eq. (3)) represents actually a mixed-basis approach and gives reliable results. (It should be noticed, however, that in calculation of the orbital magnetic moment eigenfunctions of the L_Z operator must be employed. Here, only partial mixing e.g. of spin-orbitals with opposite spins is possible and the convergency is somewhat less favorable.)

4 Outlook

In the last fifteen years tremendous progress has been achieved in the study of magnetic anisotropy of transition metals. However, Full-Potential ab-initio approaches are now limited to a small number of inequivalent atoms (currently less than ten atoms). Semi-empirical works should be able to describe larger systems. For instance a recent work[54] has been used to get orbital moments of semi-infinite transition metals. However, as in almost all calculations , the orbital moment values are expected to be underestimated. Only an ad-hoc treatment (the so-called orbital polarization),[8] brings values of the magnetic moments more in agreement with experimental observations. Clearly, in future effort should concentrate on fully ab-initio orbital magnetism determination.

Another necessity emerges: the real-space calculations within the tight-binding scheme as discussed previously and very recently within an *ab-initio* description[55] have shown their interest to describe complex systems. This effort should be amplified in the future. In some cases,[56] all the "physical" behavior is well described within the semi-empirical tight-binding scheme. In other cases the precision of an *ab-initio* method is necessary. Of course such a code has to be *Full Potential.* Recent work[57] using the FP-LMTO approach emphasizes the difficulties encountered to get a fully satisfactory description combining real-space and Full Potential descriptions. However getting a full real-space *ab-initio* formalism implemented on large supercomputers able to describe systems with few tens of inequivalent sites is a reasonable goal within the next years.

Acknowledgements

We would like to thank the COST P3 program for facilitating contact between the authors.

References

1. S. V. Beiden, W. M. Temmerman, Z. Szotek, G. A. Gehring, G. M. Stocks, Yang Wang, D. M. C. Nicholson, W. A. Shelton and H. Ebert, *Phys. Rev.* B **57**, 14247 (1998).
2. L. Kleinman, *Phys. Rev.* B **59**, 3314 (1999).
3. J. H. Van Vleck, *Phys. Rev.* **52**, 1178 (1937).
4. H. J. F. Jansen, *Phys. Rev.* B **59**, 4685, (1999).
5. R. Jost, *The General Theory of Quantized Fields*, in *American Mathematical Society*, Providence 1965, Chap. II, $ 5.
6. B. T. Sutcliffe, *Adv. Quantum Chem.* bf 31, 1 (1998); H. M. Quiney, H. Skaane and I. P. Grant, *Adv. Quantum Chem.* **32**, 1 (1998).
7. G. H. O. Daalderop, P. J. Kelly and M. F. H. Schuurmans, *Phys. Rev.* B **41**, 11919, (1990).
8. J. Trygg, B. Johansson, O. Eriksson and J. M. Wills, *Phys. Rev. Lett.* **75**, 2871 (1995).
9. S. V. Halilov, A. Ya. Perlov, P. M. Oppeneer, A. N. Yaresko and V. N. Antonov, *Phys. Rev.* B **57**, 9557 (1998).
10. K. Kyuno, J.-G. Ha, R. Yamamoto and S. Asano, *Phys. Rev.* B **54**, 1092 (1996).
11. Y.-S. Kim and S.-C. Shin, *Phys. Rev.* B **59**, R6597 (1999).
12. J. Henk, A. M. N. Niklasson and A. Johansson, *Phys. Rev.* B **59**, 9332 (1999).
13. D. Sandler, *Rep. Prog. Phys.* **62**, 809 (1999).
14. L. Néel, *J. Phys. Rad.* **15**, 225 (1954).
15. K. Kyuno, R. Yamamoto and S. Asano, *J. Phys. Soc. Japan* **61**, 2099 (1992).
16. D.-S. Wang R. Wu and A. J. Freeman, *Phys. Rev.* B **47**, 14932 (1993).
17. G. H. O. Daalderop, P. J. Kelly and M. F. H. Schuurmans, *Phys. Rev.* B **50**, 9989 (1994).
18. Š. Pick and H. Dreyssé, *Europhys. Lett.* **26**, 613 (1994), and unpublished results (1993).
19. H. Brooks, *Phys. Rev.* **58**, 909 (1940).
20. A. J. Bennett and B. R. Cooper, *Phys. Rev.* B **3**, 1642 (1971).
21. M. Kolář, *Czech. J. Phys.* B **25**, 806 (1975).
22. H. Takayama, K.-P. Bohnen and P. Fulde, *Phys. Rev.* B **14**, 2287 (1976).
23. J. G. Gay and R. Richter, *Phys. Rev. Lett.* **56**, 2728 (1987).
24. W. Karas, J. Noffke and L. Fritsche, *J. de Chimie Physique* **86**, 861 (1989).
25. L. Szunyogh, B. Újfalussy and P. Weinberger, *Phys. Rev.* B **51**, 9552

(1995).
26. I. V. Solovyev, P. H. Dederichs and I. Mertig, *Phys. Rev.* B **52**, 13419 (1995).
27. T. H. Moos, W. Hübner and K. H. Bennemann, *Solid State Commun.* **98**, 639 (1996); A. Lessard, T. H. Moos and W. Hübner, *Phys. Rev.* B **56**, 2594 (1997).
28. O. K. Andersen, O. Jepsen and D. Glötzel, in *Highlights of Condensed Matter Theory*, eds. F. Basani, F. Fumi and M. P. Tosi (North-Holland, Amsterdam, 1985), p. 59.
29. C. J. Pickard and M. C. Payne, *Phys. Rev.* B **59**, 4685 (1999).
30. D.-S. Wang, R. Wu and A. J. Freeman, *Phys. Rev. Lett.* **70**, 869 (1993).
31. J. von Neumann and E. Vigner, *Phys. Z.* **30**, 467 (1929); L. D. Landau and E. M. Lifshitz, *Quantum Mechanics, 3rd edition* (Pergamon, New York, 1977), $ 79.
32. Š. Pick, J. Dorantes-Dávila, G. M. Pastor and H. Dreyssé, *Phys. Rev.* B **50**, 993 (1994).
33. P. Bruno, *Phys. Rev.* B **39**, 865 (1989); (Doctoral Thesis, Université Paris-Sud (unpublished), 1989).
34. X. Wang, R. Wu and A. J. Freeman, *Phys. Rev.* B **54**, 61 (1996).
35. Š. Pick, *Solid State Commun.* (1999).
36. Jian Chen and J. L. Erskine, *Phys. Rev. Lett.* **68**, 1212 (1992).
37. D. S. Chuang, C. A. Ballantine and R. C. O'Handley, *Phys. Rev.* B **49**, 15084 (1994).
38. Hyuk J. Choi, R. K. Kawakami, E. J. Escoria-Aparicio, Z. Q. Qin, J. Pearson, J. S. Jiang, Dongqi Li and S. D. Bader, *Phys. Rev. Lett.* **82**, 1947 (1999).
39. H. J. Elmers, J. Hauschild and U. Gradmann, *Phys. Rev.* B **59**, 3688 (1999).
40. M. Maret, M. C. Cadeville, A. Herr, R. Pinsot, E. Beaupaire, S. Lefebvre and M. Bessière, *J. Magnet. Magnet. Mat.* **191**, 61 (1998).
41. Jian Chen, M. Drakaki and J. L. Erskine, *Phys. Rev.* B **45**, 3636 (1992).
42. H. A. Dürr, G. van der Laan, J. Vogel, G. Panaccione, N. B. Brokes, E. Dudzik and R. McGrath, *Phys. Rev.* B **58**, R11583 (1998).
43. M. Respaud, M. Gioran, J. M. Broto, F. H. Yang, T. Ould Ely, C. Amiens and B. Chaudret, *Phys. Rev.* B **59**, R3934 (1999).
44. Š. Pick and H. Dreyssé, *Phys. Rev.* B **46**, 5802 (1992); *Phys. Rev.* B **48** 13588 (1993).
45. Š. Pick, *J. Phys.* (France) **51**, 2011 (1990).
46. F. Ducastelle, *Order and Phase Stability in Alloys* (North Holland, Amsterdam, 1991).

47. R. Haydock, *Solid State Physics* **35**, 215 (1980).
48. F. Ducastelle and F. Cyrot-Lackmann, *J. Phys. Chem. Solids* **32**, 285 (1971).
49. Similar trends have been obtained recently by means of ab-initio schemes: P. James et al. , to be published, S. Blügel et al. , to be published, and in this book.
50. B. Újfalussy, L. Szunyogh, P. Bruno and P. Weinberger, *Phys. Rev. Lett.* **77**, 1805 (1996).
51. A. N. Anisimov, M. Farle, P. Poulopoulos, W. Platow, K. Baberschke, P. Isberg, R. Wäppling, A. M. N. Niklasson and O. Eriksson, *Phys. Rev. Lett.* **82**, 2391 (1999).
52. J. Dorantes-Dávila, H. Dreyssé and G. M. Pastor, *Phys. Rev.* B **55**, 15033 (1997).
53. G. M. Pastor, J. Dorantes-Dávila, Š. Pick and H. Dreyssé, *Phys. Rev. Lett.* **75**, 326 (1995); J. Dorantes-Dávila and G. M. Pastor, *Phys. Rev. Lett.* **81**, 208 (1998).
54. J. L. Rodríguez López, J. Dorantes-Dávila and G. M. Pastor, *Phys. Rev.* B **57**, 1040 (1998).
55. L. Petit, S. V. Beiden, W. N. Temmerman, Z. Szotek, G. M. Stocks and G. A. Gehring, *Phil. Mag.* B **78**, 449 (1998).
56. For the system TM/Co/Pd(111), with a tickness of Co up to 8 layers and TM is a transition metal overlayer, J. Dorantes-Dávila, H. Dreyssé and G. M. Pastor, to be published.
57. I. Galanakis, M. Alouani, J. M. Wills and H. Dreyssé, *Phil. Mag.* B **78**, 463 (1998).

CLASSICAL AND QUANTUM MAGNETIZATION REVERSAL IN SINGLE NANOMETER-SIZED PARTICLES AND WIRES

W. WERNSDORFER

Lab. L. Néel - CNRS, BP166, 38042 Grenoble Cedex 9, France
E-mail: wernsdor@labs.polycnrs-gre.fr

We discuss briefly various measuring techniques and review the basic concepts of a new micro-SQUID technique which allows one to study single nanometer- sized magnetic particles at low temperature. In the case of sufficiently small particles, the magnetization reversal occurs by uniform rotation and the measurement of the angular dependence of the magnetization reversal yields the effective magnetic anisotropy. The influence of the temperature on the magnetization reversal is discussed. Probabilities of switching, switching field distributions, and telegraph noise measurements are proposed to check the predictions of the Néel-Brown theory describing thermal activated magnetization reversal. At very low temperature, we will see how quantum effects can be revealed.

1 Introduction

Since the late 1940s, nanometer-sized magnetic particles have generated continuous interest as the study of their properties has proved to be scientifically and technologically very challenging. In particular it was recognized that the ferromagnetic state, with a given orientation of the particle moment, has a remanent magnetization if the particle is small enough. This was the starting point of huge permanent magnets and magnetic recording industries. However, despite intense activity during the last few decades, the difficulties in making nanoparticles of good enough quality has slowed the advancement of this field. As a consequence, for 50 years, these applications concentrated above and then near the micrometer scale. In the last few years, this has been no longer the case because of the emergence of new fabrication techniques which have led to the possibility of making small objects with the required structural and chemical qualities. In order to study these objects new techniques were developed such as magnetic force microscopy, magnetometry based on micro-Hall probes or micro-SQUIDs. This led to a new understanding of the magnetic behavior of nanoparticles, which is now very important for the development of new fundamental theories of magnetism and in modeling new magnetic materials for permanent magnets or high density recording.

In order to put this review article into perspective, let us consider Figure 1 which presents a scale of size which goes from macroscopic down to nanoscopic

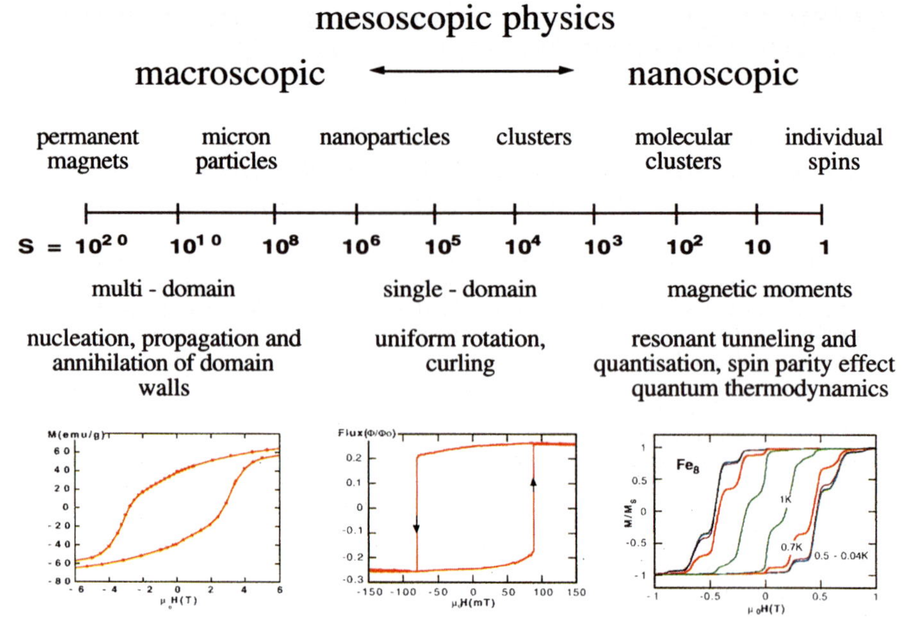

Figure 1. Scale of size which goes from macroscopic down to nanoscopic sizes. The unit of this scale is the number of magnetic moments in a magnetic system.

sizes. The unit of this scale is the number of magnetic moments in a magnetic system. At the macroscopic level, magnetism is governed by domains (Weiss 1907) and domain walls. The magnetization reversal occurs via nucleation, propagation and annihilation of domain walls. Size and width of the domain walls depend on the material of the magnetic system, on its size, shape and surface, and temperature.[1] The material dependence of the domain walls has motivated the definition of two length scales: (i) the domain wall width δ defined by $\delta = \sqrt{A/K}$ and (ii) the exchange length λ defined by $\lambda = \sqrt{A}/M_S$ where A is the exchange energy, K is the crystalline anisotropy constant and M_S is the spontaneous magnetization. Qualitatively, the first definition shows that the anisotropy energy favors a thin wall, while the exchange energy favors a thick wall. For very small crystalline anisotropy, the first definition suggests an infinite domain wall width which has a large total energy. This is due to the magnetostatic energy term which can be reduced by subdividing the ferromagnetic crystal into domains. Therefore, for very small crystalline anisotropy, the domain wall width is of the order of magnitude of the exchange length λ. Both length scales can range from sub-micrometer scales in alloys to atomic scales in rare earth systems.

When the system size is of the order of magnitude of δ and λ, the formation of domain walls requires too much energy, i.e. the magnetization remains in the so-called single domain state. Hence, the magnetization has to reverse mostly by uniform rotation. In this article, we discuss mainly this size range where the physics is rather simple.

For system sizes well below δ and λ, one must take into account the band structure of spins which are complicated by band structure modifications at the particle's boundaries.[2]

At the smallest size (below which one must consider individual atoms and spins) there are either free clusters made of some atoms[3] or molecular clusters which are molecules with a central complex containing magnetic atoms. In the last case, measurements on the Mn_{12}ac and Fe_8 molecular clusters showed that the physics can be described by a collective moment of spin $S = 10$. By means of simple hysteresis loop measurements, the quantum character of these molecules showed up by well defined steps which are due to resonance quantum tunneling between energy levels.[4] In addition, the Fe_8 molecular cluster is particularly interesting because of its biaxial anisotropy which allowed to observe for the first time[5] the existence of what in a semi-classical description is the quantum spin phase or Berry phase associated with the magnetic spin of the cluster. This could be done by measuring the tunneling matrix element Δ as a function of a magnetic field applied along the hard anisotropy axis. Oscillations of Δ were revealed which are due to constructive or destructive interference between two spin paths.[6] Furthermore, parity effects were observed when comparing the transitions between different energy levels of the system[5] which are analogous to the parity effect between systems with half integer or integer spins.[7] Hence, molecular chemistry had a large impact in the research of quantum tunneling of magnetization at molecular scales.

2 Single particle measurement techniques

The dream of measuring the magnetization reversal of an individual magnetic particle goes back to the pioneering work of Néel.[8] The first realization was published by Morrish and Yu in 1956.[9] These authors employed a quartz-fiber torsion balance to make magnetic measurements on individual micrometer sized $\gamma-$ Fe_2O_3 particles. With their technique, they wanted to avoid the complication of particle assemblies which are due to different orientations of the particle's easy axis of magnetization and particle-particle

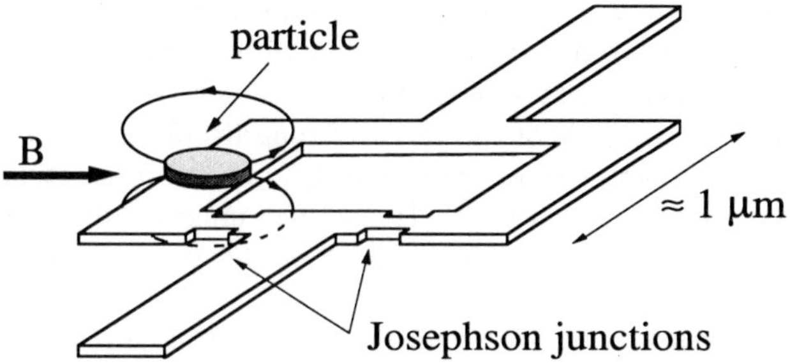

Figure 2. Drawing of a planar Nb micro-bridge-DC-SQUID on which a ferromagnetic particle is placed. The SQUID detected the flux through its loop produced by the sample's magnetization. Due to the close proximity between sample and SQUID we had a very efficient and direct flux coupling.

dipolar interaction. They aimed to show the existence of a single-domain state in a magnetic particle. Later on, other groups tried to study single particles but the experimental precision did not allow a detailed study. A first breakthrough came via the work of Knowles[10] who developed a simple optical method for measuring the switching field, defined as the minimum applied field required to reverse the magnetization of a particle. However, the work of Knowles failed to provide quantitative information on well defined particles. More recently, insights into the magnetic properties of individual and isolated particles were obtained with the help of electron holography,[11] vibrating reed magnetometry,[12] Lorentz microscopy,[13,14] and magnetic force microscopy.[15,16] Recently, most of the studies have been carried out using magnetic force microscopy at room temperature. This technique has an excellent spatial resolution but dynamical measurements are difficult due to the sample-tip interaction.

Only few groups could study the magnetization reversal of individual nanoparticles or nanowires at low temperatures. The first magnetization measurements of individual single-domain nanoparticles and nanowires at very low temperatures were presented by Wernsdorfer *et al.*[17] The detector (a Nb micro-bridge-DC-SQUID) and the studied particles were fabricated using electron-beam lithography. Coppinger *et al.*[18] indirectly investigated the magnetic properties of nanoparticles by resistance measurements. They observed the two-level fluctuations in the conductance of a sample containing self-organizing ErAs quantum wires and dots in a semi-insulating GaAs matrix. By measuring the electrical resistance of isolated Ni wires with diameters between 20 and 40 nm, Giordano and Hong studied the motion of magnetic domain walls.[19,20] Other low temperature techniques which may be adapted to single particle measurements are Hall probe magnetometry,[21] magnetometry based on the giant magnetoresistance[22,23] or spin-dependent tunneling with Coulomb blockade.[24] At the time of writing, the micro- SQUID technique allowed the most detailed study of the magnetization reversal of nanometer-sized particles.[25,26] The main principes of the micro-SQUID technique are resumed in Figure 2.

3 Mechanisms of magnetization reversal

For a sufficiently small magnetic sample it is energetically unfavorable for a domain wall to be formed at remanence. The specimen then behaves as a single domain. For extremely small particles, the magnetization should reverse by uniform rotation of magnetization. For somewhat larger particles, the curling reversal mode should be dominant.[1] For even larger particles, magnetization reversal occurs via a domain wall nucleation process starting in a rather small volume of the particle.[17,27] The crossover from coherent rotation to nucleation was also studied numerically.[28,29]

The model of uniform rotation of magnetization, developed by Stoner and Wohlfarth,[30] and Néel,[31] is the simplest classical model describing the magnetization reversal. One supposes a particle of an ideal material where exchange energy holds all spins tightly parallel to each other, and the magnetization does not depend on space. In this case, the exchange energy is constant, and does not enter the energy minimization. Therefore, there is competition only between the anisotropy energy of the particle and the effect of the applied field. The original study of Stoner and Wohlfarth assumed only uniaxial shape anisotropy which is the anisotropy of the magnetostatic energy of the sample induced by its non spherical shape. This is sufficient to describe highly symmetry cases like a prolate spheroid of revolution or an infinite cylinder. However, real systems are often quite complex and the anisotropy is a sum of mainly shape (magnetostatic), magnetocrystalline and surface anisotropy.

Shape anisotropy can be written as a biaxial anisotropy with two second order terms. Magnetocrystalline anisotropy is in most cases either uniaxial (hexagonal systems) or cubic, yielding mainly second and fourth order terms. Finally, in the simplest case, surface anisotropy is of second order. One additional complication arises because the different contributions of anisotropies are aligned in an arbitrary way to each other. All these facts motivated a generalization of the Stoner-Wohlfarth model for an arbitrary effective anisotropy which was done by Thiaville.[32] It is a geometrical method to determine the switching field for all angles of the applied magnetic field yielding the critical surface of switching fields, and to calculate the particle's energy.

The main interest of Thiaville's calculation is that measuring the critical surface of the switching field allows one to find the effective anisotropy of the nanoparticle. The knowledge of the latter is important for temperature dependent studies and quantum tunneling investigations. Knowing precisely the particle's shape and the crystallographic axis allows one to determine the different contributions of the effective anisotropy.

In order to demonstrate experimentally the uniform rotation mode, the angular dependence of the magnetization reversal has often been studied. It is important to choose a proper system which minimizes nonuniform magnetization states. This can by done by choosing a system in which one particular anisotropy contribution is dominant.

For example, the magnetization reversal in a single-crystalline nanowire, of a material with low magnetocrystalline anisotropy, is dominated by uniaxial shape anisotropy.[33] Another example are BaFeO nanoparticles which have a dominant uniaxial magnetocrystalline anisotropy.[34] Finally, nearly spherical nanoparticles are also nice model systems as ellipsoids favor uniform magnetization inside the particle. Here, magnetocrystalline and shape anisotropies are the dominant contributions.[35]

4 Experimental methods for the study of the Néel-Brown model of thermally activated magnetization reversal

The thermal fluctuations of the magnetic moment of a single-domain ferromagnetic particle and its decay towards thermal equilibrium was introduced by Néel[8] and further developed by Brown.[36] In their model of thermally activated magnetization reversal, a single domain magnetic particle has two equivalent ground states of opposite magnetization separated by an energy barrier which is due to shape and crystalline anisotropy. The system can escape from one state to the other by thermal activation over the barrier. Just as in the Stoner-Wohlfarth model, they assumed uniform magnetization and uniaxial anisotropy in order to derive a single relaxation time. Néel supposed further that the energy barrier between the two equilibrium states is large in comparison to the thermal energy $k_B T$ which justified a discrete orientation approximation. Brown criticized the model of Néel because the system is not explicitly treated as a gyromagnetic one.[36] Brown considered the magnetization vector in a particle to wiggle around an energy minimum, then jump to the vicinity of the other minimum, then wiggle around there before jumping again. He supposed that the orientation of the magnetic moment may be described by a Gilbert equation with a random field term which is assumed to

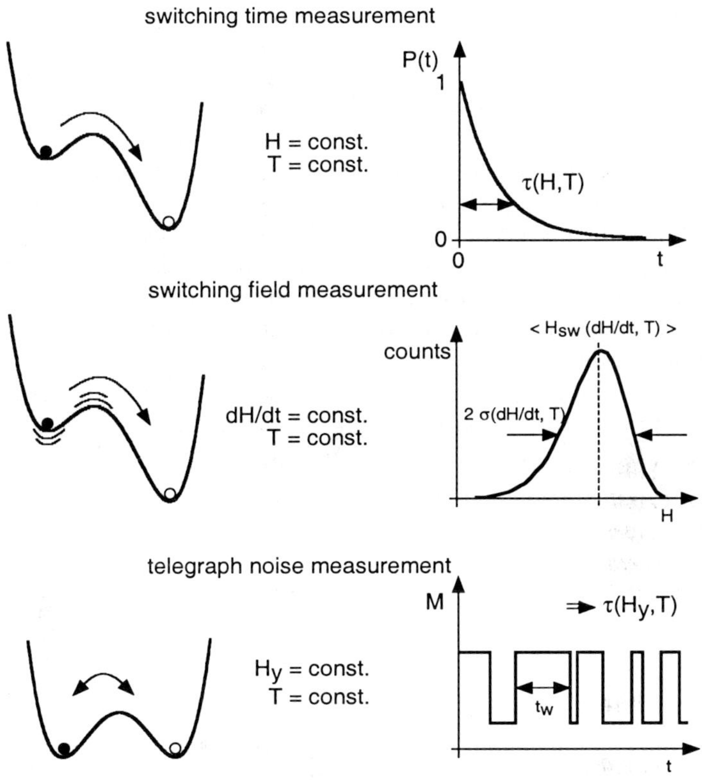

Figure 3. Schema of three methods for studies of the escape from a metastable potential well.

be white noise. On the basis of these assumptions, Brown was able to derive a Fokker-Planck equation for the distribution of magnetization orientations. Brown did not solve his differential equation. Instead he tried some analytic approximations and an asymptotic expansion for the case of the field parallel or perpendicular to the easy axis of magnetization. More recently, Coffey *et al.* found by numerical methods an exact solution of Brown's differential equation for uniaxial anisotropy and an arbitrary applied field direction.[37] They derived also an asymptotic general solution for the case of large energy barriers in comparison to the thermal energy k_BT. In the following, we consider a simplified version of the Néel-Brown model and propose three techniques to verify the validity of this model.

4.1 Waiting time measurements

The waiting time method corresponds to magnetization relaxation measurements on assemblies of particles. In this case, the decay of magnetization is often logarithmic in time which is due to a broad distribution of switching fields. For individual particle studies, waiting time measurements give direct access to the switching probability (Fig. 3). At a given temperature, the magnetic field is increased to a waiting field H_w near the switching field H^0_{sw}. Next, the elapsed time until the magnetization switches is measured. This process is repeated several hundred times, yielding a waiting time histogram. The integral of this histogram yields the switching probability. Finally, the switching probability is measured at different waiting fields H_w in order to explore several barrier heights, and at different temperatures. According to the Néel-Brown model, the probability that the magnetization has not switched after a time t is given by:

$$P(t) = e^{-t/\tau} \tag{1}$$

where τ (inverse of the switching rate) can be expressed by an Arrhenius law of the form:

$$\tau(T,H) = \tau_0 e^{E(H)/k_B T} \tag{2}$$

where τ_0 is the inverse of the attempt frequency. A simple analytical approximation for the field dependence of the energy barrier $E(H)$ is:

$$E(H) = E_0 \left(1 - H/H^0_{sw}\right)^\alpha = E_0 \varepsilon^\alpha \tag{3}$$

where E_0 is the energy barrier at zero field, and ε is defined as a reduced field difference value. It can be shown that the exponent α is in general equal to 1.5.[38] Using the Stoner-Wohlfarth analytical expressions of $E(H)$,[30] we can numerically show that α is near 1.5[26] and increases up to α value of 2 (more frequently cited in the literature) if the applied field forms an angle smaller than a few degrees with the easy axis of magnetization.

4.2 Switching field measurements

For single particle studies, it is often more convenient to study the magnetization reversal by ramping the applied field at a given rate and measuring the field value as soon as the particle's magnetization switches. In this case, the switching process is characterized by a distribution of switching fields (Fig. 3) evaluated by Kurkijärvi.[39,40] The most probable switching field H_{sw} is given by:

$$H_{sw} = H_{sw}^0 \left(1 - \left[\frac{k_B T}{E_0} ln \left(\frac{cT}{v\varepsilon^{\alpha-1}}\right)\right]^{1/\alpha}\right) \tag{4}$$

where $c = k_B H_{sw}^0/(\tau_0 \alpha E_0)$ and v is the field sweeping rate. The width of the switching field distribution is given by:

$$\sigma \approx H_{sw}^0 \frac{1}{\alpha} \left(\frac{k_B T}{E_0}\right)^{1/\alpha} \left[\frac{k_B T}{E_0} ln \left(\frac{cT}{v\varepsilon^{\alpha-1}}\right)\right]^{(1-\alpha)/\alpha} \tag{5}$$

4.3 Telegraph noise measurements

At zero applied field, a single domain magnetic particle has two equivalent ground states of opposite magnetization separated by an energy barrier. When the thermal energy $k_B T$ is sufficiently high, the total magnetic moment of the particle can thermally fluctuate, like a single spin in a paramagnetic material. Such magnetic behavior of an assembly of independent single domain particles is called superparamagnetism.

In order to study the superparamagnetic state of a single particle, it is simply necessary to measure the particle's magnetization as a function of time which is called telegraph noise measurement as stochastic fluctuations between two states are expected. According to the Néel-Brown model, the mean time τ spent in one state of magnetization is given by an Arrhenius law of the form of equations (2). As τ increases exponentially with decreasing temperature, it is very unlikely to observe an escape process at low temperature. However, applying a constant field in direction of a hard axis (hard plane) of magnetization reduces symmetrically the height of the energy barrier (Fig. 3). When the energy barrier is sufficiently small, the particle's magnetization can fluctuate between two orientations which are close to hard axis (hard plane) of magnetization.

The time spent in each state follows an exponential switching probability law as given by equations (1-3) with $\alpha = 2$. Note that for a slightly asymmetric energy potential, one switching probability can be so long that two level fluctuation becomes practically unobservable.

4.4 Experimental evidence for the Néel-Brown model

The Néel-Brown model is widely used in magnetism, particularly in order to describe the time dependence of the magnetization of collections of particles, thin films and bulk materials. However until recently, all the reported measurements, performed on individual particles, were not consistent with the Néel-Brown theory. This disagreement was attributed to the fact that real samples contain defects, ends and surfaces which could play an important, if not dominant, role in the physics of magnetization reversal. It was suggested that the dynamics of reversal occurs via a complex path in configuration space, and that a new theoretical approach is required to provide a correct description of thermally activated magnetization reversal even in single-domain ferromagnetic particles.[16,41] Similar conclusions were drawn from numerical simulations of the magnetization reversal.[28]

A few years later, Wernsdorfer *et al.* found experimentally for the first time[42,35,34,44] a very good agreement with the Néel-Brown model by using waiting time, switching field and telegraph noise measurements. They found also that sample defects, especially sample oxidation, play a crucial role in the physics of magnetization reversal. The field and temperature dependence of the exponential prefactor τ_0 was also taken into account.[44]

5 Macroscopic quantum tunneling of magnetization

Macroscopic quantum tunneling (MQT) represents one of the most fascinating phenomena in condensed matter physics.[45] MQT means the tunneling of a macroscopic variable through a barrier of an effective potential of a macroscopic system. Today, macroscopic systems showing amazing agreements with theoretical predictions are Josephson Junctions and SQUIDs. Here, quantum tunneling between two macroscopically distinct current states has been observed. After a slow evolution during the last ten years, MQT in magnetism now constitutes a new and very interesting field of research. It has been predicted that MQT can be observed in magnetic systems with low dissipation. In this case, it is the tunneling of the magnetization vector of a single-domain particle through its anisotropy energy barrier or the tunneling of a domain wall through its pinning energy. These phenomena have been studied theoretically and experimentally,[46,34] and more results are expected in the near future.

Acknowledgements

We acknowledge A. Aharoni, R. Ballou, A.-L. Barra, B. Barbara, A. Benoit, E. Bonet Orozco, I. Chiorescu, W.T.Coffey, B. Doudin, A. Garg, D. Gatteschi, D. Givord, K. Hasselbach, D.Mailly, B. Pannetier, C. Paulsen, J.-L. Porteseil, R. Sessoli, P. Stamp, A.Thiaville, and C. Thirion for very helpful discussions and experimental contributions.

References

1. A. Aharoni, *An Introduction to the Theory of Ferromagnetism* (Oxford University, London, 1996).
2. A.J. Freemann and R. Wu, *J. Magn. Magn. Mater* **100**, 497, (1991).
3. I.M.L. Billas, A. Chtelain, W.A. de Heer, *J. Magn. Magn. Mat.* **168**, 64 (1997).
4. J.R. Friedman *et al.* , *Phys. Rev. Lett.* **76**, 3830 (1996).
5. W. Wernsdorfer and R. Sessoli, *Science* **284**, 133 (1999).
6. A. Garg, *Europhys. Lett.* **22**, 205 (1993).
7. D. Loss, D.P. DiVincenzo, and G. Grinstein, *Phys. Rev. Lett.* **69**, 3232 (1992).
8. L. Néel, *Ann. Geophys.* **5**, 99 (1949) and *C. R. Acad. Science, Paris* **228**, 664 (1949).
9. A.H. Morrish and S.P. Yu, *Phys. Rev.* **102**, 670 (1956).
10. J.E. Knowles, *IEEE Trans. Magn.* **MAG-14**, 858 (1978) and *IEEE Trans. Magn.* **MAG-20**, 84 (1984).
11. A. Tonomura *et al.* , *Phys. Rev.* B **34**, 3397 (1986).
12. H.J. Richter, *J. Appl. Phys.* **65**, 9 (1989).
13. S.J. Hefferman, J.N. Chapman and S. McVitie, *J. Mag. Mag. Mat.* **95**, 76 (1991).
14. C. Salling, S. Schultz, I. McFadyen, M. Ozaki, *IEEE Trans. Mag.* **27**, 5185 (1991).
15. T. Chang and J.G. Chu, *J. Appl. Phys.* **75**, 5553 (1994).
16. M. Ledermann, S. Schultz and M. Ozaki, *Phys. Rev. Lett.* **73**, 1986 (1994).
17. W. Wernsdorfer *et al.* , *J. Magn. Magn. Mat.* **145**, 33 (1995).
18. F. Coppinger *et al.* , *Phys. Rev. Lett.* **75**, 3513 (1995).
19. K. Hong and N. Giordano, *J. Mag. Mag. Mat.* **151**, 396 (1995) and *Europhys. Lett.* **36**, 147 (1996).
20. J.E. Wegrowe *et al.* , *IEEE Trans. on Mag.* **34**, 903 (1998).
21. A.D. Kent, S. von Molnar, S. Gider, D.D. Awschalom, *J. Appl. Phys.* **76**,

6656 (1994).
22. V. Gros *et al.* , *J. Magn. Magn. Mat.* **165**, 512 (1997).
23. W.J. Gallagher *et al.* , *J. Appl. Phys.* **81**, 3741 (1997).
24. L.F. Schelp *et al.* , *Phys. Rev.* B **56**, R5747 (1997).
25. W. Wernsdorfer *et al.* , *J. Appl. Phys.* **78**, 7192 (1995).
26. W. Wernsdorfer, *Ph. D. thesis* (Joseph Fourier University, Grenoble, 1996).
27. W. Wernsdorfer *et al.* , *Phys. Rev.* B **53**, 3341 (1996).
28. H.L. Richards, S.W. Sides, M.A. Novotny and P.A. Rikvold, *J. Appl. Phys.* **79**, 5749 (1996); E.D. Boerner and H. Neal Bertram, *IEEE Trans. Mag.* **33**, 3052 (1997).
29. D. Hinzke and U. Nowak, *Phys. Rev.* B **58**, 265 (1998).
30. E.C. Stoner and E.P. Wohlfarth, *Philos. Trans. London Ser.* A **240**, 599 (1948); reprinted in *IEEE Trans. Magn.* **MAG-27**, 3475 (1991).
31. L. Néel, *C. R. Acad. Science, Paris* **224**, 1550 (1947).
32. A. Thiaville, *J. Magn. Magn. Mat.* **182**, 5 (1998) and *Phys. Rev.* B (1999).
33. W. Wernsdorfer *et al.* , *J. Appl. Phys.* **81**, 5543 (1997).
34. W. Wernsdorfer *et al.* , *Phys. Rev. Lett.* **79**, 4014 (1997).
35. W. Wernsdorfer *et al.* , *Phys. Rev. Lett.* **78**, 1791 (1997).
36. W.F. Brown, *J. Appl. Phys.* **30**, 130S (1959); *J. Appl. Phys.* **34**, 1319 (1963); *Phys. Rev.* **130**, 1677 (1963); *IEEE Trans. Mag.* **15**, 1196 (1979).
37. W.T. Coffey *et al.* , *Phys. Rev.* B **52**, 15951 (1995); L.J. Geoghegan, W.T. Coffey and B. Mulligan, *Adv. Chem. Phys.* **100**, 475 (1997); W.T. Coffey, *Adv. Chem. Phys.* **103**, 259 (1998).
38. R.H. Victora, *Phys. Rev. Lett.* **63**, 457 (1989).
39. J. Kurkijärvi, *Phys. Rev.* B **6**, 832 (1972).
40. A. Garg, *Phys. Rev.* B **51**, 15592 (1995).
41. W. Wernsdorfer *et al.* , *J. Magn. Magn. Mat.* **151**, 38 (1995).
42. W. Wernsdorfer *et al.* , *Phys. Rev. Lett.* **77**, 1873 (1996).
43. W. Wernsdorfer *et al.* , *Phys. Rev.* B **55**, 11552 (1997).
44. W.T. Coffey *et al.* , *Phys. Rev. Lett.* **80**, 5655 (1998).
45. A.J. Leggett *et al.* , *Rev. Mod. Phys.* **59**, 1 (1987).
46. *Quantum Tunneling of Magnetization - QTM'94*, ed. L. Gunther and B. Barbara (NATO ASI, Serie E: Applied Sciences, Vol. 301, 1995).

PHASE DIAGRAM OF A CLASSICAL, ANISOTROPIC HEISENBERG ANTIFERROMAGNET IN A FIELD

M. ACHARYYA, U. NOWAK AND K. D. USADEL

Theoretische Tieftemperaturphysik, Gerhard-Mercator-Universität-Duisburg, 47048-Duisburg, Germany

E-mail: uli@thp.uni-duisburg.de

We study a classical, anisotropic Heisenberg antiferromagnet in a field by Monte Carlo methods. We calculate the transverse and longitudinal antiferromagnetic order parameters as function of temperature and axial magnetic field in order to obtain the phase diagram in the field-temperature plane consisting of antiferromagnetic-, spin-flop-, and saturated paramagnetic phase. For zero temperature the critical fields can be calculated analytically. Crossing the spin flop phase by varying the temperature, the magnetic specific heat shows two peaks.

1 Introduction

The *metamagnetic*[1] behavior of various antiferromagnetic materials has drawn much attention in modern research.[1,2,3,4,5,6,7] The field induced phase transition and the multicritical phenomena are the major part of interest. In recent experiments[2,3,4] on $FeBr_2$, the temperature variation of the specific heat shows the appearance of a second anomalous peak at lower temperature in addition to the usual critical peak at the Néel temperature. A possible explanation of this anomalous peak was first given by Pleimling and Selke[5] who performed extensive Monte Carlo calculations of an Ising-metamagnet considering a hexagonal lattice structure. It was speculated that due to the positive axial field small clusters of positive spins in the negative sea may form a stable phase. Nevertheless, a detailed characterization of this "intermediate phase" is missing in the literature.

However, recent experimental observations[4] of a transverse spin ordering, associated with a weak first order transition and a sharp peak of the specific heat, cannot be explained by a simple Ising model. A model with transverse spin components is necessary. This prompted us to study a classical anisotropic Heisenberg model.

Here, we report briefly on our Monte Carlo results of an anisotropic Heisenberg model. In the next section we present the model and the Monte Carlo simulation scheme is discussed. In section III numerical results are shown and compared with exact analytical calculation given in section IV. The paper ends with a summary.

2 The Classical anisotropic Heisenberg Model and the Monte Carlo Simulation

The classical anisotropic Heisenberg model with competing interactions, in presence of a magnetic field, can be represented by the Hamiltonian

$$\begin{aligned}\mathcal{H} = & - J/2 \sum_{<ij>} \vec{S}_i \cdot \vec{S}_j - J'/2 \sum_{<ij>'} \vec{S}_i \cdot \vec{S}_j \\ & - D \sum_i (S_i^z)^2 - \vec{H} \cdot \sum_i \vec{S}_i, \end{aligned} \tag{1}$$

where the $\vec{S}_i$ represent classical spin vectors of unit length at site i of the lattice. The spin vectors may point into any direction in spin space continuously. For simplicity, we have chosen a tetragonal lattice of linear size L and the range of interaction is limited to nearest neighbors only. D is the uniaxial anisotropy constant favoring the spin to be aligned either parallel or antiparallel to the z-axis and $\vec{H}$ is the external, uniform magnetic field which throughout this article will be parallel to the easy axis of the system. The exchange interaction of spins within planes perpendicular to the z-axis is ferromagnetic ($J > 0$) while the inter-planar interaction is antiferromagnetic ($J' < 0$). We use periodic boundary condition in all directions.

We have performed Monte Carlo simulations[8] (using the Metropolis rate for the single spin change from $\vec{S}$ to $\vec{S}'$ where $\vec{S}'$ is chosen randomly) for the system described above where we used a linear system size of $L = 20$. L^3 random updates of spins defines one Monte Carlo Step per site (MCSS).

Measuring all energetical quantities in units of the ferromagnetic inter-planar interaction J we set the antiferromagnetic intra-planar interaction to $J' = -0.5J$) and the anisotropy to $D = 0.3J$. It should be noted here, that very large values of D will yield Ising-like behavior. To have transverse ordering one has to choose lower values for the anisotropy. We have suitably chosen the parameter values in such a way that the anisotropy is high enough to yield a longitudinal antiferromagnetic phase at low axial field and low enough to show a spin-flop spin structure with transverse spin components for a larger axial field.

We used two different procedures to investigate the phase diagram. One of those is a cooling procedure: starting from an initially random configuration (corresponding to a high temperature phase) we equilibriate the system up to 4×10^4 MCSS and calculated thermal averages and fluctuations from further 4×10^4 MCSS. Hence, the total length of the simulation for one fixed temperature T is 8×10^4 MCSS. Then we decrease the temperature and use the last spin configuration of the simulation before as the initial configura-

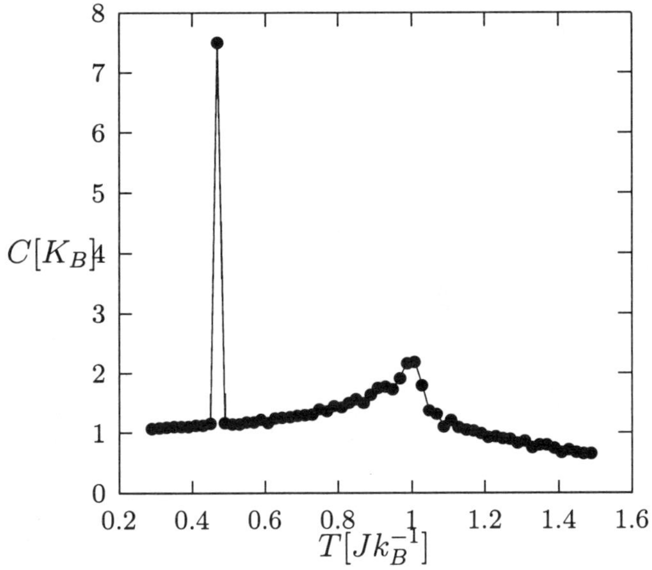

Figure 1. Monte Carlo results for the temperature variation of the specific heat ($C = dE/dT$) for $H_z = 0.68$. The solid line is just çonnecting the data points.

tion for the calculation for the next temperature. In this way we simulate a cooling procedure which is closer to equilibrium compared to starting with a random configuration for each temperature. The other procedure is a field cycle for constant temperature. Here, we start with an antiferromagnetically ordered configuration at zero field. Then we increase the field stepwise until the system reaches a saturated paramagnetic state. Then we decrease the field again. This cycle gives the chance to observe hysteresis in case of a first order phase transition.

3 Simulation results

We measure the total energy (per spin) $E = \langle \mathcal{H} \rangle / L^3$ of the system from which the specific heat can be calculated as the derivative, i.e., $C = dE/dT$.

Fig. 1 shows the temperature variation of the specific heat for $H_z = 0.68$. The high temperature peak is identified as a signature of the continuous transition from saturated paramagnetic (SP) to a spin-flop (SF) phase and the low temperature sharp peak corresponds to a discontinuous transition

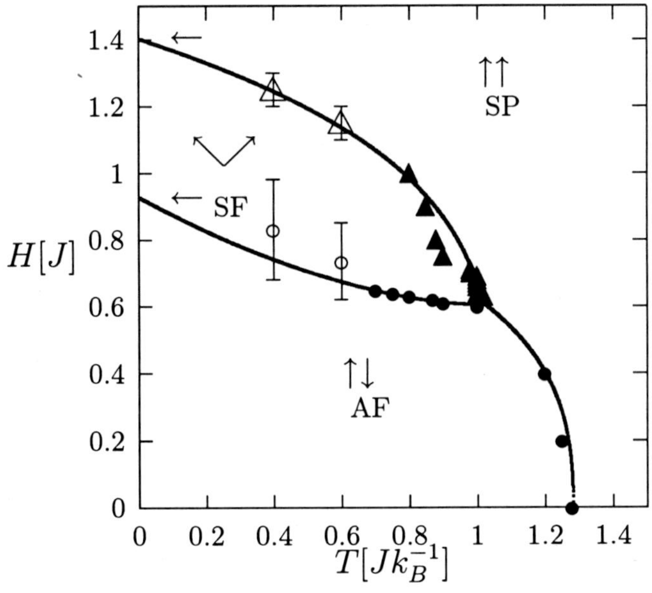

Figure 2. The Monte Carlo results for the phase diagram in the field-temperature plane. The horizontal arrows indicate the critical fields h_1 and h_2. The solid lines are guides to the eye.

from a spin-flop phase to a longitudinal antiferromagnetic (AF) phase. To demonstrate this, we have also calculated the longitudinal and transverse antiferromagnetic order parameters. From the temperature variations of these two order parameters we estimated the transition point as a function of the axial field shown as filled symbols in the phase diagram in Fig. 2. We also used field cycling procedures to investigate the phase diagram. The corresponding transition points are shown as open symbols in Fig. 2. During field cycling at the lower transition from AF to SF a broad hysteresis can be observed the width of which is shown as error bars. The occurrence of hysteresis is a hint for a first order phase transition.

Comparing this phase diagram with the specific heat data in Fig. 1 we find that the high temperature peak corresponds to a continuous transition from a saturated paramagnetic phase to a spin-flop phase while the low temperature sharp peak stems from the latent heat of the discontinuous transition from a spin-flop phase to a longitudinal antiferromagnetic phase.

4 The exact critical fields at zero temperature

One can easily calculate[9] the critical fields at zero temperature. In a ground state all spins within one sublattice can be assumed to be parallel oriented. Let θ be the angle between the easy axis and the sublattice magnetization vector (or spin). For the saturated paramagnetic phase the angle 2θ between two spins in different sublattices is 0 and the corresponding energy is

$$E_{SP} = -2J + J' - D - h.$$

Within a spin-flop ground state this angle can take any value. For any general angle θ the total energy (per site) of the system following from the Hamiltonian above, is

$$E_{SF} = -2J + J' \cos 2\theta - D \cos^2 \theta - h \cos \theta.$$

By minimizing this energy with respect to θ one can find the stable angle between two spins from the equation

$$\theta = \arccos \frac{h}{2(2J' - D)}$$

In the longitudinal antiferromagnetic ground state the energy is:

$$E_{AF} = -2J - J' - D$$

By comparing E_{AF} and E_{SF}) one finds the critical field

$$h_1 = 2\sqrt{D(2J' - D)}$$

for the transition from the antiferromagnetic phase to the spin-flop phase and by comparing E_{SF} and E_{SP} one can get the critical field

$$h_2 = 2(2J' - D)$$

for the transition from spin-flop to saturated paramagnetic phase. One can compare our Monte Carlo estimates with $h_1 \cong 0.92$ and $h_2 \cong 1.4$ for $J' = 0.5$ and $D = 0.3$ (see Fig.2). It is interesting to note that the region of spin-flop phase disappears for $D \geq J'$, which follows from the condition $h_1 = h_2$. Also, the values of the critical fields are independent of the ferromagnetic interaction strength.

5 Summary

We have studied the classical, anisotropic Heisenberg model with competing intra-layer and inter-layer interaction by Monte Carlo simulations. We have calculated the transverse and longitudinal antiferromagnetic order parameters and studied them as a function of temperature axial fields. In the field-temperature plane we have observed three different phases, namely the longitudinal antiferromagnetic phase, the spin-flop phase and the saturated paramagnetic phase. The zero temperature critical fields are obtained from exact calculations. They agree roughly with the Monte Carlo results.

Acknowledgements

We would like to thank Ch. Binek, W. Kleemann, O. Petracic, M. Pleimling and W. Selke for various discussion. MA would like to acknowledge the financial support from the Graduiertenkolleg "Struktur und Dynamik heterogener Systeme" at the University of Duisburg.

References

1. E. Stryjewski and N. Giordano, *Advances in Physics* **26**, 487 (1977).
2. M.M.P. de Azevedo, Ch. Binek, J. Kushauer, W. Kleemann and D. Betrand, *J. Mag. Mag. Mat.* **140**, 1557 (1995).
3. H. Aruga Katori, K. Katsumata and M. Katori, *Phys. Rev.* B **54**, R9620 (1996); H. Aruga Katori, K. Katsumata and M. Katori, *J. Appl. Phys.* **81**, 4396 (1997); K. Katsumata, H. Aruga Katori, S.M. Shapiro and G. Shirane, *Phys. Rev.* B **55**, 11466 (1997).
4. O. Petracic, Ch. Binek and W. Kleemann, *J. Appl. Phys.* **81**, 4145 (1997); O. Petracic, Ch. Binek, W. Kleemann, U. Neuhausen and H. Leuken, *Phys. Rev.* B **57**, R11051 (1998).
5. M. Pleimling and W. Selke, *Phys. Rev.* B **56**, 8855 (1997) and the references therein.
6. M. Pleimling and W. Selke, *Phys. Rev.* B (1999), in press.
7. M. Pleimling, *Eur. Phys. J.* B, (1999), in press.
8. K. Binder and D. Heermann, *Monte Carlo Simulation in Statistical Physics: An Introduction* (Springer, Berlin, 1992).
9. W. Gebhardt and U. Krey, *Phasenübergänge und kritische Phänomene*, (Vieweg, 1980), Kapitel 4.

TWO-DIMENSIONAL BEHAVIOR OF THE SUBLATTICE MAGNETIZATION IN THREE DIMENSIONAL ISING ANTIFERROMAGNETS

O. PETRACIC, CH. BINEK AND W. KLEEMANN

Laboratorium für Angewandte Physik, Gerhard-Mercator-Universität Duisburg, D-47048 Duisburg, Germany

A three-dimensional layered Ising-Antiferromagnet with a ferromagnetic intra-layer coupling to z neighbors, $zJ > 0$, and an antiferromagnetic interlayer coupling to z' neighbors, $z'J' < 0$, is investigated by Monte Carlo simulations on a hexagonal lattice. The physical nature of the anomalous temperature bahavior of the sublattice magnetizations, which is found for certain values of $r = zJ/z'J'$ and z' in magnetic fields is explained in terms of successive phase transitions. They take place on the ferromagnetic 2-dimensional spin-down sublattice at $T \approx T_c^{2d}$, smeared by a finite stabilizing molecular field, and on both antiferromagnetically coupled sublattices at $T_c^{3d} > T_c^{2d}$.

1 Introduction

Anisotropic three-dimensional antiferromagnets (AFs) are still an interesting subject of theoretical investigations. There exist many investigations e.g. on their magnetic phase diagram within the framework of Ising or anisotropic Heisenberg models.[1–11] It is found, that layered Ising AFs with two competing interaction parameters (Fig. 1) may exhibit two rather different phase diagrams (Fig. 2) depending only on two parameters, the ratio $r = zJ/z'J'$ and z',[9] where J and J' are the coupling constants of the intra-layer and of the inter-layer exchange, respectively; z and z' are the coordination numbers of the couplings. The Ising Hamiltonian is of the form:

$$\mathcal{H} = -J \sum_{<i,j>} S_i S_j - J' \sum_{<i,j>} S_i S_j - H \sum_i S_i, \tag{1}$$

where H is the applied magnetic field acting on all spins S_i, with $S_i = \pm 1$. The two kinds of phase diagrams are continuously transformed into one another by changing the crucial parameters r and z'. For large values of $|r|$, i.e. $|r| > 0.6$,[3] and small values for z', i.e. $z' < 10$,[9] the $FeCl_2$-like phase diagram[12] is found (Fig. 2 (b)), whereas for low values of $|r|$ and large values for z' the case (b) in Fig. 2 appears. This second case is characterized by three interesting features. On the one hand, a possible decoupling of the tricritical point (TCP) into a critical endpoint (CEP) and a bicritical endpoint (BCE)

Figure 1. Schematic drawing of a layered 3-dimensional antiferromagnet with a ferromagnetic coupling in the layers, $J_{A-A} = J_{B-B} = zJ$, and an antiferromagnetic coupling between adjacent layers, $J_{A-B} = J_{B-A} = z'J'$. The magnetic field, H, is applied along the antiferromagnetic stacking direction.

is encountered. From previous investigations it follows, that this decoupling is only observed in mean field calculations,[2–9] while in Monte Carlo simulations only one multicritical point is found.[10] On the other hand, the second-order phase line has a balloon-like shape and extends even above the limiting field value H_{c0} (spin-flip field at $T = 0$). This means, that for some fixed field values $H > H_{c0}$ it is possible to cross this phase line twice with increasing temperature T. Furthermore, above and below the critical line anomaly lines are found, where the magnetization exhibits an additional inflection point and the specific heat shows an additional broad maximum.[9,10]

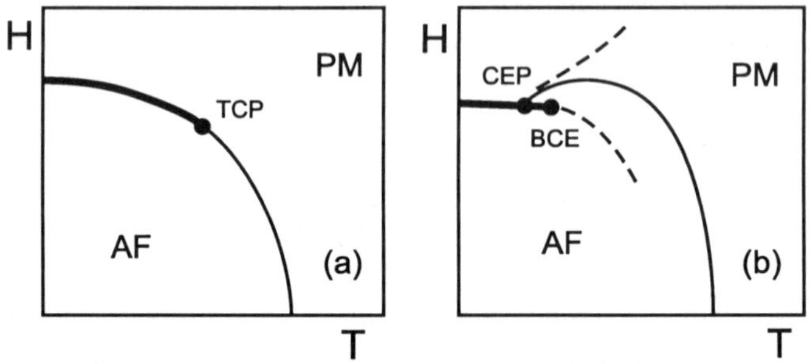

Figure 2. Schematic magnetic phase diagrams of 3-dimensional layered Ising AFs (see Figure 1) choosing (a) $|r|$ large and z' small (e.g. $|r| = 1.0$ and $z' = 4$), and (b) vice versa (e.g. $|r| = 0.4$ and $z' = 20$). Bold, thin and broken lines refer to first- and second-order phase transitions and anomalies, respectively. Transitions occur between the antiferromagnetic (AF) and paramagnetic (PM) phase. TCP, CEP, BCE denote tricritical, critical end- and bicritical endpoints, respectively.

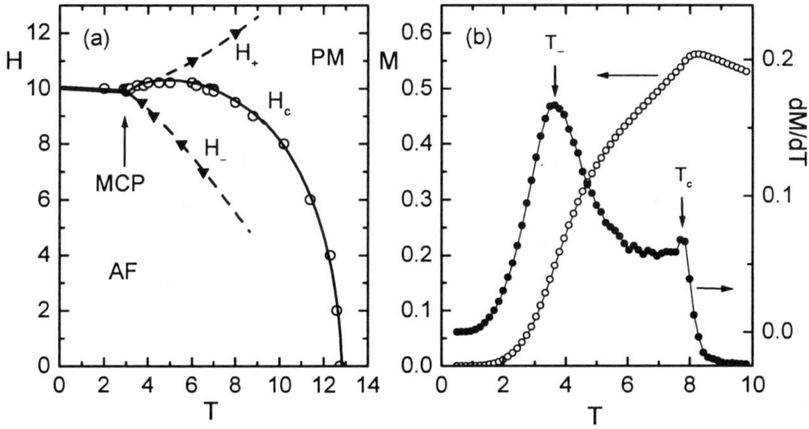

Figure 3. (a) H-T phase diagram of a 3-dimensional Ising antiferromagnet on a hexagonal lattice (24x24x18) with coupling constants $zJ = 6 \cdot 0.7$ and $z'J' = -20 \cdot 0.5$ ($k_B = 1$). Periodic boundary conditions for all axes were used. For $T < T_{MCP} \approx 3$ the phase transition is discontinuous, while for $T > T_{MCP}$ a second-order phase transition occurs at the line $H_c(T)$. The broken lines denoted as H_- and H_+ are the anomaly lines. (b) Calculated magnetization M (open circles) and its derivative dM/dT (full circles) vs temperature T in a field $H = 9.5$.

2 Comparison between theory and experiment

Fig. 3 (a) shows the H-T phase diagram of a 3-dimensional layered Ising AF obtained from Monte-Carlo simulations on a hexagonal lattice with periodic boundary conditions. All quantities like T, H, J and J' are considered to be dimensionless. For the simulation the Metropolis algorithm was used with $k_B = 1$. The parameters z, z', J, J' are chosen such as to reproduce case (b) of Fig. 2. One observes only one multicritical point (MCP), where a first-order phase line ($T < T_{MCP}$) and a second order phase line, $H_c(T)$ meet, and where the two anomaly lines, H_- and H_+, originate.

Fig. 3 (b) shows one magnetization curve, referring to the phase diagram of Fig. 3 (a), as a function of temperature for $H = 9.5$ and its derivative, dM/dT. The magnetization is defined to be $M = \sum_i S_i/N$, where N is the number of lattice sites. M vs T clearly shows an anomalous curvature for $T < T_c$, which manifests itself in the derivative as an additional broad maximum.

The phase diagram shown in Fig. 3 (a) resembles that of $FeBr_2$, an insulating uniaxial antiferromagnet with $T_N = 14.2$ K[13–16] (Fig. 4). Both

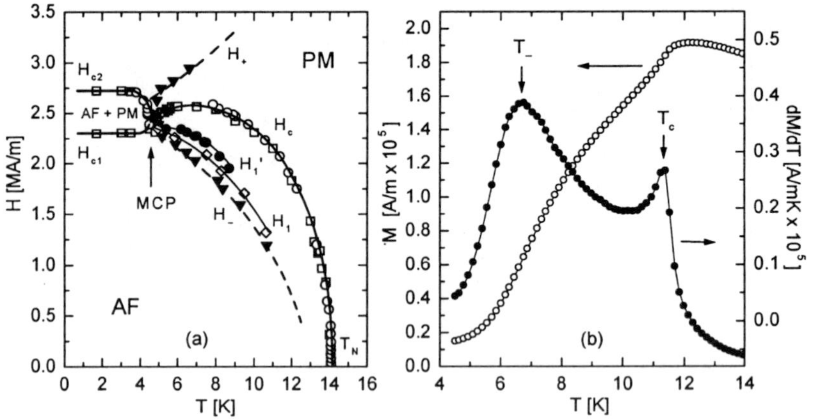

Figure 4. (a) H-T phase diagram of $FeBr_2$, where H is the applied axial magnetic field. The lines H_c, H_{c1}, H_{c2}, H_- and H_+ were obtained by magnetization or susceptibility measurements. H_1 and H_1' denote the positions of the spikes found by specific heat (open diamonds) (Ref. 14) and off-axis magnetization measurements vs temperature (solid circles), respectively. (b) Magnetization M vs temperature T and its derivative for a field $H = 2.07$ MA/m (field parallel to the c-axis). The phase lines H_1 or H_1' are not seen in this configuration (see Ref. 16). The broad peak at $T_- = T(H_-)$ is due to strong non-critical fluctuations (Ref. 13).

in Fig. 3 and Fig. 4 lines of non-critical fluctuations (or anomaly lines) do appear. In this article the attention is focused on these non-critical fluctuations at $T_- = T(H_-)$. Although this phenomenon is well investigated both in experiments[13,15,16] on $FeBr_2$ and in theory[8–10] no clear explanation for the occurrence of these fluctuations yet exists. Especially one wonders, how it is possible that two completely different types of phase diagrams are found by varying only two parameters, r and z'.

3 Non-critical fluctuations

In order to gain insight into the origin of the non-critical fluctuations at $H_-(T)$, we performed systematic Monte Carlo simulations of the sublattice magnetizations with the same parameters as in Fig. 3, $zJ = 4.2$ and $z'J' = -10.0$. The exchange constants and especially the number of coupled neighbors are comparable to those in $FeBr_2$.[17] However, here we used only two exchange couplings. The different intra-planar exchange parameters found in $FeBr_2$ were absorbed in one ferromagnetic effective intra-planar cou-

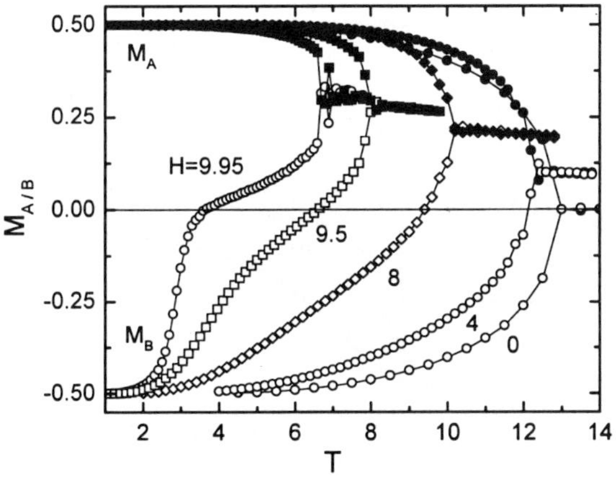

Figure 5. Sublattice magnetization of an Ising AF with an hexagonal lattice (24x24x18) and $zJ = 6 \cdot 0.7$ and $z'J' = -20 \cdot 0.5$ ($k_B = 1$). For low values of H the sublattice magnetization M_A and M_B are as expected for an uniaxial antiferromagnet. By increasing the field value the sublattice magnetizations become strongly asymmetric.

pling constant. Fig. 5 shows the temperature dependences of the sublattice magnetizations, M_A and M_B (see Fig. 1), for $H = 0$, 4, 8, 9.5 and 9.95, respectively. While M_A (parallel) and M_B (antiparallel to $\mathbf{H}$) are symmetric for zero field, $M_A(T) = -M_B(T)$, they become more and more inequivalent with increasing field. For fields coming close to the spin-flip one, $H_{c0} = 10$, anomalous bumps appear at $T \approx T_-$ in $M_B(T)$, whereas $M_A(T)$ is virtually constant up to that temperature. Obviously all of the non-critical fluctuations observed at $T \approx T_-$ happen to occur merely on the B-sublattice. In other words, these fluctuations are essentially constrained to 2-dimensional (2d) layers separated by magnetically saturated up-spin layers of the A-sublattice. It is, hence, tempting to compare the B-sublattice with an ensemble of 2d ferromagnets (FMs) with the same intra-layer parameters, $zJ > 0$, but subjected to a field $H_{eff} = H - H_{c0}$, where H is the field applied to the corresponding 3d AF.

Fig. 6 shows the sublattice magnetization curves for $H = 9.5$ and 9.95 as before and the magnetization curves of a 2d FM with $zJ = 4.2$ in fields $H_{eff} = -0.5$ and -0.05, respectively. One observes, that the magnetization curve of the 2d FM fits well with the spin-down sublattice magnetization M_B up to the inflection point at T_- (arrows) . It seems, that the magnetization of the spin-down sublattice behaves like a 2d FM in an effective mean field,

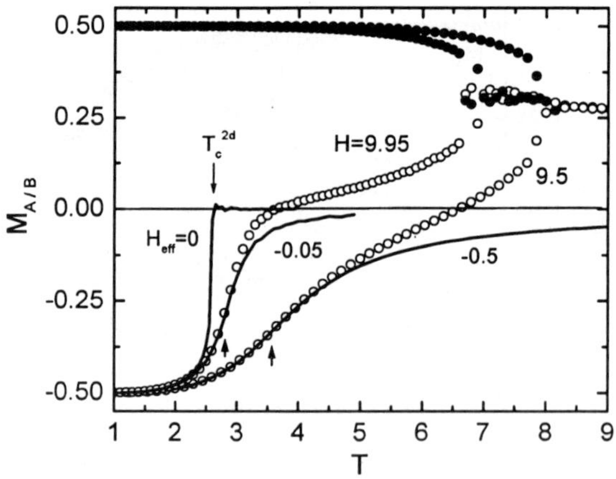

Figure 6. Two examples of sublattice magnetizations (circles) of the model as in Fig. 5 ($H_{c0} = -z'J' = 10.0$) and in addition magnetization curves of a 2d FM (solid lines) with $zJ = 6 \cdot 0.7$ in fields $H_{eff} = H - H_{c0} = 0, -0.05$, and -0.5, respectively. Arrows denote the inflection point, which is associated with T_-.

which is the sum of the external applied field H and the field, produced by the fully magnetized spin-up sublattice (A-sublattice), $H_{eff} = H + H_A$. Since M_A is completely magnetized, $H_A = z'J' = -H_{c0}$, one has $H_{eff} = H - H_{c0}$.

If the effective field H_{eff} becomes zero ($H_{eff} = 0 \Leftrightarrow H = H_{c0}$) we have the case of a 2d FM in zero field, which undergoes a phase transition at T_c^{2d} and becomes paramagnetic for $T > T_c^{2d}$ (see Fig. 6). Although this case cannot be obtained here, because of the spin-flip occurring in the range $H_{MCP} < H < H_{c0}$, the anomaly temperatures T_- can be associated with the points of inflection of the $M(H_{eff})$ vs T curves of the 2d Ising FM (arrows in Fig. 6). Therefore one can conclude, that the anomaly of M_B signifies the thermal destruction of 2d ferromagnetic order on the quasi-decoupled B-sublattice layers, which precedes the global 3d phase transition of both sublattices. In other words, the anomaly we find in the magnetization is due to the finite temperature range lying between T_c^{2d} and T_c^{3d}. If this splitting is reduced by increasing the intra-planar ferromagnetic interaction and thus increasing the 2d transition temperature, the anomaly decreases and vice versa.

4 Conclusion

The occurrence of anomalies, which are observed in certain 3d Ising AFs is due to the separation of the smeared 2d phase transition on one sublattice from the 3d global phase transition of both sublattices. This is typical of antiferromagnets, whose inter-planar antiferromagnetic coupling is strong compared with the intra-planar exchange ($|r| = |zJ/z'J'| < 0.6$).[3] A high value of the inter-planar exchange gives rise to a high value of the spin-flip field $H_{c0} = -z'J'$. This results in a strong stabilization of the spin-up sublattice, which acts only as a mean-field $H_{MF} = z'J'$ on the spin-down sublattice. Hence, the spin-down sublattice behaves like a 2d FM in an effective field $H_{eff} = H + H_{MF}$. The quality of the mean-field approximation crucially depends on a high coordination number z', which is needed for integrating out thermal spin-fluctuations.

Acknowledgements

We would like to thank M. Acharyya and U. Nowak for helpful discussions and the Deutsche Forschungsgemeinschaft (Graduiertenkolleg "Struktur und Dynamik heterogener Systeme") for financial support.

References

1. F. Harbus and H.E. Stanley, *Phys. Rev.* B **8**, 1141 (1973).
2. A.R. Fert, J. Gelard, and P. Carrara, *J. Phys. Chem. Sol.* **35**, 261 (1974).
3. J.M. Kincaid and E.G.D. Cohen, *Phys. Rep.* C **22**, 57 (1975).
4. H.J. Herrmann, E.B. Rasmussen, and D.P. Landau, *J. Appl. Phys.* **53**, 7994 (1982).
5. I. Vilfan and S. Galam, *Phys. Rev.* B **34**, 6428 (1986).
6. Y.L. Wang and J.D. Kimel, *J. Appl. Phys.* **69**, 6176 (1991).
7. H.J. Herrmann and D.P. Landau, *Phys. Rev.* B **48**, 239 (1993).
8. W. Selke and S. Dasgupta, *J. Magn. Magn. Mater.* **147**, L245 (1995).
9. W. Selke, *Z. Phys.* B **101**, 145 (1996).
10. M. Pleimling and W. Selke, *Phys. Rev.* B **56**, 8855 (1997).
11. M. Pleimling, *Eur. Phys. J.* B (in print, cond-mat/9902237).
12. C. Vettier, H.L. Alberts, and D. Bloch, *Phys. Rev. Lett.* **31**, 1414 (1973).
13. M.M.P. de Azevedo, Ch. Binek, J. Kushauer, W. Kleemann, and D. Bertrand, *J. Magn. Magn. Mater.* **140-144**, 1557 (1995).
14. H. ArugaKatori, K. Katsumata, and M. Katori, *Phys. Rev.* B **54**, R9620 (1996).

15. O. Petracic, Ch. Binek, and W. Kleemann, *J. Appl. Phys.* **81**, 4145 (1997).
16. O. Petracic, Ch. Binek, and W. Kleemann, U. Neuhausen, and H. Lueken, *Phys. Rev.* B **57**, R11051 (1998).
17. W.B. Yelon, C. Vettier, *J. Phys.* C **8**, 2760 (1975).

THERMALLY ACTIVATED MAGNETIZATION REVERSAL IN CLASSICAL SPIN CHAINS

D. HINZKE, U. NOWAK AND K. D. USADEL
Theoretische Tieftemperaturphysik, Gerhard-Mercator-Universität-Duisburg, 47048 Duisburg /Germany
E-mail:denise@thp.uni-duisburg.de, uli@thp.uni-duisburg.de, usadel@thp.uni-duisburg.de

We investigate the thermally activated magnetization switching in a classical Heisenberg spin chain driven by an external magnetic field. For small system sizes we expect that the magnetic moments rotate coherently, while in the case of larger system sizes the magnetization reversal is proposed to be due to soliton-antisoliton nucleation. We compare Monte Carlo simulations with the direct integration of the Landau-Lifshitz-Gilbert equation of motion with Langevin dynamics as well as with asymptotic solutions for the escape rates following from the Fokker-Planck equation, finding agreement for low temperatures and high damping. We also discuss deviations in the intermediate temperature regime.

Magnetic particles which are small enough to be single-domain are proposed to have good qualities for magnetic recording. The ultimate limit for the density of magnetic storage is given by that size of the particles where superparamagnetism sets in.[1] Therefore, the role of thermal activation for the magnetization reversal of single-domain particles is very actively studied at present analytically,[2,3,4,5,6,7] experimentally[7,8,9] and numerically.[10,11,12]

In order to investigate thermal activation further, we study a chain of classical magnetic moments. This system can be considered as a model for thin ferromagnetic wires[3] which are also investigated in recent experiments.[9] Furthermore, this model is very useful as a test tool for numerical methods since it was also treated analytically.[3] Depending on the system size, given anisotropies and the strength of the magnetic field different reversal mechanisms may appear. For small system sizes the magnetic moments are expected to rotate coherently,[13] while for sufficiently large system sizes the so-called soliton-antisoliton nucleation is proposed.[3]

In the following we consider a chain of magnetic moments of length L with periodical boundary conditions. Our system is defined by a classical Heisenberg Hamiltonian,

$$E = -J \sum_{\langle ij \rangle} \mathbf{S}_i \cdot \mathbf{S}_j - d_x \sum_i (S_i^x)^2 + d_z \sum_i (S_i^z)^2 - \mu_s \mathbf{H} \cdot \sum_i \mathbf{S}_i. \qquad (1)$$

where the $\mathbf{S}_i = \mu_i/\mu_s$ are three dimensional magnetic moments of unit length.

The first sum which represents the exchange of the magnetic moments is over nearest neighbor interactions with the exchange coupling constant J. The second and third sum represent uniaxial anisotropies of the system where the first one favors the x-axis as easy axis (anisotropy constant $d_x > 0$) and the other one the z-axis as hard axis of the system ($d_z > 0$). The last sum is the coupling of the moments to an external magnetic field $\mathbf{H}$.

On the one hand we investigate our system using the Landau-Lifshitz-Gilbert (LLG) equation of motion with Langevin dynamics. This equation has the form

$$\frac{(1+\alpha^2)\mu_s}{\gamma}\frac{\partial \mathbf{S}}{\partial t} = \mathbf{S} \times \left(\zeta(t) - \frac{\partial E}{\partial \mathbf{S}}\right) - \alpha\, \mathbf{S} \times \left(\mathbf{S} \times \left(\zeta(t) - \frac{\partial E}{\partial \mathbf{S}}\right)\right), \tag{2}$$

with the gyromagnetic ratio $\gamma = 1.76 \times 10^{11} (\mathrm{Ts})^{-1}$ and a damping constant α. The first part of Eq. 2 describes the spin precession while the second part includes the relaxation of the moments. In both parts of this equation the thermal noise $\zeta(t)$ represents thermal fluctuations. Eq. 2 is solved numerically using the Heun method.[12]

On the other hand we simulate the system using Monte Carlo methods[14] with a heat-bath algorithm. We use a new Monte Carlo technique proposed recently[15] in order to obtain a quantified time step. The trial step of our algorithm is a random movement of a magnetic moment within a cone with a size that is chosen in such a way that one Monte Carlo step represents a certain time interval of the LLG equation in the high damping limit (for details see[15]). Throughout this work we will use $\alpha = 4$, which – as we will show – is large enough so that our Monte Carlo simulation and the numerical solution of Eq. 2 yield identical time scales.

All our simulations, using the two different methods above, start with a configuration where all magnetic moments point into the x-direction, antiparallel to the external magnetic field $\mathbf{H} = -H_x\hat{\mathbf{x}}$. The time when the x-component of the magnetization changes its sign is the characteristic time τ which is the reciprocal of the escape rate λ which has been calculated in certain limits asymptotically from the Fokker-Planck equation (for details see[3,4,5]).

In the case of small chain lengths the magnetic moments rotate coherently while overcoming the energy barrier[13]

$$\Delta E_{cr} = L d_x (1-h)^2 \tag{3}$$

with $h = \mu_s H_x/(2d_x)$. The corresponding characteristic time calculated from

Figure 1. Snapshots of a spin chain for various times during coherent rotation. The x-component (up-down) of the magnetic moments is also color coded. $L = 20$, $h = 0.75$, $d_x = 0.1$, $d_z = 1.0$ and $k_BT = 0.0019$.

Figure 2. Snapshots of a spin chain for various times during soliton-antisoliton nucleation. $L = 80$, $h = 0.75$, $d_x = 0.1$ and $d_z = 1.0$, $k_BT = 0.0019$.

the Fokker-Planck equation[4] is for low temperatures $T \ll \Delta E_{cr}$

$$\tau = \tau_0 \exp \frac{\Delta E_{cr}}{k_B T} \tag{4}$$

with the prefactor

$$\tau_0 = \frac{\pi \mu_s (1+\alpha^2) \sqrt{\frac{d_z(1+h)}{d_x(1-h)+d_z}}}{\gamma\left(\alpha(d_x(1-h^2)-d_z) + \sqrt{\alpha^2\left(d_z + d_x(1-h^2)\right)^2 + 4d_x d_z(1-h^2)}\right)}. \tag{5}$$

In Fig. 1 snapshots of such a reversal process are shown. Since the energy barrier ΔE_{cr} for coherent rotation is proportional to L with increasing length of the chain a different reversal mechanism becomes energetically favorable.

Fig. 2 shows snapshots during the reversal of a system with length $L = 80$ but with the same parameters as before. Due to thermal fluctuations a nucleus is originated and two domain walls pass the system. This is the so-called soliton-antisoliton nucleation proposed by Braun.[3] The corresponding energy barrier ΔE_{nu} which has to be overcome during this nucleation process is

$$\Delta E_{nu} = \sqrt{2Jd_x}\left(4\tanh R - 4hR\right), \tag{6}$$

with $R = \text{arcosh}(\sqrt{1/h})$. The corresponding characteristic time follows once again a thermal activation law (see Eq. 4) but with a different prefactor which in the overdamped limit is given by

$$\tau_0 = \frac{\mu_s(1+\alpha^2)}{16\gamma\alpha L} \sqrt{\frac{k_B T J \pi^3 \sqrt{2}}{d_x^3 \sqrt{Jd_x}}} \frac{1}{|E_0(R)| \tanh R^{3/2} \sinh R}. \tag{7}$$

The eigenvalue $E_0(R)$ has been calculated numerically in.[5] In the limit $h \to 1$ it is $|E_0(R)| \approx 3R^2$. In the following, our numerical results for the characteristic times τ for soliton-antisoliton nucleation as well as for coherent rotation are compared with the equations above.

In Fig. 3 we show the temperature dependence of the characteristic time τ for two different system sizes where each data point is an average over 100 independent runs. Note, that since the Langevin dynamics simulations are much more time consuming, here we have only data for higher temperatures. As is evident from Fig. 3, the data of the Monte Carlo simulations and the Langevin dynamics simulations agree in both cases, for coherent rotation as well as for soliton-antisoliton nucleation. This confirms that our Monte Carlo method indeed yields results with a realistic time scale. Also, in the limit of low temperatures our data confirm the asymptotic solutions above. For

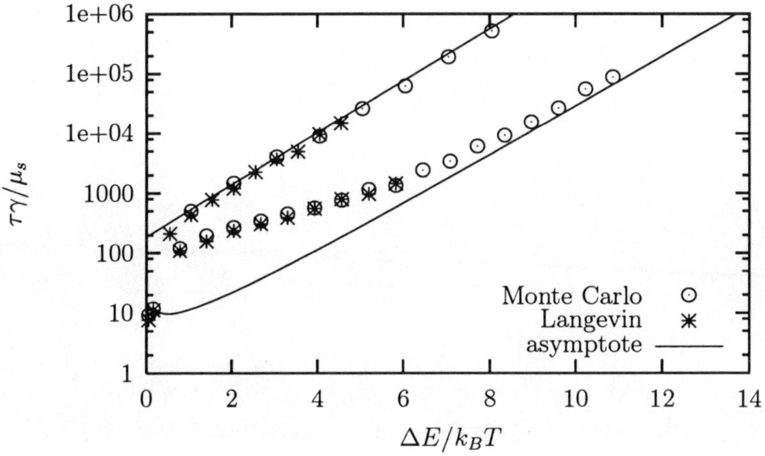

Figure 3. Reduced characteristic time $\tau\gamma/\mu_s$ vs. $\Delta E/k_B T$ where ΔE is the appropriate energy barrier of nucleation and coherent rotation, respectively. The data are from MC and Langevin dynamics simulations. The solid lines are the asymptotic formulae for the inverse escape rates. $d_x = 0.1$, $d_z = 1.0$ and $h = 0.75$. Lower curve: nucleation, $L = 80$. Upper curve: coherent rotation, $L = 8$.

the case of coherent rotation a very good agreement is already obtained for temperatures $\Delta E_{cr}/k_B T > 1$. In the case of nucleation the asymptotic formula is confirmed only for very low temperatures $\Delta E_{nu}/k_B T > 8$. In order to understand the deviations of our data from the asymptote in the intermediate temperature range we show here Fig. 4 where one can see several nuclei simultaneously. Obviously, depending on temperature (and also other quantities, like system size and field) with a certain probability many nuclei can grow at the same time. This is the so-called multidroplet nucleation which until now was investigated mainly in Ising models.[16,11] Due to multidroplet nucleation the effective energy barrier is reduced leading to the deviations in the intermediate temperature regime.

To summerize, we presented numerical results on the characteristic switching time in a classical spin chain. The model exhibits different reversal mechanisms. The energy barriers and the characteristic times for coherent rotation and soliton-antisoliton nucleation which have been calculated asymptotically from the Fokker-Planck equation[3,4] are confirmed with Monte Carlo simulations as well as Langevin dynamics simulations. For the case of nucleation, in an intermediate temperature range multidroplet nucleation leads to

Figure 4. Snapshots of a chain for various times during multidroplet nucleation. $L = 120$, $h = 0.95$, $d_x = 0.1$ and $d_z = 1.0$ and $k_B T = 0.038$.

deviations from the asymptotic formula.

Acknowledgements

We thank H. B. Braun for providing numerical results for the eigenvalue $|E_0(R)|$ and for helpful discussions. This work was supported by the Deutsche Forschungsgemeinschaft through the Graduiertenkolleg "Struktur und Dynamik heterogener Systeme".

References

1. R. W. Chantrell and K. O'Grady, in *Applied Magnetism*, edited by R. Gerber, C. D. Wright, and G. Asti (Kluwer Academic Publishers, Dordrecht, 1994), p. 113.
2. I. Klik and L. Gunther, *J. Stat. Phys.* **60**, 473 (1990).
3. H. B. Braun, *Phys. Rev. Lett.* **71**, 3557 (1993); H. B. Braun, *J. Appl. Phys.* **85**, 6127 (1999).
4. H. B. Braun, *J. Appl. Phys.* **76**, 6310 (1994).
5. H. B. Braun, *Phys. Rev.* B **50**, 16485 (1994)
6. D. A. Garanin, *Phys. Rev.* B **55**, 3050 (1997)
7. W. T. Coffey, D. S. F. Crothers, J. L. Dorman, Yu. P. Kalmykov, E. C.

Kennedy, and W. Wernsdorfer, *Phys. Rev. Lett.* **80**, 5655 (1998);

8. W. Wernsdorfer, E. Bonet Orozco, K. Hasselbach, A. Benoit, B. Barbara, N. Demoncy, A. Loiseau, H. Pascard, and D. Mailly, *Phys. Rev. Lett.* **78**, 1791 (1997).
9. W. Wernsdorfer, B. Doudin, D. Mailly, K. Hasselbach, A. Benoit, J. Meier, J. -Ph. Ansermet, and B. Barbara, *Phys. Rev. Lett.* **77**, 1873 (1996) and *Phys. Rev.* B **55**, 11552 (1997).
10. D. Hinzke and U. Nowak, *Phys. Rev.* B **58**, 265 (1998).
11. U. Nowak and D. Hinzke, *J. Appl. Phys.* **85**, 4337 (1999).
12. J. L. Garcí-Palacios and F. J. Lázaro, *Phys. Rev.* B **58**, 14937 (1998).
13. L. Néel, *Ann. Geophys.* **5**, 99 (1949), W. F. Brown, *Phys. Rev.* **130**, 1677 (1963).
14. K. Binder and D. W. Heermann, *Monte Carlo Simulation in Statistical Physics* (Springer-Verlag, Berlin, 1988), p. 21.
15. U. Nowak, R. W. Chantrell, and E. C. Kennedy, cond-mat/9906089
16. D. Stauffer, *Int. J. Mod. Phys.* C **3**, No.5, 1059 (1992); P. A. Rikvold and B. M. Gorman, in *Annual Reviews of Computational Physics I*, edited by D. Staufer (World Scientific, Singapore, 1994), p. 149; M. Acharyya and B. Chakrabarti, *Phys. Rev.* B **52**, 6550 (1995)

DYNAMICS OF DRIVEN INTERFACES IN HETEROGENOUS MAGNETIC SYSTEMS

L. ROTERS, A. HUCHT, S. LÜBECK, U. NOWAK, AND K. D. USADEL
Theoretische Tieftemperaturphysik, Gerhard-Mercator-Universität Duisburg, 47048 Duisburg, Germany
E-mail:{lars,fred,sven,uli,usadel}@thp.uni-duisburg.de

We analyze interface motion in the random field Ising model with quenched disorder by means of Monte Carlo simulations. In the absence of thermal fluctuations a depinning transition occurs at some critical threshold field. We study the interface motion in the vicinity of the threshold field by varying the temperature as well as the driving field. It turns out that thermal fluctuations yield a rounded transition which can be characterized by a critical exponent.

1 Introduction

Driven interfaces in systems with quenched disorder display with increasing driving force a transition from a phase where no interface motion takes place to a phase with a finite interface velocity. This so-called depinning transition is caused by a competition of driving force and quenched disorder which hinders the interface motion (see e.g.[1]). Depinning transitions are found in a large variety of physical problems, like fluid invasion in porous media,[2] depinning of charge density waves,[3,4] or field-driven motion of domain walls in ferromagnets.[5]

With the assumption that the lattice constant is substantially smaller than characteristic fluctuations of the interface position a continuum theory has been proposed.[5] This continuum theory yields the conclusion that the motion of domain walls can be described by an Edwards-Wilkinson equation[6] with quenched disorder. However, by describing the time evolution of a single valued function the Edwards-Wilkinson equation takes no overhangs into account. Overhangs occur in magnetic systems where interfaces separate, for instance, regions of different spin orientations. While the interface motion in a system with quenched disorder near the depinning transition is theoretically investigated often in the absence of thermal fluctuations, these fluctuations affect the experimental study of the depinning transition.[7,8,9] The crucial point is that energy barriers which are responsible for a trapping of the interface in a metastable state at zero temperature can always be overcome due to thermal fluctuations. In the vicinity of the transition point this yields a rounded transition (for experimental evidence see, for instance, Fig. 2 in[9]). To characterize this rounding quantitatively a scaling ansatz has been proposed.[4] This ansatz is based on an equation of motion for sliding charge density waves and predicts its characteristic velocity to be a power law of temperature at the critical threshold.

2 Interfaces in the random field Ising model

We investigate interface motion in the $3d$ random field Ising model (RFIM) with quenched disorder on a simple cubic lattice. The Hamiltonian of the RFIM is given by

$$\mathcal{H} = -\frac{J}{2} \sum_{\langle i,j \rangle} S_i \, S_j - \sum_i (H + h_i) \, S_i \,, \tag{1}$$

where we restrict the exchange energy $J = 1$ to nearest-neighbors. The magnetic moments S_i are coupled both to a driving field H and to quenched random fields h_i which are uniformly distributed within an interval $[-\Delta, \Delta]$.

An initially flat interface is built into the system separating regions of up- and down oriented spins corresponding to $S_i = \pm 1$. One orientation of the magnetic moments is favored by the driving field by which a magnetization reversal process is started. Interface motion is caused by flips of magnetic moments adjacent to the interface. The dynamics of this process is simulated by a Monte-Carlo based single-spin-flip dynamics with transition probabilities $p(S_i \rightarrow -S_i, T)$ according to a heat-bath-algorithm[10] at finite temperatures T. At zero temperature these transition probabilities reduce to

$$p(S_i \rightarrow -S_i, 0) = \begin{cases} 1 & : \delta\mathcal{H} < 0 \\ 1/2 & : \delta\mathcal{H} = 0 \\ 0 & : \delta\mathcal{H} > 0 \end{cases} \tag{2}$$

with $\delta\mathcal{H} = \mathcal{H}(-S_i) - \mathcal{H}(S_i)$ denoting the corresponding change in $\mathcal{H}$. Beside interface motion, also nucleation may occur, i.e. a magnetic moment initially parallel to all of its neighbors turns and remains turned during the time evolution of the system. Since we are interested in studying interface motion it is essential that within the observation time nucleation does not occur. The minimum energy needed for isolated spin flips is $2(zJ - H - \Delta)$. As long as this quantity is large as compared to the temperature the time scales of nucleation processes and interface motion are separated.[11] In particular, there is no need to suppress artificially nucleation or isolated spin flips during the simulation.

The analysis of interface motion on simple cubic lattices usually considers [10]-interfaces for $d = 2$ and [100]-interfaces for $d = 3$, respec-

Figure 1. Interface in the RFIM at $H \approx H_c$. For reasons of simplicity a two dimensional interface is shown. Different colors indicate different orientations of the magnetic moments, corresponding to $S_i = \pm 1$.

tively (see e.g.[12,13,14,15,16]). However, investigating these interfaces in the limit of vanishing disorder means that interface motion is restricted to driving fields $H/J > z - 2$ where z denotes the number of next neighbors.[12] To avoid this, we consider [111]-interfaces in the $3d$-RFIM which move in the absence of disorder at arbitrarily small driving fields. Hereby, the separation of time scales for interface motion and nucleation is even further increased.[11]

3 Depinning transition

First we consider the depinning transition of the interface which occurs for $T = 0$. This situation is illustrated in a $2d$ sketch in Fig. 1. If the driving field exceeds some critical value H_c interface motion takes place. Recent investigations on the depinning transition of diagonal interfaces in the RFIM show that characteristics of this transition depend crucially on the existence of overhangs.[17] For $\Delta/J < 1$ there are no overhangs near the transition point $H_c = \Delta$. Under these circumstances, interface motion is restricted to driving fields exceeding each random field and the interface velocity is neither affected by the driving field nor by the strength of the random fields.[17] This behavior changes for $\Delta/J > 1$ where overhangs are created. Here, the depinning transition takes place at $H_c < \Delta$ leading to a disorder affected interface motion.

We analyse the disorder averaged interface velocity in the limit of large times which can be interpreted as the order parameter of the depinning transition. Figure 2 shows the interface velocity for a fixed value of Δ and its

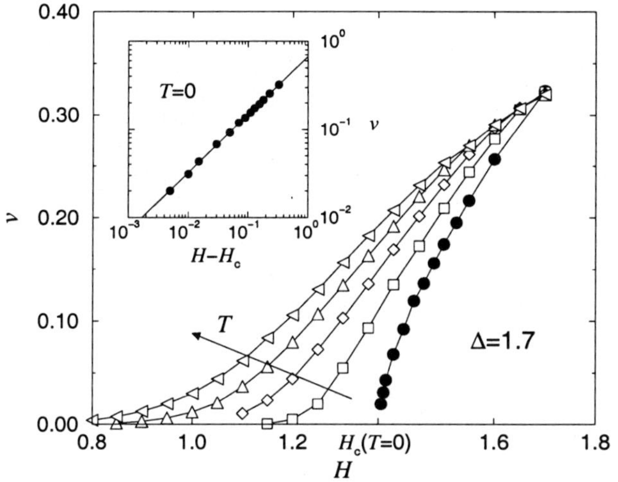

Figure 2. Dependence of the interface velocity on temperature and driving field in the thermodynamic limit. Filled and opened symbols denote velocities found at $T = 0$ and $T \in \{0.05, 0.10, 0.15, 0.20\}$, respectively. The inset shows a logarithmic plot of the interface velocities found at $T = 0$ which can be fitted according to Eq. (3) (solid line).

dependence on the driving field, both, for $T = 0$ and $T > 0$. Since in the vicinity of the transition point finite size effects may become important, we calculated the interface velocities in systems of different size L. We found that for sufficiently large L the interface velocity is almost unaffected by finite-size effects. Therefore, the values of the interface velocity shown in Fig. 2 correspond within negligible errors with those of the limit $L \to \infty$.

We first focus our attention to $T = 0$. The corresponding interface velocities are denoted by filled symbols in Fig. 2. These data suggest that without thermal fluctuations the interface velocity vanishes continuously at the transition point, i.e.

$$v(H) \sim (H - H_c)^{\beta}\,. \tag{3}$$

The inset of Fig. 2 shows a logarithmic plot of the interface velocities which confirms this assumption. Fitting the data according to Eq. (3) yields $\beta = 0.66 \pm 0.04$ and $H_c = 1.37 \pm 0.01$. In the presence of thermal fluctuations, however, Eq. (3) is no longer valid since pinning does not occur. This is due to the fact that a pinned interface is trapped in a metastable state which can always be overcome by thermal fluctuations. As can be seen from Fig. 2

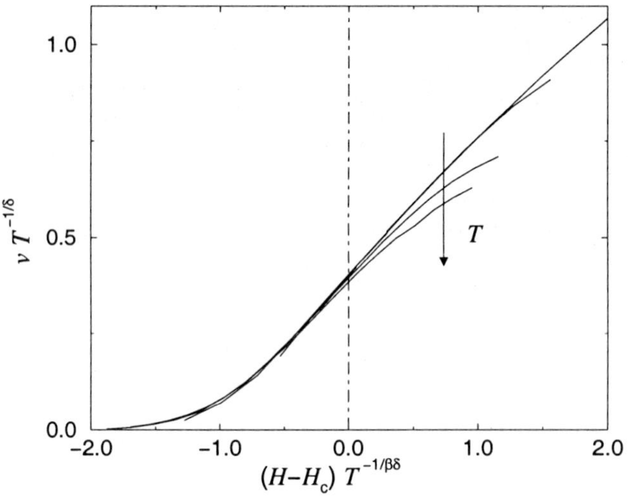

Figure 3. Interface velocities for $\Delta = 1.7$. The data are rescaled according to Eq. (4) and correspond with those shown in Fig. 2 for $T > 0$. From the data collapse one obtains $\beta = 0.63 \pm 0.05$, $\delta = 2.38 \pm 0.2$, and $H_c = 1.375 \pm 0.01$.

the transition is rounded with increasing temperature. To characterize this rounding quantitively a scaling ansatz has been proposed,[4]

$$v(T,H) = T^{1/\delta} f[(H-H_c)\,T^{-1/\beta\delta}]\,. \tag{4}$$

The asymptotes of the scaling function $f(x)$ are given by $f(x \to \infty) \sim x^{\beta}$ and $f(x \to 0) = \text{const}$, respectively. Figure 3 shows the dependence of the interface velocity on the driving field for various values of T. The data are rescaled according to the scaling ansatz, Eq. (4), and are identical with those shown in Fig. 2 for $T > 0$. As can be seen the scaling ansatz is fulfilled and one obtains $\delta = 2.38 \pm 0.2$, $\beta = 0.63 \pm 0.05$, and $H_c = 1.375 \pm 0.01$. Both, the values of H_c and β correspond within the error-bars with the values found for $T = 0$. The scaling ansatz, Eq. (4), has also been shown to be a valid description for an [11]-interface in the $2d$-RFIM with quenched disorder.[11] Thus, the interface velocity evolves both for $d = 2$ and for $d = 3$ according to a power law with increasing temperature, $v \sim T^{1/\delta}$ at $H = H_c$. This coincides with properties of an order parameter at continuous equilibrium phase transitions (see, for instance,[19] and references therein) and leads us to the conclusion that the influence of temperature on the depinning transition can be understood in well-established terms of criticality.

4 Summary

We studied numerically the motion of an [111]-interface in the random field Ising model with quenched disorder which is a magnetization reversal process. An investigation of the interface motion at finite temperatures is possible since within the time scale which is used to investigate the interface no nucleation processes occur.[11] We analyzed the interface velocity near the critical threshold by both varying temperature and driving field. It turns out that a scaling ansatz [Eq. (4)] found in the context of sliding charge density waves is a valid description for the depinning transition observed in the RFIM. This scaling ansatz has two meaningful consequences: (i) without thermal fluctuations the interface velocity vanishes at the transition point according to a power law which suggests that critical behavior takes place. (ii) For $H = H_c$ the dependence of the interface velocity on the temperature is also given by a power law.

Acknowledgements

This work was supported by the Deutsche Forschungsgemeinschaft through the Graduiertenkolleg *Struktur und Dynamik heterogener Systeme* at the University of Duisburg, Germany.

References

1. H. Leschhorn, T. Nattermann, S. Stepanow, and L.-H. Tang, *Ann. Physik* **6**, 1 (1997).
2. M. Cieplak and M. Robbins, *Phys. Rev. Lett.* **60**, 2042 (1988).
3. D. S. Fisher, *Phys. Rev. Lett.* **50**, 1486 (1983).
4. D. S. Fisher, *Phys. Rev.* B **31**, 1396 (1985).
5. R. Bruinsma and G. Aeppli, *Phys. Rev. Lett.* **52**, 1547 (1984).
6. S. F. Edwards and D. R. Wilkinson, *Proc. R. Soc. London*, Ser. A **381**, 17 (1982).
7. F. Ladieu, M. Sanquer, and J. P. Bouchaud, *Phys. Rev. B.* **53**, 973 (1996).
8. U. Nowak, J. Heimel, T. Kleinefeld, and D. Weller, *Phys. Rev.* B **56**, 8143 (1997).
9. S. Lemerle, J. Ferré, C. Chappert, V. Mathet, T. Giamarchi, and P. Le Doussal, *Phys. Rev. Lett.* **80**, 849 (1998).

10. K. Binder and D. W. Heermann, *Monte Carlo Simulation in Statistical Physics*, in *Springer Series in Solid-State Sciences 80* (Springer, Heidelberg, 1997).
11. U. Nowak and K. D. Usadel, *Europhys. Lett.* **44**, 634 (1998).
12. B. Koiller, H. Ji, and M. O. Robbins, *Phys. Rev.* B **46**, 5228 (1992).
13. N. Martys, M. O. Robbins, and M. Cieplak, *Phys. Rev.* B **44**, 12294 (1991).
14. H. Ji and M. O. Robbins, *Phys. Rev.* B **46**, 14519 (1992).
15. L. A. N. Amaral, A. L. Barabási, and H. E. Stanley, *Phys. Rev. Lett.* **73**, 62 (1994).
16. L. A. N. Amaral, A. L. Barabási, H. A. Makse, and H. E. Stanley, *Phys. Rev.* E **52**, 4087 (1995).
17. L. Roters, A. Hucht, S. Lübeck, U. Nowak, and K. D. Usadel, (accepted for publication in *Phys. Rev.* E); see also cont-mat/9905324.
18. K. Binder, in *Phase Transitions and Critical Phenomena*, Vol. 5B, edited by C. Domb and M. S. Green, (Academic Press London Ltd., London, 1976).
19. H. E. Stanley, *Introduction to Phase Transitions and Critical Phenomena*, (Clarendon Press, Oxford, 1971).